HIGH-PERFORMANCE ENERGY-EFFICIENT MICROPROCESSOR DESIGN

SERIES ON INTEGRATED CIRCUITS AND SYSTEMS

Anantha Chandrakasan, Editor
Massachusetts Institute of Technology
Cambridge, Massachusetts, USA

Published books in the series:

A Practical Guide for SystemVerilog Assertions
Srikanth Vijayaraghavan and Meyyappan Ramanathan
2005, ISBN 0-387-26049-8

Statistical Analysis and Optimization for VLSI: Timing and Power
Ashish Srivastava, Dennis Sylvester and David Blaauw
2005, ISBN 0-387-25738-1

Leakage in Nanometer CMOS Technologies
Siva G. Narendra and Anantha Chandrakasan
2005, ISBN 0-387-25737-3

Thermal and Power Management of Integrated Circuits
Arman Vassighi and Manoj Sachdev
2005, ISBN 0-398-25762-4

High-Performance Energy-Efficient Microprocessor Design

Edited by

VOJIN G. OKLOBDZIJA
Integration Corp., Berkeley, California and University of California

and

RAM K. KRISHNAMURTHY
Microprocessor Research Laboratory, Intel Corp., Hillsboro, Oregon

Springer

A C.I.P. Catalogue record for this book is available from the Library of Congress.

ISBN-10 0-397-28594-6 (HB)
ISBN-10 0-387-34047-5 (e-book)

Published by Springer,
P.O. Box 17, 3300 AA Dordrecht, The Netherlands.

www.springer.com

Printed on acid-free paper

Printed in the Netherlands.

TABLE OF CONTENTS

INTRODUCTION

Microprocessor design is a discipline and an art. Since the introduction of the first microprocessor in 1971 containing 2108 transistors embodied in Intel's 4004, the complexity of the design has increased several orders of magnitude with contemporary multi-core processors containing over two billion transistors. Yet microprocessor design is still driven by the inspiration, passion and vision of individuals involved.

This book deals with energy efficiency in microprocessors. As the complexity increased and the number of transistors integrated on the chip skyrocketed, power became the single most important issue limiting the otherwise unlimited progress. Not only does power determine the maximal speed at which we can allow a microprocessor to run, but it also determines the form factor, packaging and price. This is particularly important as computers migrated into consumer electronics characterized by mobility, portability and battery operation. Microprocessor design is a centerpiece of contemporary electronics, computer engineering and almost every complex electronic endeavor. Today microprocessors are found in almost every product, from the personal computer to iPod, in personal digital assistants, cell phones, games consoles, digital electronic cameras and television sets. They became an indispensable part of our everyday life which is increasingly dependent on them and their sustained and reliable functioning. In almost all of these applications energy efficiency is a must.

The importance of energy-efficient processor design is also recognized in courses taught at universities, industrial courses or seminars. Several conferences have been dedicated to energy-efficient design and several workshops have augmented important conferences in attempts to highlight this particular aspect. This book describes and teaches important topics of energy-efficient processor design starting from circuits to architecture, test and design for testability. It is targeted toward engineers, practitioners and researchers, as well as graduate students who want to learn about energy-efficient microprocessor design. It is suitable for advanced undergraduate and graduate courses in electrical engineering where the subject of low-power design is taught. The book is divided into chapters that can be covered one per week, thus being suitable for universities adhering to the quarter system courses. In the next edition we plan to introduce exercises and problems at the end of each chapter to make it more suitable for teaching. The chapters are written by the world's top experts in the field highlighting a particular aspect of their expertise.

The book starts with a chapter "Ultra-Low Power Processor Design", describing the ways of designing for energy efficiency in low- and ultra-low-power processor design. It contrasts the ways of achieving low power with the flexibility of design which is often of more importance. The techniques widely used for the power reduction of microcontrollers and DSP processors are reviewed in this chapter. They include basic CPI (clocks per instruction) reduction, gated-clock mechanisms, optimal pipeline length, hardware accelerators, reconfigurable units and techniques to reduce leakage power. This is augmented by several examples such as RISC 8-bit and 32-bit microcontrollers and DSP cores. They describe the necessary tradeoffs between flexibility and energy efficiency.

Energy-efficient design of digital circuits is discussed in the second chapter of this book. In particular this chapter describes how transistor sizing affects the energy and delay of digital circuits. It examines design methodology based on the logical effort, and shows its limitations when designing for energy efficiency. The chapter presents a new methodology for the design and analysis of digital circuits in the energy-delay space which allows for energy reduction without performance penalty.

Clocking, which is one of the most critical aspects of processor design, is described in the third chapter of this book. Clocking strategy determines processor performance and largely impacts its power consumption. Conventional clocking strategies and circuit techniques descriptions, augmented with an overview of the state-of-the art clocked storage elements used in modern microprocessors, are contained in this chapter. Emerging methods aimed at handling incoming challenges in microprocessor design are also described.

The fourth chapter is dedicated to static memory design and issues related to the design of memory peripherals. This chapter presents design techniques to reduce SRAM dynamic and static power. It explores the design of static memory structures starting with a description of the single-port six-transistor SRAM cell operation and subsequently addressing voltage-mode and current-mode differential reads, single-ended reads, and control logic operation. The chapter covers issues pertaining to SRAM reliability, testing, and yield. Subsequently, implementation of efficient multi-port register file storage cells and peripherals is described.

The fifth chapter addresses the problem of placing design blocks of widely varying sizes and shapes and interconnecting them. Stability of placement methods is now a key concern. To achieve timing closure it is essential that gate sizing, buffer insertion, and routing can be completed without large disruptions to the overall physical structure of a circuit. This chapter surveys modern techniques for circuit placement, with an emphasis on how placement interacts with logic synthesis and routing.

The choice of the right algorithm and a corresponding adder topology is discussed in the sixth chapter. This depends on many factors closely related to the technology of implementation. With the transition to deep-submicron CMOS technologies further complexity has been introduced. Thus, it has become even more difficult to make the right selection of appropriate design topology when power consumption is included. In this chapter this complex relationship is addressed and the important factors that influence the right selection of algorithm, circuit topology, operating conditions and power consumption are explained.

Fast 32-bit and 64-bit ALU with single-cycle latency and throughput are described in Chapter 7. They are one of the most performance-limiting units within the integer and floating-point execution clusters. ALU also contribute to one of the highest power-density locations on the processor, resulting in thermal hotspots and sharp temperature gradients within the execution core. This strongly motivates energy-efficient ALU designs that satisfy the high-performance requirements, while reducing peak and average power dissipation. The chapter, describes a single-cycle 64-bit integer execution ALU fabricated in 90 nm dual-Vt CMOS technology.

Chapter 8 deals with algorithms and implementation details used in today's floating-point units. It shows the implementation of the different parts of the fused multiply-add dataflow including the counter tree, suppression of sign extension encoding, leading zero anticipation, and end around carry adder design which has a huge performance advantage over a separate multiplier and adder. With one compound operation, effectively two dependent operations per cycle can be achieved. This type of design has a huge performance advantage over a separate multiplier and adder.

Chapter 9 addresses microarchitectural techniques for avoiding defects. Error detection and correction microarchitecture can significantly reduce the probability of failure and enhance the yield and the reliable operation of a microprocessor. The concept and methods of error detection and correction are discussed, followed by a description of microarchitecture and logic design error detection and recovery techniques. The failure mechanisms of nanometer-class semiconductor VLSI circuits and commercial microprocessors using error detection and recovery techniques are presented.

Chapter 10 deals with the issue of microprocessor bandwidth. It discusses its effects on the performance of an individual processor as well its impact on the overall system. The current trends in system evolution are examined, and the implication of those trends on bandwidth is discussed. Finally, the chapter explores the technologies that are likely to emerge and satisfy those trends.

Chapter 11 explores common methods and circuit architectures used to transmit and receive data through off-chip links. It discusses the most prevalent of these techniques, focusing on the chip-to-chip communication topologies common for microprocessors, namely access to memory, processor-to-processor communication for parallel computing, and processor-to-chipset communication.

The final chapter summarizes the approaches presented in this book through an example of a system-on-chip (SOC) design where low power is imperative. The chapter is based on the processor core implementing SuperH™ architecture in a 130-nm CMOS process. The processor is suited for a wide range of usage in consumer, low-power digital appliances such as cellular phones, digital still/video cameras, and car navigation systems.

Finally we would like to thank the people who helped with this project, Mark de Jongh of Springer in particular, for his diligence in executing this project and patience and understanding when things did not go as expected. The students Milena Vratonjic, Mandeep Singh and Christophe Giacomotto provided invaluable help in reviewing the chapters.

Berkeley, California
Hillsboro, Oregon
March 30, 2006

Chapter 1

ULTRA-LOW-POWER PROCESSOR DESIGN

Christian Piguet
CSEM & EPFL

Abstract: Processor energy efficiency is a major issue in a majority of products; however, it is difficult to achieve it, as it is in contradiction with the main characteristic of processors, i.e. flexibility provided by embedded software. A given function implemented in random logic could consume 100–1000 times less energy than the same function implemented in a processor and corresponding embedded software. However, flexibility is more and more required, and ultra-low-power processors are mandatory. Only a few techniques have been widely used for the power consumption reduction of microcontrollers and DSP processors. This chapter will review these techniques, which are basically CPI (clocks per instruction) reduction, gated-clock mechanisms, optimal pipeline length, hardware accelerators, reconfigurable units and techniques to reduce leakage power. Several examples will be described in more details, such as some RISC 8-bit and 32-bit microcontrollers as well as some DSP cores. The latter are a good example of the necessary tradeoffs between flexibility and energy efficiency, as many random logic-based accelerators are very often used.

Key words: Ultra-low power; CPI; gated-clock; pipeline; accelerators; reconfigurable; leakage; microcontrollers; DSP cores.

1. Introduction

For any application, several embedded processors such as microcontrollers and DSP cores, as well as a large number of embedded memories and specialized peripheral circuits, are today integrated on the same die called systems on chip (SoC). The two main constraints are computation power and low-power design. In any portable application for the consumer market, low power is generally the most dramatic issue. Furthermore, the use of deep

Vojin G. Oklobdzija and Ram K. Krishnamurthy (eds.),
High-Performance Energy-Efficient Microprocessor Design, 1–30.

submicron technologies implies solving new problems, such as very low supply voltages, leakage, wire delays, networks on chip, signal input slopes, noise and crosstalk effects.

Memories generally consume most of the power. However, the principles to reduce power consumption of memories are basically well known: the memory has to be cut in small pieces and only one piece is addressed to fetch or to store data (cache, hierarchical, divided wordline and divided bitline). However, this chapter is focused on low-power processors even if memories are larger consumers. The principles to reduce power consumption of processors are more difficult: they have to deliver more and more computation power, they have to be totally flexible, and of very low power. These requirements are fundamentally contradictory; so basically, designers have to trade off these requirements. They have to choose the right processors with more or less flexibility to achieve the required goals, which are also application-dependent.

2. Energy Efficiency Versus Flexibility

Figure 1 shows that the flexibility [1], i.e. to use a general-purpose processor or a specialized DSP processor, has a large impact on the energy required to perform a given task compared to the execution of the same given task on dedicated hardware.

2.1. The Basic Choices

At the system level there are many important choices, as shown in Figure 1. Specialized architectures in random logic are obviously better in performances

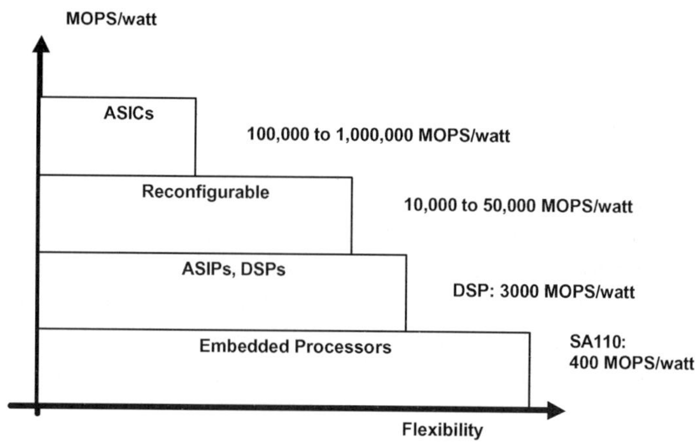

Figure 1. MOPS/watt versus flexibility.

(speed, power consumption) but are fixed without any flexibility. That is why today microprocessor-based architectures are generally selected; however, with reduced performances but much better flexibility.

It should be noted that a µP-based implementation results in very high sequencing. This is due to the µP structure that is based on reuse of the same operators and registers (ALU, accumulators). For instance, only one clock is necessary to increment a hardware counter. For its software counterpart the number of clock cycles is much higher, while executing several instructions with many clocks each. This simple example shows that the number of clock cycles executed for the same task can be very different depending on the architecture. For an electronic wristwatch 2000 instructions were necessary to update the time versus one clock for a random logic circuit.

Assuming applications presenting control tasks as well as DSP tasks, one can choose different architectures:

- a single microprocessor for control and DSP tasks
- a microcontroller with co-processors
- a microcontroller (8 or 32-bit) with a DSP processor
- configurable processors.

For the microprocessor-based approach the processor type is an important issue. As shown in Figure 1, there is a trade-off between performance and flexibility. This choice is obviously dependent on the application and required performances. Reuse is mandatory, and the designer has to choose IP cores for the microcontroller, DSP or co-processors. Configurable processors are quite interesting, as specialized instructions and execution units can be configured to the specified application.

2.2. Processor Types

There are several points to fulfill in order to save power. The first point is to adapt the data width of the processor to the required data. This results in a relatively increased sequencing to manage, for instance, 16-bit data on an 8-bit microcontroller. As shown in Table 1, for a 16-bit multiply, 30 instructions are required (add-shift algorithm) on a 16-bit processor, while 127 instructions are required on a 8-bit machine (double precision). However, a better architecture would be to have a hardware 16 $*$ 16 bit parallel-parallel multiplier with only one instruction to execute a multiplication.

Another point is to use the right processor for the right task. For control tasks do not use DSP processors; they are largely inefficient. Conversely, do not use 8-bit microcontrollers for DSP tasks! For instance, to perform a JPEG compression on an 8-bit microcontroller requires about 10 million executed instructions for a 256 $*$ 256 image (CoolRISC, 10 MHz, 10 MIPS, 1 second per

Table 1. Number of executed instructions in 8-bit microcontrollers

	No. of instructions		No. of instructions	
	CoolRISC 88: in the code	CoolRISC 88: executed	PIC 16C5x: in the code	PIC 16C5x: executed
8-bit multiply linear	30	30	35	37
8-bit multiply looped	14	56	16	71
16-bit multiply linear	127	127	240	233
16-bit multiply looped	31	170	33	333

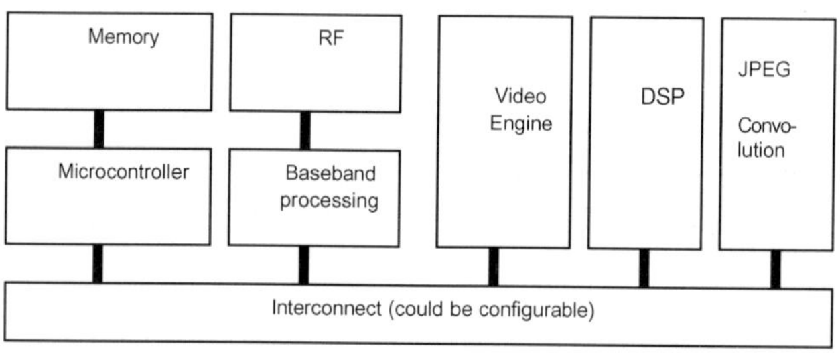

Figure 2. Heterogeneous architectures with many different processors.

image). It is very inefficient! Factor 100 in energy reduction can be achieved with JPEG dedicated hardware.

Figure 2 shows an interesting architecture to save power. For any applications there is some control that is performed by a microcontroller (the best machine to perform control). But in most applications there is also a main task to execute, that could be a DSP task, convolution, JPEG or another task. The best architecture is to design a specific machine (co-processor) to execute this task. So this task is executed by the smallest and most energy-efficient machine. Most of the time both microcontroller and co-processors are not running in parallel.

3. Power Reduction Techniques

This section will review power reduction techniques, which are basically CPI (clocks per instruction) reduction, gated-clock mechanisms, optimal pipeline length, hardware accelerators, reconfigurable units and techniques to reduce leakage power. At the architecture level most of the improvements in single processor performance are likely to come from lowering CPI.

3.1. Overview of Power Reduction Techniques

Future SoC will contain several different processor cores on a single chip. This results in parallel architectures, which are known to be less dynamically power-hungry than fully sequential architectures based on a single processor [2]. The design of such architectures has to start with very high-level models in languages such as System C, SDL or MATLAB. The very difficult task is then to translate such very high-level models in application software in C and in RTL languages (VHDL, Verilog) to be able to implement the system on several processors. One could think that many tasks running on many processors require a multitask but centralized operating system (OS), but regarding low power, it would be better to have tiny OS (2K or 4K instructions) for each processor [3], assuming that each processor executes several tasks. Obviously, the latter solution is easier to implement, even if performances could be reduced due to the inactivity of a processor that has nothing to do at a given time frame.

One has to note that most of the dynamic power can be saved at the highest levels. At the system level partition, activity, number of steps, simplicity, data representation and locality (cache or distributed memory instead of a centralized memory) have to be chosen (Figure 3). These choices at high level are, however, strongly application-dependent.

At the architecture level many low-power techniques aiming at dynamic power reduction have been proposed (Figure 3). The list could be: gated clocks,

	Dynamic Power			Static Power
High-level	Reduction of the number of executed tasks, steps and instructions. Processor types. Processor versus random logic. Reconfigurability			Remove units that do nothing or nearly nothing
Archi-tecture	Asynchro-nous Encoding	Parallel Pipeline	Simplicity	Architectures with less inactive gates
Circuit Layout	Gated Clock	Sub 1V. DVS Low V_T	Low-power Library and Basic cells	Gated Vdd, MTCMOS, VTCMOS, DTMOS stacked transistors
	Activity reduction	Vdd reduction	Capacitance reduction	

Figure 3. Power reduction techniques.

pipelining, parallelization, very low Vdd, several Vdd, variable Vdd (DVS or dynamic voltage scaling) and V_T, activity estimation and optimization, low-power libraries, reduced swing, asynchronous and adiabatic. Some are used in industry, but some are not, such as adiabatic and asynchronous techniques. At the lowest levels, for instance a low-power library, only a moderate factor (about 2) in dynamic power reduction can be reached. At the logic and layout level the choice of a mapping method to provide a netlist and the choice of a low-power library are crucial. At the physical level layout optimization and technology have to be chosen.

However, leakage or static power becomes more and more important in very deep submicron technologies, and low-power design methodologies have to take into account the total power, i.e. dynamic and static power [4].

3.2. Leakage Reduction at Architecture Level

Many techniques have been proposed to reduce leakage power, such as gated Vdd, multi-V_T technologies, DTMOS, VTCMOS and static or dynamic SATS [4]. These techniques are effective at the cost of more or less complex circuits or technologies. Another technique already proposed is weak inversion logic for which transistors work in a weak inversion regime [5]. Despite the use of these techniques, reducing static power in very deep submicron technologies has to be addressed at all design levels, including system and architecture levels. This section addresses this problem at architecture level.

3.2.1. Activity, logic depth and number of gates

By analyzing the ratio of dynamic over static power it is observable that microprocessors presenting a low or very low activity will present a too large static power compared to dynamic power. Performing a logic function with a very small activity can be considered as far from the optimum, as the microprocessor is idle most of the time. In other words, if a transistor or a logic gate is not switching for a very long period it is not very efficient, as the ratio between the switching time (related to dynamic power) and the idle time (for which the transistor or the logic gate is leaky) is very small. This is obviously the case when the microprocessor is in sleep mode, as no dynamic power is present. Hence the only consumed power is the static one. However, in that case nothing can be done at the architecture level and only circuit techniques [4] can be used to reduce leakage.

The presented design methodology aims at searching an optimum in which one has better use of switching transistors or gates in the reference period of time, i.e. an increased activity in such a way that dynamic power is not too small compared to static power. Obviously, this relative increase of activity has to be understood as useful, and not, for instance, by suppressing gated clocks or increasing glitches.

The conventional definition of activity is the factor a in the dynamic power formula: $P = a * f * C * Vdd^2$. This means that a is the ratio between the number of switching gates in a clock period over the total number of gates. Combinational circuits generally present activities around 1–5%. One has also to consider idle gates when they are connected in series. If there are 20 gates in series for a given pipeline stage or logic block (logic depth LD = 20), these 20 gates have to switch in series in a clock period. So only 1/20 of the clock period is used for switching; the rest of the time the considered gate is only a leaky gate. In running modes, with more and more leaky transistors, it seems that it could be better to avoid designing very inactive gates and therefore to search for an optimum ratio of dynamic power over static power. Leakage energy could be considered roughly proportional to the number of gates and to the duration of the clock period. It is not the case of the dynamic energy that is only proportional to the number of switching gates. To have a better balance of dynamic versus static energy, it could be mandatory to decrease significantly the total number of gates, generally resulting in increase of global activity.

3.2.2. Optimal total power

This optimum of total power (dynamic + static) is roughly obtained with similar amount of static and dynamic power [6, 7]. To reduce the total power consumption an attractive goal could be to have fewer, but more active, transistors or gates to perform the same logic function. If a given logic function requires 10,000 gates for its implementation and presents an activity factor of 1%, this means that on average 100 gates are switching in a clock period. If the same logic function could be implemented with only 1000 gates, keeping the same number of switching gates (100), the activity will be 10%, with the same dynamic power but with a leakage reduced by a factor of 10 due to the reduction of the total number of gates [7].

Figure 4 shows at the left such a theoretical situation with the same dynamic power and a significant reduced leakage power for the architecture containing fewer gates. This is however a naive situation, as nothing is said about the speed of the two architectures. The architecture with the smaller number of gates could be slower and would therefore require a larger Vdd to achieve the same speed, impacting the dynamic power.

Figure 4 (at the right) shows a practical situation comparing a 16 × 16 multiplier automatically synthesized (Synopsys) with two different architectures. Architecture A is realized with four RCA multipliers working in parallel (3250 gates) and the second architecture B is implemented with two Wallace

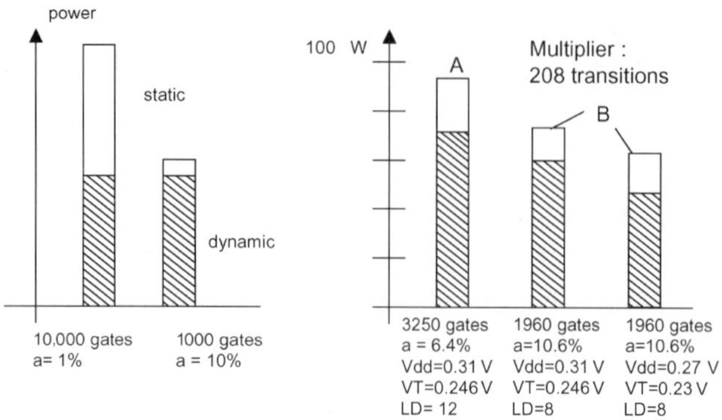

Figure 4. Various architectures with the same number of transitions for a given function (white rectangle: static; filled rectangle: dynamic).

multipliers working in parallel (1960 gates). Both present 208 transitions for executing a 16×16 multiply (3250 gates $* 6.4\% = 1960$ gates $* 10.6\% = 208$ transitions), showing that various architectures with a quite different number of gates can achieve their function with the same number of transitions. Architecture B (second bar, at the same Vdd and V_T) with fewer gates than A presents a larger activity, a smaller leakage but also a smaller dynamic power (smaller transistors, due to improved delay slack). However, at the optimum of total power consumption this architecture B can be supplied with smaller Vdd and V_T (third bar) to meet the same speed performances as A, with better results in total power consumption. This shows, however, that the speed performances are a very important parameter in such comparisons.

The goal of designing architectures with more global activity is not to reduce only leakage power with same dynamic power (Figure 4 at left), but to find an optimum of the total power, as shown at the right of Figure 4. This optimum has to be searched for the same speed constraints, and looking at first and last bars at the right of Figure 4, architecture B with less gates and more activity is better in terms of optimal total power; however, with a similar ratio of dynamic over static power of respectively 3.2 for A and 2.5 for B. This example shows very clearly that the reduction of the total number of gates and the increase of the activity could result in the same or even smaller dynamic power and optimal total power. It is the paradigm shift, not increasing the number of switching gates, but reducing the total number of gates. In this respect the resulting activity, although increased, can be considered as a useful activity. However, as already pointed out, the reduction in the total number of gates and the increase of activity cannot be a goal *per se*, as some architectures for a given logic function with significantly less gates could be extremely slow. So one has to dramatically increase the supply voltage and decrease the V_T

for satisfying the speed constraints, resulting in a much larger total power. It could be the case, for instance, for a sequential multiplier using an add-shift mechanism compared to a parallel multiplier.

3.2.3. Design methodologies based on optimal total power

In refs 7 and 10, it is shown that at optimal total power, the ratio Ion/Ioff of a single device in a given technology is proportional to K1*(LD/a), i.e. proportional to architectural parameters such as logical depth (LD) and activity (a). K1 is not fixed as shown in refs 8 and 9, but goes from 3 to 6 depending on the architecture. Dynamic over static power is also equal to K1, i.e. dynamic power is 3 to 6 times larger than static power at the optimum of the total power. It turns out that Ion/Ioff is quite low in very deep submicron technologies, showing that leakage tolerant architectures do have to present small LD and large activities. So pipelined architectures, reducing LD proportionally to the number of stages with the same activity, have to be preferred to parallel architectures that reduce LD and activity. In other words, parallel architectures do reduce LD, but the transistor count is also largely increased, having too much impact on the number of inactive gates and therefore the leakage.

Obviously, what is searched for is design methodologies [10, 11] starting from a given logic function architecture and generating a new architecture in which the number of cells is reduced, the number of transitions for this given logic function is constant, the activity is increased, but the logic depth is also constant (or even smaller). If the logic depth is increased, as is generally the case by increasing sequencing to reduce the gate count, the speed constraints have to be satisfied by increasing Vdd and going down with V_T, resulting in a larger optimal total power.

4. Low-power Microcontrollers

4.1. Eight-bit Embedded Microcontrollers

The most well-known 8-bit microcontrollers are Intel 8051-based (many versions), Motorola 68K, Microchip PIC, Zilog eZ80Acclaim! and many Japanese microcontrollers. Basically, all these 8-bit microcontroller architectures are based on old CISC (complex instruction set computer) architectures, with instruction formats containing several bytes. This means that several memory accesses are required to execute a single instruction, and it is impossible to execute an instruction in a single clock cycle. So this results in a CPI (clock per instruction) of 4–20. To provide good MIPS performances (million of instructions executed per second), according to the following formula: MIPS $= f/\text{CPI}$, they have to be clocked at relatively high f frequencies. As the dynamic power consumption is proportional to f, they present a figure MIPS/watt that is not reduced as it could be.

However, for power consumption or for energy consumed per instruction, RISC microcontrollers are much better. As the energy per clock cycle is roughly the same in RISC and CISC microcontrollers (about the same number of transistors), RISC-like microcontrollers with a single clock cycle per instruction provide a significant advantage regarding energy per instruction. Only two RISC 8-bit microcontrollers are commercially available:

- Xemics/Semtech-CSEM CoolRISC (www.coolrisc.com)
- Atmel AVR (www.atmel.com)

4.2. CoolRISC 8-bit Microcontroller

The CoolRISC is a three-stage pipelined core [12, 13]. The branch instruction is executed in only one clock. In that way no load or branch delay can occur in the CoolRISC core, resulting in a strictly CPI $= 1$ (Figure 5). For each instruction the first half clock is used to precharge the program memory. The instruction is read and decoded in the second half of the first clock. As shown in Figure 5, a branch instruction is also executed during the second half of this first clock, which is long enough to perform all the necessary transfers. For a load/store instruction, only the first half of the second clock is used to store data in the RAM memory. For an arithmetic instruction the first half of the second clock is used to read an operand in the RAM memory or in the register set, the second half of this second clock to perform the arithmetic operation and the first half of the third clock to store the result in the register set.

Furthermore, to reduce the energy per clock cycle, the gated clock technique has been extensively used in the design of the CoolRISC core (Figure 6). The ALU, for instance, has been designed with input and control registers that are loaded only when an ALU operation has to be executed. During the execution of another instruction (branch, load/store), these registers are not clocked, thus

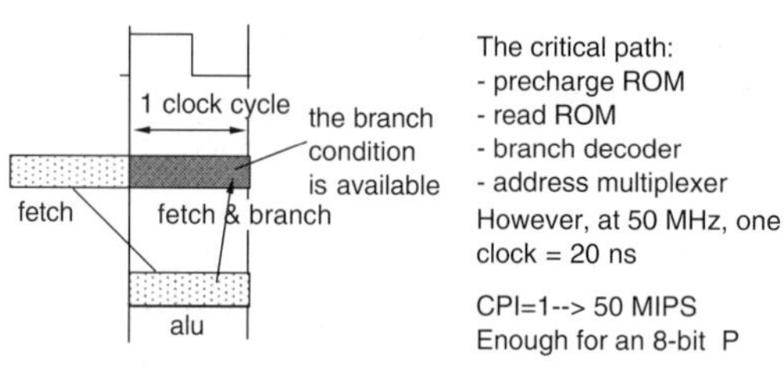

Figure 5. No branch delay.

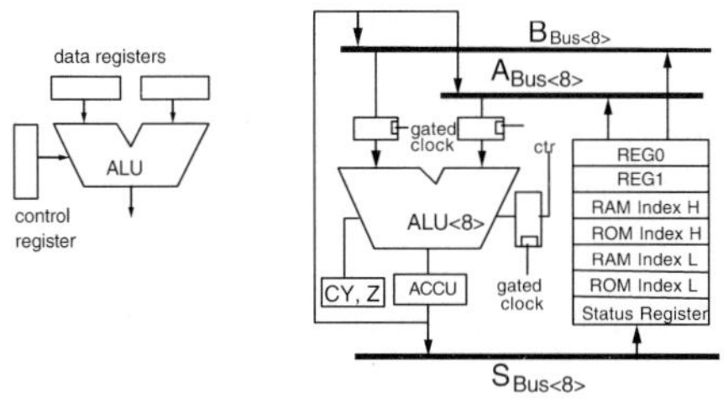

Figure 6. Gated-clock ALU.

no transition occurs in the ALU (Figure 6). A similar mechanism is used for the instruction register; thus, in a branch, which is executed only in the first pipeline stage, no transition occurs in the second and third stages of the pipeline. It is interesting to see that gated clocks can be advantageously combined with the pipeline architecture; the input and control registers implemented to obtain a gated clocked ALU are naturally used as pipelined registers. Automatic gated clocks insertion is proposed today in many papers, such as ref. 14 and CAD tools [15].

A main issue in the design of processor cores [16] is the design of the clock tree with the smallest possible clock skew and avoiding any timing violations. In deep submicron technologies this design becomes more and more difficult as wire delays are larger and larger compared to gates delays. Generally, clock trees do have very large buffers, resulting in a significant increase in power consumption. Today, most processor cores are based on a single-phase clock and are based on D-flip-flops. As clock input slopes are a key factor in D-flip-flop reliability, as clock input capacitances of D-flip-flops in standard cell libraries are significant, as clock skew is an issue, the design of clock trees becomes more and more difficult in deep submicron technologies.

Another approach than the conventional single-phase clock with D-flip-flops (DFF) has been shown to be more effective. This is based on using only latches with two non-overlapping clocks. This clocking scheme has been used for the 8-bit CoolRISC microcontroller [17] as well as for other DSP cores such as ref. 18. Figure 7 shows this latch-based clocking scheme that has been chosen to be more robust to clock skew, flip-flop failures and timing problems at very low voltage [17]. The clock skew of Ø1 (respectively Ø2) has to be shorter than half a period of CK, i.e. Ø1 can be delayed for half a period of CK before Ø1 and Ø2 become active at the same time. However, this clocking scheme requires two clock cycles of the master clock CK to execute a single instruction. This means 100 MHz is necessary to generate 50 MIPS (CoolRISC with CPI = 1), but the two Øi clocks and the two clock trees are at 50 MHz. Only a very small logic block is clocked at 100 MHz to generate two 50 MHz clocks.

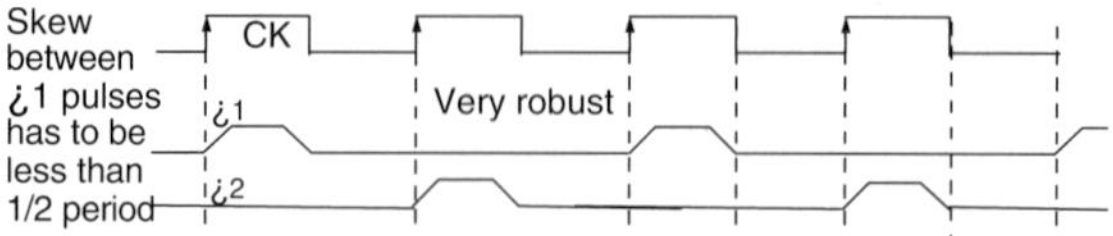

Figure 7. Latch-based clocking schemes.

The design methodology using latches and two non-overlapping clocks has many advantages over the use of D-flip-flop methodology. Due to the non-overlapping of the clocks, and the additional time barrier caused by having two latches in a loop instead of one D-flip-flop, latch-based designs support greater clock skew before failing than a similar D-flip-flop design (each targeting the same MIPS). This allows the synthesizer and router to use smaller clock buffers and to simplify the two clock trees generation, which will reduce the power consumption of the two clock trees. Despite the fact that this clocking scheme requires two different clock trees, each of them sees a much smaller clock capacitance, and due to clock skew relaxed constraints it is likely that the total power consumption is smaller that the power of a single clock tree using D-flip-flops.

Furthermore, if the chip has clock skew problems at the targeted frequency after integration, it is possible with a latch-based design to reduce clock frequency. The result is that the clock skew problem will disappear, allowing the designer to test the chip functionality and eventually to detect other bugs or to validate the design functionality. Another advantage with a latch design is time borrowing [17]. The latch design has additional time barriers, which stop the transitions and avoid unneeded propagation of signal and thus reduce power consumption (Figure 8). Using latches can also reduce the number of MOS in a design. For example, a microcontroller has 16 $*$ 32-bits registers, i.e. 512 DFF or 13,312 MOS (using DFF with 26 MOS). With latches the master part of the registers can be common for all the registers, which gives 544 latches or 6528 MOS (using latches with 12 MOS). In this example the register area is reduced by a factor of 2.

Using latches for pipeline registers significantly reduces power consumption when using such a scheme in conjunction with clock gating. The clock gating of each stage (latch register) of the pipeline with individual enable signals, reduces the number of transitions in the design compared to the equivalent DFF design, where each DFF is equal to two latches clocked and gated together.

The latch-based design allows a very natural and safe clock gating methodology. Figure 8 shows a simple and safe way of generating enable signals for clock gating. This method gives glitch-free clock signals without the adding of memory elements, as is needed with DFF clock gating. Synopsys handles very nicely the proposed latch-based design methodology. It performs nicely

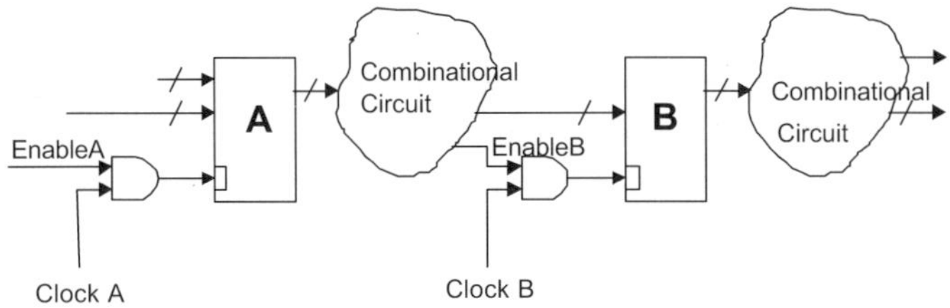

Figure 8. Latch-based clock gating.

Figure 9. CoolRISC core and SRAM memories in 0.18 μm technology.

the time borrowing and correctly analyzes the clocks for speed optimization. So it is possible to use this design methodology with Synopsys, although there are a few points of discussion linked with the clock gating.

This clock gating methodology cannot be inserted automatically by Synopsys. The designer has to write the description of the clock gating in his VHDL code. This statement can be generalized to all designs using the above latch-based design methodology. Synopsys can do automatic clock gating for a pure double-latch design (in which there is no combinatorial logic between the master and slave latch), but such a design causes a loss of speed over a similar DFF design. The most critical problem is to prevent the synthesizer from optimizing the clock gating AND gate with the rest of the combinatorial logic. To ensure a glitch-free clock this AND gate has to be placed as shown in Figure 8. This can be easily done manually by the designer by placing these AND gates in a separate level of hierarchy of his design or placing a "don't touch" attribute on them.

It is interesting to note that the CoolRISC core in a quite old TSMC 0.25 μm consumes about 10 μW/MIPS, resulting in 100,000 MIPS/W at 1.05 V. This is roughly better for a factor of 2 over a similar design using D-flip-flops [17]. Figure 9 shows a CoolRISC core with SRAM memories embedded in a SoC chip integrated in 0.18 μm TSMC technology.

4.3. 32-bit Microcontrollers

Embedded 32-bit RISC microcontrollers are used in portable applications for which low power consumption is an issue [19]. These 32-bit cores are quite small, from 30,000 MOS to about 100,000 MOS. They are very power-efficient. The main characteristics are as follows:

- load/store architectures (RISC);
- instructions of 32 bits (ARM, MIPS, StrongArm) or 16 bits (SH, Thumb, TinyRISC);
- 32 registers (ARM, MIPS) or 16 registers (SH);
- small number of generic instructions;
- single pipeline, with a peak CPI = 1;
- the number of pipeline stages was generally three for the first 32-bit cores (ARM7, Thumb, MiniRISC, TinyRISC) some years ago;
- for most recent 32-bit cores the number of pipeline stages has been significantly increased to achieve larger frequencies, from five (MIPS R3000, Hitachi SH, StrongArm, X-Scale, ARM9) to six pipeline stages (ARM10) or even eight pipeline stages (MIPS32 24K).

The basic principle of a pipelined μP is to execute N-cycles instructions in an overlapped fashion in a N-stages pipeline. At each cycle an instruction is provided to the pipeline, and an instruction is completed. This results in the execution of one instruction per cycle. However, there are some drawbacks, such as pipeline stalls. Branch hazards occur when the target instruction after a branch is determined in the second, third or xth stage of the pipeline. This can be solved by the insertion of load delay or branch delay (reducing the throughput) or by using a bypass technique: the preceding instruction is able to transfer its result directly to the next instruction if needed.

The resolution of pipeline hazards can be achieved while using several different approaches:

- Static approach, for which it is the compiler that is responsible for re-organizing the code and/or inserting NOP instructions. Generally the code size is increased [20, 22].
- Dynamic approach, for which the processor hardware is in charge to solve the pipeline bubbles at run-time. Generally load and branch delays are inserted, but code could also be reorganized dynamically, resulting in out-of-order execution [20, 22].
- Pipelined multithreaded architecture.
- Short pipeline with branch instructions executed in one clock (Cool-RISC [12]).

Prediction techniques can be used to guess the target instruction and to start to fetch (sometimes to execute) it before the branch is resolved. Obviously, in case of misprediction, the processor has to go back. The cost of a branch and of mispredicted branches becomes so high in terms of clock cycles that prediction techniques become very sophisticated. A given branch instruction is generally not 50% taken and 50% not taken. More than 90% of branch instructions are taken (or not taken) at 90% to 99%. Prediction techniques are based on this fact, i.e. based on the prediction that the branch will do what it did last time [21]. There are several techniques to implement a prediction mechanism. For instance, the first ARM having prediction is the the ARM10 (static prediction: backward taken, forward not taken).

The most well-known 32-bit microcontrollers are as follows: ARM, Thumb, Atmel AT91, StrongArm, X-Scale, MIPS, ColdFire, Hitachi SH SuperH, Embedded X86, Transmetta Crusoe, XAP3, NECVR4 and many others. Just to illustrate the energy efficiency of some 32-bit RISC cores, let us see some well-known cores. ARM7 was a three-stage pipeline core. ARM8 and ARM9 have been designed while extending the pipeline to five stages with no prediction (all taken) and branches executed in two clock delays. The ARM10 has six stages with a static prediction. With this prediction logic the CPI will be around 1.5. ARM11 has eight pipeline stages. In some deep submicron technologies one has the following numbers:

- In 0.18 μm and 1.8 V, ARM9 core achieves 330 MHz, 365 MIPS, 120 mW, 3000 MIPS/watt.

- In 0.13 μm ARM1026 with caches achieves 475 MHz, 500 MIPS, 250 mW, 2000 MIPS/watt.
- The ARM11 processor family delivers up to 550 MHz of performance with power as low as 0.24 mW/MHz using a 0.13 μm foundry process, so 4000 MIPS/watt.

XScale is Intel's implementation of the fifth generation of the ARM architecture. It is the successor to the Intel StrongARM microcontrollers. The number of pipeline stages was extended to seven or eight stages. The most interesting feature of the XScale architecture is what Intel calls "Dynamic Voltage Scaling". In 2001, XScale was dissipating 450 mW at 600 MHz and 1.3 V and much less at reduced supply voltage [23]. In 0.18 μm and 1.3 V, XScale with caches achieves 600 MHz, 660 MIPS, 750 mW, 880 MIPS/watt.

MIPS microcontrollers have been designed following the original MIPS and MIPS-X of Stanford [24]. These first RISC machines were based on the "delayed branch" mechanism that was handled by the compiler (called Reorganizer). In the MIPS design the instruction following a branch is always executed. The compiler is responsible for filling this delay slot with a useful instruction or, if it is not possible, by a NOP. The original pipeline provided five stages. The number of transistors was about 115,000. Performances today are as follows:

- MIPS32 4 Kp: 1 mW/MHz or 1 mW/MIPS (1000 MIPS/watt).
- MIPS32-R2 4KEc Pro: 255-300 MHz, five pipe stages, 0.12–0.37 mW/MHz (8500 MIPS/watt for 255 MHz and 0.12 mW/MHz) [25].
- MIPS32-R2 24K Pro: 400–550 MHz, eight pipe stages, 0.5 mW/MHz [25] (2000 MIPS/watt).

As the pipeline becomes deeper and deeper due to increased frequencies, branch prediction is used for MIPS, such as the MIPS32 24K with eight-stage, dynamic prediction (four clocks penalty in case of mispredicted branch).

Adding DSP instructions to a RISC core was one of the first ideas to provide flexible cores for both control and DSP tasks (see Section 5). Many companies are offering such combinations, such as ARM, ARC and Tensilica. However, while provided DSP extensions are meaningful, the resulting performances are impacted. Generally, the resulting core is larger, maximum frequency is reduced of 30% and power consumption is increased. Control as well as DSP tasks are not energy efficiently executed on these cores. If some companies like MIPS claim no performance degradation (MIPS32 24KE), it means that the DSP extensions are very limited or that MAC-like operations are executed in two clock cycles [26].

There is a clear trend for multicore and multithreaded architecture for high-performance microprocessors [27]. For embedded microcontrollers into SoC, as applications are more specific, generally the multicore approach is implemented with different processors including DSP cores and co-processors. A variety of chip vendors have recently announced multicore SoC designs, including PMC-Sierra (MIPS), Broadcom (MIPS), RMI Raza Microelectronics Inc. (MIPS), Freescale (PPC), Cavium (MIPS), and ARM Ltd. (ARM). Multicore advantages include a better ratio of performance to power usage, less heat dissipation, and a smaller physical footprint. One prime market for multicore SoC appears to be networking equipment such as firewalls that deeply inspect packets or perform compute-intensive spam filtering.

Most microprocessors are single-threaded computers, i.e. they are executing a single flow of instructions stored in a program. This means that the dependency between sequential instructions is quite high. For multi-threaded computers or multiple-context processors, several independent tasks are executed at the same time by the processor. Instructions of different tasks are completely independent. So pipeline bubbles as well as limited parallel execution could be avoided, providing excellent throughput. A multi-threaded scheme is generally used for very-high performance microprocessors, but it has also been used for small, 8-bit microcontrollers like the low-power CSEM Punch [28]. The main problem in multi-threaded computers is the number of active tasks. The processor is fully utilized only if all the tasks are active. However, if only one task is active the throughput is drastically reduced. SUN is also designing multi-threaded microprocessor architectures called Niagara [29]. So all branch and load delays are removed, all the very sophisticated prediction mechanisms about branch instructions are removed, this "new" microprocessor seems so simple! But the idea is quite old.

5. Low-power DSP Architectures

DSP microprocessors are being used more and more in high-volume applications, such as portable phones, hard-disk drives, audio and image processing, hearing aids, wired and wireless communication devices. DSP processors are different from RISC processors: they are dedicated for executing arithmetic operations and have to be very energy-efficient in executing DSP algorithms [30].

5.1. Main Features of DSP Architectures

DSP architectures are specialized in the execution of digital signal processing (DSP) algorithms. The basic DSP operation is the multiply-accumulate (MAC), i.e. a sum of multiplications used, for instance, in digital filters, correlation and Fast Fourier transforms. The goal is to execute the MAC in one clock ($CPI = 1$) while using a pipeline. The accumulator provides extra bits to accommodate growth of the accumulated result (for 16-bit data the accumulators are 40 bits, i.e. 32-bit multiplication result + 8 extra bits for the accumulation).

Another main feature of a DSP processor is to complete several memory accesses in a single clock. Simultaneously, instruction fetch from the program memory, as well as two operands fetch from two data memories and one result store, have to be performed in a single clock. DSP architectures are either load/store (RISC) architectures (input registers to the datapath) or memory-based architectures in which the operands are directly fetched from the data memories to be processed. However, load/store architectures are capable of executing in parallel an arithmetic operation and register-memory transfers (to fetch operands for the next arithmetic operation).

The third basic DSP feature is specialized addressing modes. Both data RAM are addressed through two banks of pointers with pre- or post-incrementation/decrementation as well as circular addressing (modulo). These addressing modes provide efficient management of arrays of data, to which a repetitive algorithm is applied. These operations are performed in a specialized address generation unit.

The fourth basic feature is the capability to perform efficiently loops with zero overhead. Loop or repeat instructions are able to repeat 1 to N instructions without loop counter (no need to initialize and to up-date a loop counter). These instructions are fetched from the memory and stored in a small cache memory in which they are read during the execution of the loop.

All these features have been introduced in DSP cores to increase performances, mainly to reduce the number of clock cycles to execute some tasks. Compared to microcontrollers, what is executed in one clock cycle in a DSP core could take 10–20 clock cycles in microcontrollers.

5.2. Single MAC DSP Cores

Figure 10 shows a typical DSP datapath (in this case, 24-bit Motorola 56,000) producing a single MAC per clock cycle. It contains two data buses that are connected to four 24-bit input registers. It is therefore a load/store architecture in which operands have to be loaded into these few input registers before processing. These registers contain two operands that can be multiplied and accumulated in two 56-bit accumulators with 8 guard bits. The guard bits allow the DSP to perform a series of MAC operations without arithmetic overflow. The extra bits in the accumulators are capable of accommodating the bit growth resulting from the repeated additions. Memory-register transfers can be executed in parallel with an arithmetic operation, for instance:

$$A \leftarrow X0 * Y0, X0 \leftarrow RAM(R0), RAM(R1) \leftarrow Y1;$$

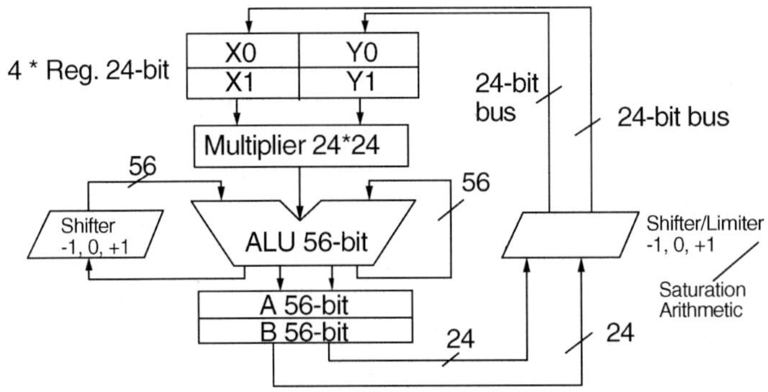

Figure 10. Typical single MAC DSP datapath.

Shifters are used to scale the operands or the results to avoid overflows. In Figure 10 there are two shifters with limited capabilities. One is used in the MAC and the other is used to scale the results that have to be stored in the data memory. Overflows are generally solved by using saturation arithmetic, i.e. an overflow value is replaced by the largest number that can be represented. It is called a limiter in Figure 10.

Many single MAC DSP cores are still offered on the market, such as Ceva-TeakLite [31] running at 200 MHz in 0.13 μm and consuming, for instance, 0.1 mA/MHz in an audio platform dedicated to MP3 stereo decoding.

5.3. RISC Cores with DSP Enhancements

As mentioned above, adding DSP instructions to a RISC core was one of the first ideas to provide flexible cores for both control and DSP tasks. Many companies have designed such cores. There are two options: adding these instructions definitively, such as ARM, MIPS and SH3-DSP, or giving the possibility to customers to add the desired instructions (customizable ARC, Hyperstone and Tensilica).

Cores having control and DSP instructions result in lower performances, as the cores are 30% bigger, maximum frequency reduced by 30% and power consumption increased. Neither the control nor the DSP tasks are really energy-efficiently executed on these cores. If some companies claim no performance degradation, it means that the DSP extensions are very limited [26]. For instance, in ARM DSP extensions there are no X/Y memories, no zero overhead looping or no special addressing modes [32].

ARC is offering synthesizable soft macros that can be configured to meet specific performances. In a four-stage pipeline the instruction set contains 32 separate instructions; the first 16 are predefined and the last 16 are available for customers. They may be instructions from an ARC library or completely new instructions defined by customers. This is a much better approach than the previous one as the customer is free to add only the required DSP instructions for a given task. ARC 605 is a five-stage pipeline with static branch prediction. The 605 is the smallest one [33] in 0.13 μm, 0.36 mm^2, 250 MHz maximum frequency and 0.06 mW/MHz. This means 15 mW at 250 MHz, 250 MIPS, so 17,000 MIPS/W. Tensilica Xtensa proposed an instruction set containing 78 basic immutable commands. Customers are also free to add their special-purpose instructions by defining them using the provided hardware RTL language. Not only new instructions can be specified, but also I/O ports can be modified depending on the memories used, as well as the pipeline number of stages [34]. The RTL code includes gated clocks for powering down various blocks and can be synthesized with any library. The process of synthesis could be as short as 8 hours. Tensilica Xtensa LX has five to seven pipeline stages and occupies 0.2 mm^2 in 0.13 μm (19,000–25,000 gates). It consumes 0.038 mW/MHz, or 15 mW at 390 MHz [33]. This means 390 MIPS, so 26,000 MIPS/W.

5.4. More Computation Power Required

Various DSP architectures can be and have been proposed to reduce power consumption significantly while keeping the largest throughput. Furthermore, many portable applications require a significantly increased throughput, due to new, quite sophisticated DSP algorithms and to wireless communication. Beyond the single MAC DSP core of 5–10 years ago, it is well known that parallel architectures with several MAC working in parallel allow designers to reduce supply voltage and power consumption at the same throughput. That is why many VLIW or multitask DSP architectures have been proposed and used, even for hearing aids. The key parameter to benchmark these architectures is the number of simple operations executed per clock cycle, up to 50 or more. Another key point is the design of specialized execution units, such as Viterbi and Turbo code. These dedicated execution units in DSP are mandatory to speed up these algorithms while significantly reducing power consumption [35].

However, there are some drawbacks regarding very parallel architectures such as VLIW or superscalar DSP processors. The very large instruction words of VLIW of up to 256 bits significantly increase the energy per memory access. Some instructions in the set are still missing for new, better algorithms. Finally the growing core complexity and transistor count (roughly 2 million transistors for the cores in a hearing-aid circuit) become a problem because leakage is roughly proportional to transistor count.

To be significantly more energy-efficient there are basically two ways however, impacting either flexibility or the ease of programming. The first one is to use reconfigurable DSP cores in which configuration bits allow the user to modify the hardware in such a way that it can fit much better to the executed algorithms. The second one is to have specific and very small DSP hardware accelerators for each task (Figure 2). In that way each DSP task is executed in the most energy-efficient way on the smallest piece of hardware.

5.5. VLIW and Superscalar DSP Cores

Texas Instruments provides an eight-issue VLIW 16-bit fixed-point TMS320C6x with two execution units including two MAC and six ALU units. As many as eight instructions can be executed in parallel (peak) with 32-bit data. All instructions can be conditional, thus eliminating branches and pipeline delays. In 2005, in 90 nm, the chip reaches 1000 MHz (so 8000 MIPS peak) and contains some hardware accelerators (Viterbi, Turbo). The TMS320C6414T consumes 673 mW at 600 MHz and 1.1V (7000 MOPS/W). But this peak performance can never be reached, as only three or four instructions over eight are executed in a single clock on average. Another drawback is the energy which is required to fetch VLIW 256-bit words into the program memory. It is significantly more than 32-bit instruction words of superscalar DSP cores. StarCore SC 140 is also a VLIW DSP core with 128-bit instructions.

Superscalar DSP cores contain parallel execution units (dual MAC or quad MAC) with a 32-bit instruction set. Energy savings occur due to the small 32-bit instruction word fetched from the memory. Ceva-X offers a scalable architecture, i.e. dual MAC, four and even eight MAC architectures. Many new cores are dual cores, i.e. a limited parallelism also due to the small instruction width. These cores are, for instance, offered by 3DSP, LSI Logic and Philips [36].

Philips has introduced the dual MAC CoolFLUX DSP, 43,000 gates, 0.1 mW/MHz at 1.2 V, 0.13 μm, 175 MHz, about 1000 MOPS. This gives about 57,000 MOPS/W. The SP3 of 3DSP is also a dual MAC DSP core five pipeline stages, 0.13 μm, maximum frequency 320 MHz, dissipating 8 mW at 1.0 V at reduced frequency. The ZSP400 of LSI LOGIC is a low-power dual MAC DSP core consuming about 0.1 mW/MHz in 0.13 μm running at 225 MHz, achieving about 900 MOPS (40,000 MOPS/W). Table 2 shows the number of clock cycles necessary for a FIR filter and a 256 point FFT.

Table 2. Number of clock cycles for a FIR and FFT256

Algorithm	MACGIC 4MAC	CoolFlux	SP-3	LSI403LP
FIR	N/4	N/2	N/2	N/2
FFT 256	1500	5500	n/a	5000

5.6. Reconfigurable DSP Cores

As mentioned above, reconfigurable DSP architectures are between power-hungry FPGA (reconfigurable at the bit-level) and programmable DSP processors (not at all reconfigurable). So reconfigurable DSP are reconfigurable at the functional level, i.e. execution units (Figure 11) and interconnection networks are reconfigurable. It is interesting to note [37] that an energy-efficient architecture is the one in which the main sources of power dissipation are operators. In a general-purpose processor there is a waste of power due to load/store, branch, prediction, etc. This is even more dramatic in FPGA in which the main source (and waste) of power is due to interconnect (60–70%).

Reconfigurable DSP architectures are much more power-efficient than FPGA. The key point is to reconfigure only a limited number of units within the DSP core, such as some execution units and addressing units [38]. The latter are interesting, as the operands fetch from memory is generally a severe bottleneck in parallel machines for which 8–16 operands are required each clock cycle. So sophisticated addressing modes can be dynamically reconfigured depending on the DSP task to be executed. Figure 12 shows an example in which several addressing modes can be reconfigured depending on the user's algorithms.

An interesting reconfigurable architecture is DART cluster [37], in which configuration bits allow the user to modify the hardware in such a way that it can much better fit to the executed algorithms. Figure 11 shows an example. The DART architecture has been used for a WCDMA receiver in 0.18 μm [37]. The power consumption was at 79% in the operators (15% in a general-purpose microprocessor [39]). The energy efficiency was 39,000 MOPS/W. Compared to FPGA (Virtex) achieving for the same application 3000 MOPS/W, and to

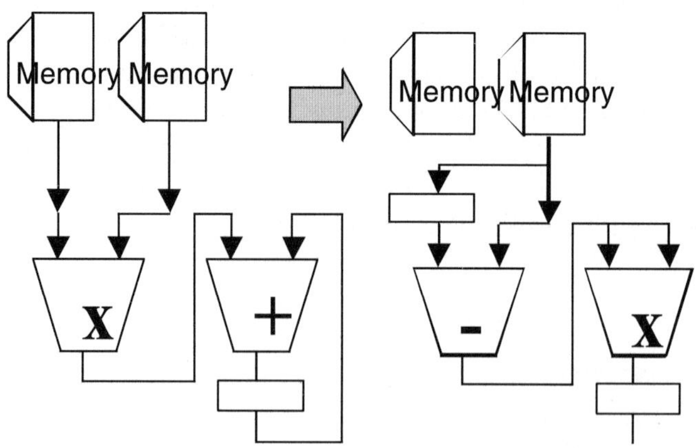

Figure 11. Hardware reconfiguration example.

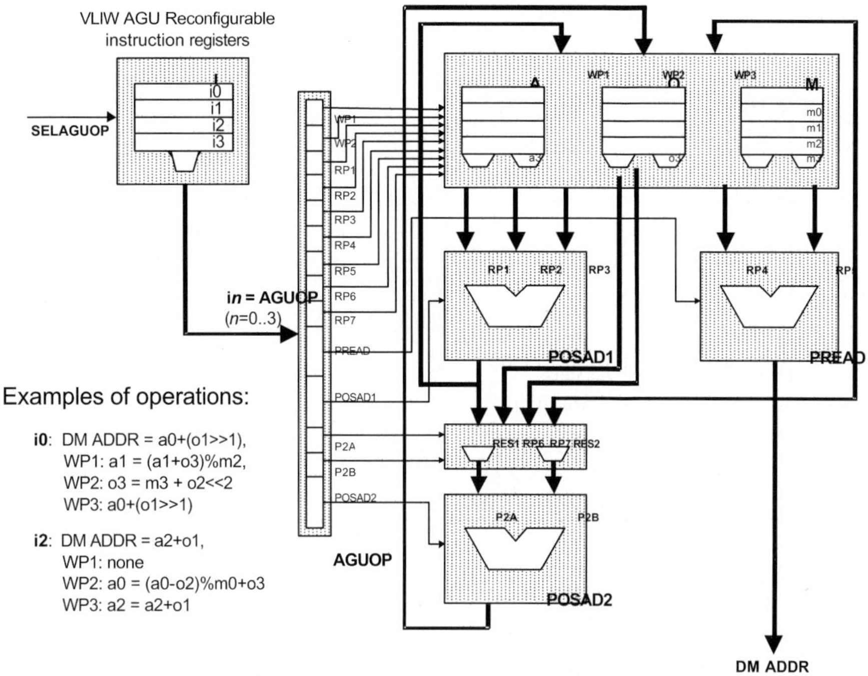

Figure 12. Addressing modes reconfiguration example (MACGIC DSP).

the TMSC64x achieving 2000 MOPS/W, the DART reconfiguration platform outperforms these two FPGA and conventional DSP by a factor of 13–19.

However, generally speaking the power consumption of reconfigurable hardware is necessarily increased due to the relatively large number of reconfiguration bits that have to be loaded in the configuration registers. Similarly, the reconfigurable units are necessarily more complex that non-reconfigurable units in terms of transistor count and therefore consume more power. Software issues are also difficult, as users can define new instructions or new addressing modes that are difficult to support by the development tools [40].

5.7. MACGIC Reconfigurable DSP Core

The MACGIC DSP is implemented as a customizable, synthesizable, VHDL software intellectual property (IP) core [38]. The core is assumed to be used in systems on chip (SoC), either as a stand-alone DSP or as a co-processor for any general-purpose microprocessor.

The DSP microprocessor is made of four units (Figure 13):

- Program sequencing unit (PSU).

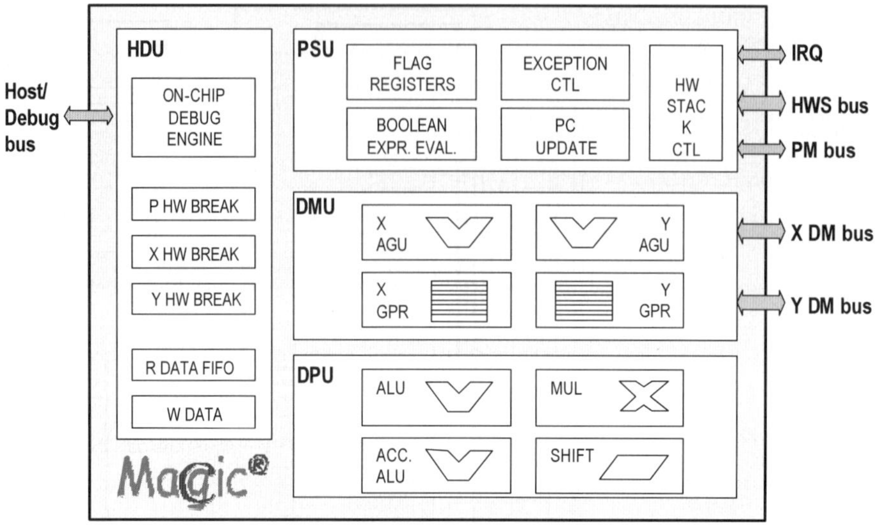

Figure 13. MACGIC architecture.

- Data move unit (DMU).
- Data processing unit (DPU).
- Host and debug unit (HDU).

The PSU handles the fetching of instructions as well as branches, subroutine calls, hardware loops, instruction repeats and external interrupts requests. The PSU dispatches operations to the DMU and to the DPU. The DMU is responsible for the handling of the data moves among DSP registers, and between data memories and DMU's general-purpose registers. The DMU contains customizable address generation units (AGU) that allow the DSP core to best fit the algorithm needs in term of data memory address computation (described in detail in the previous paragraph). The DMU is a pure load/store unit that implements simultaneous accesses to the two data memories. Up to eight data words can be transferred between the memory and the DMU during each clock cycle.

The DPU implements the signal processing operations (such as add, subtract, multiply and accumulate instructions). It can be customized to best match the class of algorithms and data to be processed. As an example, the FFT butterfly kernel is entirely implemented in hardware and provides a set of specific instructions to the software developer. DPU operations read their operands from the DMU's general-purpose registers. The wide data bandwidth available between these units allows the parallelization of many algorithms (such as: FIR, IIR, FFT, DCT, FEC, etc.). Up to 16 data words can be read from the DMU registers and up to eight X and Y data memory words can be written back to the DMU during each clock cycle.

Two versions of the MACGIC DSP core have been designed. The first one is a simple core containing a single MAC unit. The second version is a large core with four parallel multipliers and six adders. There are four categories of DPU instructions:

- Standard: MAC, MUL, ADD, CMP, MAX, AND, ...
- SIMD (Single Instr. Multiple Data): MAC4, MUL4, ... capable of performing 4 independent MAC, ... in parallel.
- Vectorized: MACV, ADDV, ... capable of performing, for instance, a MAC with 4 operands providing a single result.
- Specialized operations, targeted to specific algorithms, such as FFT, DCT, IIR, FIR, Huffmann, Viterbi, mainly "butterfly" operations.

Figure 12 shows the AGU (address generation unit). There are three classes of addressing modes, the last two being reconfigurable:

- Seven basic addressing modes (fixed): indirect, ± 1, $\pm$ offset, modulo.
- Predefined addressing modes capable of combining three operations over 48 available, coded in the Cx configuration register associated to each Ax index register, for instance the complex mode: an $<$-(an $+$ on)%mn $+$ OFFA.
- Extended addressing modes, four operations coded in the Ix configuration register, for instance, an $<$-(an $+$ om)%mp $+ 2 *$ mq (very complex, impossible to find in other DSP).

The HDU allows the use of this DSP as a co-processor for any general-purpose microprocessor. This interface allows full control of the DSP from the host processor. The HDU's debug logic helps the software designer during the software development phase by allowing the placement of breakpoints and the step-by-step execution of the DSP program.

The clock frequency for the DSP is 50 MHz in a 0.18 μm TSMC technology (slow), at 1.8 V, for a 24-bit data word size. It is 100 MHz in a 90 nm process. A FIR filter needs $\frac{1}{4}$ clock cycle per tap, an IIR biquad one clock cycle per stage and FTT radix 4 size 64 needs 222 clock cycles. The last result is similar to large VLIW DSP cores with very large instructions of 128–256 bits, while the MACGIC DSP core has 32-bit instructions. This provides a significant advantage in program memory energy accesses. The estimated number of transistors for a 24-bit powerful version is 750,000 while it is around 500,000 for a 16-bit core. Table 3 shows some performances in energy and MOPS/W for some instructions. One should note that performances are much better if the parallelism of the DSP core is used. It is useless to use a single adder or multiplier (ADD, MAC) in a four adder/multiplier datapath unit. Figure 14 shows the TSMC 0.18 μm test chip of MACGIC in July 2005; it was running nicely.

Table 3. Power consumption for a 24-bit, 8 MHz synthesis at 0.9 V, 0.18 μm TSMC "G" (generic process).

4-MAC 16-bit MACGIC	8 MHz synthesis, 0.9 V	TSMC 0.18 "G"
NOP	27 μA/MHz	—
ADD 16-bit (7 op)	98 μA/MHz	79,000 MOPS/W
MAC 16-bit/40-bit (9 op)	114 μA/MHz	88,000 MOPS/W
4 * ADD 16-bit (16 op)	155 μA/MHz	115,000 MOPS/W
4 * MAC 16-bit/40-bit (24 op)	212 μA/MHz	126,000 MOPS/W
CBFY4 radix-4 FFT (31 op)	205 μA/MHz	166,000 MOPS/W

Figure 14. MACGIC test chip in TSMC 0.18 μm.

5.8. DSP Multicore Architecture

There is a clear trend for multicore architecture for high-performance microprocessors. In this case the architecture is homogeneous, i.e. several identical cores are integrated on the same chip. This trend is also significant in embedded SoC, in which several cores are integrated on the same chip; however, with a completely different heterogeneous architecture: the various cores are completely different (microcontroller, DSP core, co-processor and accelerator).

For instance, regarding heterogeneous architectures, TI offers the OMAP2 platform which integrates an ARM11 processor with a TMS320C55x DSP and two other hardware accelerators. There are many other SoC having a microcontroller core, a DSP core and accelerators, such as Freescale Jupiter MXC275 (ARM11 and StarCore), Intel Bulverde PXA270 (XScale and

WMMX extensions), Sony PSP (MIPS and proprietary DSP core) and finally Zoran Panda ER4225 (ARM9 and only accelerators) [41]. Another ARM-based architecture is the OptimoDE acquired from Adelante. The OptimoDE is a VLIW DSP engine [42]; it can also be used as a DSP core controlled by an ARM microcontroller. Finally, many IP core vendors propose such combinations based on microcontroller, DSP core and hardware accelerators.

It is however not so clear if arrays of identical DSP cores form a valid trend for DSP. During recent years much of research has been performed on arrays of identical parallel DSP processors, but no commercial chip resulted. It is an open question whether this trend for embedded processors will also occur for DSP. Arrays of identical DSP cores could be an answer to leakage increase. Pushing too high frequencies implies lowering the V_T, resulting in increased leakage power. So arrays of identical DSP cores with high V_T could provide the same computation power with less leakage, provided that increased transistor count will not impact too much leakage. Some recently announced chips do have hundreds of identical DSP cores, such as the PicoChip 102 with 344 processors, most of them capable of executing MAC operations. It is a massive parallelism running at low frequencies (160 MHz) in 0.13 μm but delivering a huge power computation [43].

5.9. A set of DSP Co-processors

For ultra-low-power applications, the architecture consisting of a small DSP and co-processors is the most natural architecture, allowing programming the DSP and putting heavy tasks in dedicated hardware co-processors (Figure 2). Each DSP task uses the minimal number of transistors and transitions to perform its work. The control code unavoidable in every application is also efficiently executed on the microcontroller or on the simple DSP, and some unexpected DSP tasks can be executed on the simple DSP if no accelerator is available. It is certainly very efficient in terms of energy. An example of such architecture is Zoran Panda ER4225: controlled by an ARM9, all DSP functions are accomplished via hardware cores [41]. However, such architectures present some drawbacks:

- How to perform the software mapping of a given application onto so many heterogeneous processors and co-processors.
- What the development tools are, how the software and hardware designers cooperate, how they use a "programming" language for both DSP and co-processors.
- It could be possible to have more hardware if several co-processors use multipliers which are repeated in these co-processors and not shared by a single processor.

- There is some memory issue as to how DSP and co-processors share memories.
- Regarding leakage, unused engines have to be cut off from the supply voltages, resulting in complex procedures to start/stop them.
- The time to market of a new chip is longer, as the decision to add some co-processors is taken at the system level and not at the core level, as shown below.

The other architecture, based on a customizable DSP core, is basically the same. However, hardware accelerators can be embedded in its datapath. So to each hardware accelerator corresponds a specific instruction (or some specific instructions). One has the following advantages:

- It is straightforward to add supplementary instructions to the development tools. Furthermore, the software designer is not confused with something different and the validation of the system is easier.
- These hardware accelerators can share some basic resources, such as multiplier.
- The development effort to add some hardware accelerators is limited through the concept of "customizable" DSP.
- Energy is not significantly higher than the original architecture.

Consequently, it is likely that customizable and reconfigurable DSP architectures will be preferred in the future to architectures consisting of a single core (microcontroller or simple DSP) with many co-processors.

6. Conclusion

This chapter shows that power reduction techniques for low-power processors are well-known (CPI reduction, parallelism, gated clock, etc.), regarding dynamic power. However, regarding leakage reduction in very deep submicron technologies, many circuit and architectural techniques have been proposed but they are not so used in industrial circuits. Looking at the best embedded processor architectures, it turns out that for 8-bit cores, RISC-like architectures are the best; however, huge investments in software for old CISC machines drastically limit the use of these new 8-bit RISC-like processors. In 32-bit embedded cores the trend is clearly toward better performances with the same instruction sets but pushing frequencies using deeper technologies and deeper pipelines requiring sophisticated prediction mechanisms. But the architectural concepts remain the same. It is different for DSP cores, for which many very different architectures have been designed and used in various industrial chips for various applications. One can find single MAC cores, customizable cores, superscalar dual MAC and quad MAC, VLIW, hardware accelerators with a controller, multicores with various DSP and controller cores and finally reconfigurable architectures. The list is quite long and it is not yet clear which architecture is the best for which application.

References

[1] Rabay, J.M. "Managing power dissipation in the generation-after-next wireless systems", FTFC'99, June **1999**, Paris.

[2] Piguet, C. "Parallelism and low-power", Invited talk, SympA'99, Symposium Architectures de Machines, Rennes, France, 8 June **1999**.

[3] Jerraya, A. "Hardware/software codesign", Summer Course, Orebro, Sweden, 14–16 August, **2000**.

[4] Anis, M.; Elmasry, M. *Multi-threshold CMOS digital circuits*, Kluwer Academic Publishers, **2003**.

[5] Vittoz, E. "Weak inversion for ultimate low-power logic", in *Low Power Electronics Design*, edited by C. Piguet. CRC Press, **2004**, chapter 16.

[6] Heer, C. *et al.* "Designing low-power circuits: an industrial point of view", PATMOS 2001, Yverdon, 26–28 September **2001**.

[7] Piguet, C.; Schuster, C.; Nagel, J-L. "Optimizing architecture activity and logic depth for static and dynamic power reduction", *Proc. 2nd Northeast Workshop on Circuits and Systems*, NewCAS'04, 20–23 June **2004**, Montréal, Canada.

[8] Brodersen, R.W. *et al.* "Methods for true power minimization", *Proc. Int. Conf. on Computer Aided Design*. San Jose, California, November **2000**, 35–42.

[9] Nose, K.; Sakurai, T. "Optimization of Vdd and Vth for low-power and high-speed applications", ASPDAC, January **2000**, 469–474.

[10] Schuster, C.; Nagel, J-L.; Piguet, C.; Farine, P-A. "Leakage reduction at the architectural level and its application to 16 bit multiplier architectures", Patmos'04, Santorini Island, Greece, 15–17 September **2004**.

[11] Schuster, C.; Piguet, C.; Nagel, J-L.; Farine, P-A. "An architecture design methodology for minimal total power consumption at fixed Vdd and Vth", *J. Low-Power Electronics*, **2005**, *1*, 1–8.

[12] Piguet, C. *et al.* "Low-power design of 8-bit embedded CoolRISC microcontroller cores", *IEEE JSSC*, **1997**, *32*(7), 1067–1078.

[13] Masgonty, J-M. *et al.* "Low-power design of an embedded microprocessor", ESSCIRC'96, 16–21 September **1996**, Neuchâ tel, Switzerland.

[14] Oh, J.; Pedram, M. "Gated clock routing for low-power microprocessor design", *IEEE Trans. Computer Aided Design of ICs and Systems*, **2001**, *20*(6), 715–722.

[15] PowerChecker, www.bulldast.com.

[16] Keating, M.; Bricaud, P. *Reuse methodology manual*, Kluwer Academic Publishers, **1999**.

[17] Arm, C.; Masgonty, J-M.; Piguet, C. "Double-latch clocking scheme for low-power I.P. cores", PATMOS 2000, Goettingen, Germany, 13–15 September **2000**.

[18] Mosch, Ph. *et al.* "A 72 μW, 50 MOPS, 1V DSP for a hearing aid chip set", ISSCC'00, San Francisco, 7–9 February, **2000**, Session 14, paper 5, 238–239, **2000**.

[19] Schlett, M. "Trends in embedded microprocessor design", *IEEE Computer*, August **1998**, 44–49.

[20] Kuga, M. *et al.* "DSNS (dynamically-hazard-resolved, statically-code-scheduled, nonuniform superscalar): yet another superscalar processor architecture", *Computer Architecture*, **1991**, *4*, 14–29.

[21] Michaud, P. "La prédiction de branchement", SympA'99, Symposium Architectures de Machines, Rennes, France, 8 June **1999**.
[22] Omondi, A.R. *The microarchitecture of pipelined and superscalar computers*. Kluwer Academic Publishers, **1999**.
[23] Clark, L.T. *et al.* "A scalable performance 32b microprocessor", *Proc. ISSCC'2001*, San Francisco, 6 February **2001**, 230–231.
[24] Rowen, C. *et al.* "A pipelined 32b NMOS microprocessor", *Proc. ISSCC'84*, **1984**.
[25] Halfhill, T.R. "ARC 700 Secrets Revealed", *Microprocessor Rep.*, 21 June **2004**, 1–6.
[26] Tran, C. *et al.* "The MIPS32 24KE core family", *Microprocessor Rep.*, 31 May **2005**, 1–9.
[27] Krewell, K. "Multicore showdown", *Microprocessor Rep.*, 31 May **2005**, 1–5.
[28] Perotto, J-F; Lamothe,C.; Arm, C. *et al.* "An 8-bit multitask micropower RISC core", *JSSC*, **1994**, *29*(18), 986–991.
[29] "Sun's Big Splash", *IEEE Spectrum*, January **2005**, 50–54.
[30] Frantz, G. "Digital signal processor trends", *IEEE Micro*, November–December **2000**, 52–59.
[31] http://www.ceva-dsp.com/
[32] Halfill, T.R. "MIPS24KE: better late than never", *Microprocessor Rep.*, 31 May **2005**.
[33] Halfhill, T.R. "ARC's preconfigured cores", *Microprocessor Rep.*, 14 March **2005**, 1–6.
[34] Halfhill, T.R. "Tensilica tackles bottlenecks", *Microprocessor Rep.*, 31 May **2004**.
[35] Verbauwhede, I.; Nicol, Ch. "Low power DSP's for wireless communications", *Proc. ISLPED'00*, **2000**, Rapallo, Italy, 303310.
[36] Cravotta, R. "Targeted DSPs take aim" *EDN*, 28 April **2005** www.edn.com.
[37] David, R. *et al.*, "Low-power reconfigurable processors", in *Low power electronics design*, ed. C. Piguet. CRC Press, **2004**, Chapter 20.
[38] Rampogna, F. *et al.*, "MACGIC, a low-power, re-configurable DSP", in *Low power electronics design*, ed. C. Piguet. CRC Press, **2004**.
[39] Rabaey, J.M. "Reconfigurable processing: the solution to low-power programmable DSP", Int. Conf. on Acoustics, Speech and Signal Processing (ICASSP), April **1997**.
[40] Verbauwhede, I.; Piguet, C.; Schaumont, P.; Kienhuis, B. "Architectures and design techniques for energy-efficient embedded DSP and multimedia processing", Embedded Tutorial, *Proc. DATE'04*, Paris, 16–20 February **2004**, Paper 7G, 988–995.
[41] Baron, M. "2004: top features, low power", *Microprocessor Rep.*, 18 January **2004**.
[42] Cravotta, R. "2004 DSP directory", *EDN*, 29 April **2004**, 49–67.
[43] Halfhill, T.R. "PicoChip makes a big MAC", *Microprocessor Rep.*, 14 October **2003**.

Chapter 2

DESIGN OF ENERGY EFFICIENT DIGITAL CIRCUITS

Bart R. Zeydel and Vojin G. Oklobdzija
ACSEL Laboratory, University of California, Davis

Abstract: Recent technology advances have resulted in power being the major concern for digital design. In this chapter we address how transistor sizing affects the energy and delay of digital circuits. The state of the art in circuit design methodology (Logical Effort) is examined and we identify its limitations for design in the energy-delay space. We examine how to explore the entire energy-delay space for a circuit and present an approach for the design and analysis in the energy-delay space which allows for energy reduction without performance penalty. Finally, we present techniques for the design of energy-efficient digital circuits.

Key words: digital circuits; energy-delay optimization; energy-delay space; performance optimization; power optimization; transistor sizing

1. Introduction

Advances in CMOS technology have led to dramatic improvements in performance while maintaining constant power density. However, as device dimensions continue to decrease traditional constant field scaling can no longer be applied [1–3]. The problem with this trend is that performance and power no longer scale proportionally across technology nodes leading to increasing power density. Further adding to this problem has been the drive to produce chips operating at higher and higher clock frequencies, which has caused circuit designers to focus solely on optimizing circuits and implementations for delay regardless of energy.

In this chapter we present models to examine the energy and delay characteristics of digital circuits and relate these characteristics to the physical

Vojin G. Oklobdzija and Ram K. Krishnamurthy (eds.),
High-Performance Energy-Efficient Microprocessor Design, 31–56.

dimensions of transistors. Using these models we will analyze Logical Effort (LE) [4, 5], the state of the art design methodology for digital circuits. The location of the LE solution in the energy-delay space is then examined to determine its applicability to energy-efficient design. The analysis demonstrates that LE does not guarantee an energy-efficient circuit. To address this we examine the entire energy-delay space for a circuit that can be obtained through transistor sizing. From this we present a simplified approach for the high-level exploration of the energy-delay characteristics of a circuit. Based on this analysis we present guidelines for the design of energy-efficient digital circuits.

2. RC Modeling of Gate Delay

Delay modeling techniques for evaluating large circuits have historically involved the simplification of current based delay modeling. The most common simplification assumes a step input, allowing for the current to be approximated over the time of interest [6–10].

2.1. Logic Gate Characteristics

In this section the physical characteristics of a CMOS logic gate are related to its delay characteristics. The layout of a CMOS inverter is shown in Figure 1. The physical parameters are W_n, W_p, L_n, and L_p which represent the widths and channel lengths of the nMOS and pMOS transistors respectively. Understanding the dependence of gate capacitance, parasitic capacitance and effective channel resistance on these physical parameters is essential to the use of RC modeling for the optimization of CMOS logic gates.

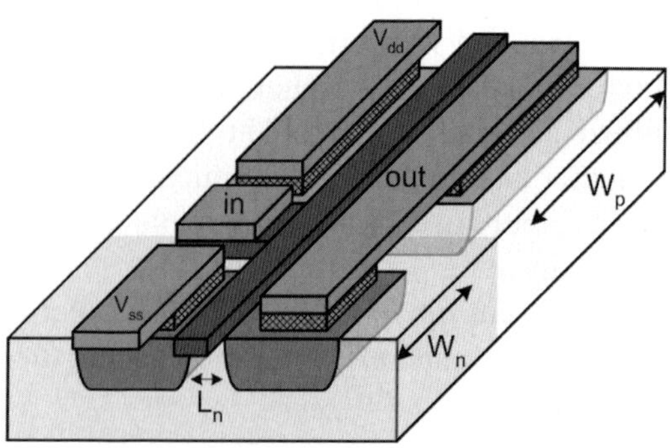

Figure 1. CMOS Inverter.

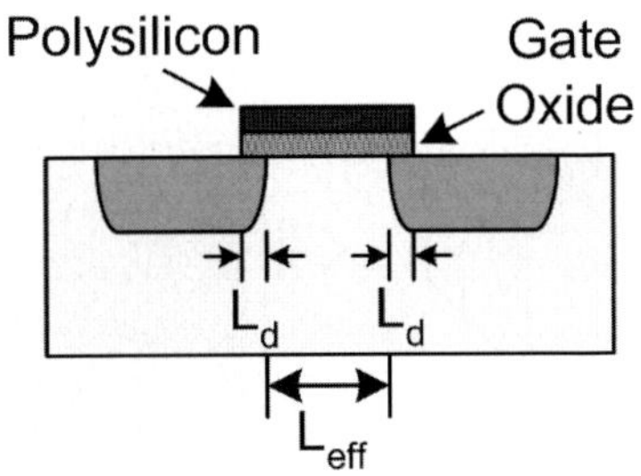

Figure 2. MOSFET Gate Capacitance.

2.1.1. *Gate capacitance*

Gate capacitance, C_{gate}, is a function of the effective channel length, L_{eff}, and the width of the transistor, W. The effective channel length can be calculated from the drawn transistor length as $L_{\text{eff}} = L_{\text{drawn}} - 2L_d$, as seen in Figure 2, where L_d refers to the lateral diffusion length of the source or drain into the channel. To simplify notation L_{eff} will be referred to as L.

The gate capacitance of each transistor can be calculated from the width and length of the transistor and the per area capacitance of the gate, C_{ox}.

$$C_{\text{gate}} = W \cdot L \cdot C_{\text{ox}}$$

The gate capacitance is directly proportional to the width of the transistor. Thus, as the width changes by a factor α the gate capacitance also changes by the same factor α.

$$C_{\text{gate}} = \alpha \cdot W \cdot L \cdot C_{\text{ox}}$$

The capacitance of an input to a gate, C_{in}, is the sum of the gate capacitances attached to the input. For example, the input capacitance of an inverter is:

$$C_{\text{in}} = (W_n \cdot L_n + W_p \cdot L_p) \cdot C_{\text{ox}}$$

Scaling the width of each transistor in the inverter by a factor α causes C_{in} to also scale by α.

2.1.2. *Parasitic capacitance*

The parasitic capacitance of a transistor has two components. The junction capacitance, C_{ja}, expressed in F per area in μm^2, and the periphery capacitance, C_{jp}, expressed in F per μm of the periphery length. These components are shown in Figure 3.

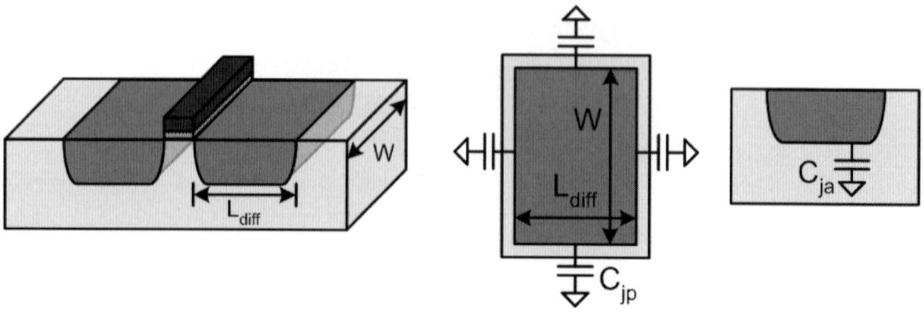

Figure 3. MOSFET Parasitic Capacitance.

The parasitic capacitance of each transistor can be computed directly from layout as:

$$C_p = C_{\text{ja}} \cdot W \cdot L_{\text{diff}} + C_{\text{jp}} \cdot (2 \cdot W + 2 \cdot L_{\text{diff}})$$

Parasitic capacitance is only roughly proportional to changes in gate width by α, due to its constant term $2C_{\text{jp}}L_{\text{diff}}$. To simplify analysis this term is often ignored allowing for the parasitic capacitance to be proportional to α.

2.1.3. *Resistance*

The channel resistance, R_{channel}, in a MOSFET is dependent on its region of operation, transistor width, and channel length. In saturation R_{channel} can be expressed as follows, where μ is the mobility of the channel and λ is the Early effect:

$$R_{\text{channel(sat)}} = \frac{\partial V_{\text{ds}}}{\partial I_{d(\text{sat})}} = \frac{2 \cdot L}{W \cdot \mu \cdot C_{\text{ox}} \cdot (V_{\text{gs}} - V_t)^2 \cdot \lambda}$$

In linear or triode, R_{channel} can be expressed as:

$$R_{\text{channel(lin)}} = \frac{\partial V_{\text{ds}}}{\partial I_{d(\text{lin})}} = \frac{L}{W \cdot \mu \cdot C_{\text{ox}} \cdot (V_{\text{gs}} - V_t - V_{\text{ds}})}$$

The resistance of the channel is inversely proportional to the width of the transistor in both saturation and linear regions of operation. Thus, by changing the width of the transistor by a factor α, the resistance of the transistor changes by $1/\alpha$.

2.2. RC Delay Model

The propagation delay of a CMOS logic gate can be represented using a RC-model [6]. The model can be derived assuming a step input (Figure 4),

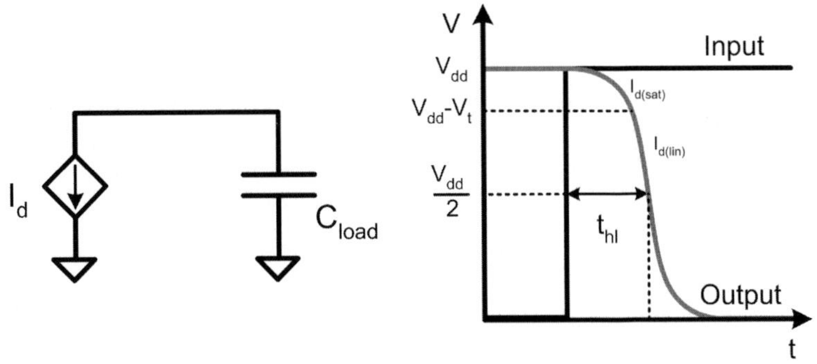

Figure 4. Step input response of a CMOS logic gate.

and related to gate capacitance, parasitic capacitance and channel resistance. The load, C_{load}, consists of the output load, C_{out}, and the parasitic load at the output of the gate, C_p. The derivation will only be shown for the high-to-low propagation delay, t_{hl}, however a similar derivation can be performed for the low-to high propagation delay, t_{lh}.

The propagation delay, t_{hl}, can be calculated from:

$$-I_d = C_{\text{load}} \cdot \frac{\partial V_{\text{out}}}{\partial t}$$

where t_{hl} is given by:

$$t_{\text{hl}} = -\int_{V_{\text{dd}}}^{V_{\text{dd}}/2} \frac{C_{\text{load}}}{I_d} \partial V_{\text{out}}$$

For a step input, the transistor will be in saturation for V_{out} from V_{dd} to $V_{\text{dd}} - V_t$. In the saturation region, the drain current is given by:

$$I_{d(\text{sat})} = \mu_n \cdot C_{\text{ox}} \cdot \frac{W}{L} \frac{(V_{\text{dd}} - V_t)^2}{2}$$

The transistor will be in the linear region for V_{out} from $V_{\text{dd}} - V_t$ to $V_{\text{dd}}/2$. In the linear region, drain current is given by:

$$I_{d(\text{lin})} = \mu_n \cdot C_{\text{ox}} \cdot \frac{W}{L} \left((V_{\text{dd}} - V_t) \cdot V_{\text{out}} - \frac{V_{\text{out}}^2}{2} \right)$$

Substituting into the integration for t_{hl}:

$$t_{\text{hl}} = -\int_{V_{\text{dd}}}^{V_{\text{dd}}-V_t} \frac{C_{\text{load}}}{I_{d(\text{sat})}} \partial V_{\text{out}} - \int_{V_{\text{dd}}-V_t}^{V_{\text{dd}}/2} \frac{C_{\text{load}}}{I_{d(\text{lin})}} \partial V_{\text{out}} = t_{\text{hl(sat)}} + t_{\text{hl(lin)}}$$

Integrating gives:

$$t_{hl(sat)} = -\frac{C_{load}}{\mu_n \cdot C_{ox} \cdot \frac{W}{L}\frac{(V_{dd}-V_t)^2}{2}} \int_{V_{dd}}^{V_{dd}-V_t} \partial V_{out} = \frac{2 \cdot V_t \cdot C_{load}}{\mu_n \cdot C_{ox} \cdot \frac{W}{L}(V_{dd} - V_t)^2}$$

$$t_{hl(lin)} = -\frac{C_{load}}{\mu_n \cdot C_{ox} \cdot \frac{W}{L}} \int_{V_{dd}-V_t}^{V_{dd}/2} \left(\frac{1}{(V_{dd} - V_t) \cdot V_{out} - \frac{V_{out}^2}{2}} \right) \cdot \partial V_{out}$$

$$= \frac{C_{load}}{\mu_n \cdot C_{ox} \cdot \frac{W}{L}(V_{dd} - V_t)} \cdot \ln\left(3 - 4\frac{V_t}{V_{dd}}\right)$$

Substituting $t_{hl(sat)}$ and $t_{hl(lin)}$ into t_{hl}:

$$t_{hl} = \frac{C_{load}}{\mu_n \cdot C_{ox} \cdot \frac{W}{L}(V_{dd} - V_t)} \cdot \left(\frac{2 \cdot V_t}{V_{dd} - V_t} + \ln\left(3 - 4\frac{V_t}{V_{dd}}\right)\right)$$

The channel resistance is physically dependent on W, L, μ_n, and C_{ox}. These terms can be grouped to describe the effective resistance of the channel, $R_{channel}$:

$$R_{channel} = \frac{L}{\mu_n \cdot C_{ox} \cdot W \cdot (V_{dd} - V_t)}$$

The remaining terms can be grouped into a constant determined from V_{dd} and V_t:

$$\kappa = \left(\frac{2 \cdot V_t}{V_{dd} - V_t} + \ln\left(3 - 4\frac{V_t}{V_{dd}}\right)\right)$$

The resulting delay of a gate can be expressed as:

$$t_{hl} = \kappa \cdot R_{channel} \cdot C_{load} = \kappa \cdot R_{channel} \cdot (C_{out} + C_p)$$

In this form, delay is seen to be linear with respect to C_{load}. A graphical representation of this model is shown in Figure 5. R_{up} and R_{down} denote the equivalent pull-up and pull-down resistance of a gate.

We would like to observe the delay dependence as transistor widths are scaled by a factor α. The original resistances and capacitances will be referred to as the template. The resistance of a gate changes inversely with α, as:

$$R_{channel} = R_{template}/\alpha$$

The input capacitance and parasitic capacitance of the gate both change directly with α:

$$C_{in} = C_{template} \cdot \alpha$$

$$C_p \approx C_{p(template)} \cdot \alpha$$

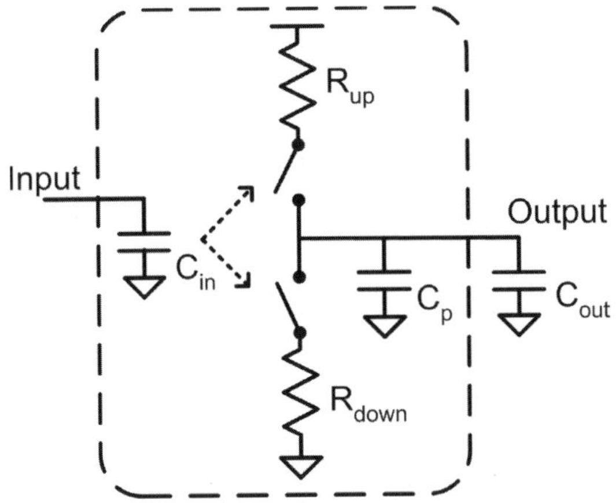

Figure 5. RC Model for a CMOS gate.

Plugging the scaled values for resistance and capacitance into the RC delay model yields:

$$t_d = \kappa \cdot \left(\frac{R_{\text{template}}}{\alpha}\right)(C_{\text{out}} + \alpha \cdot C_{p(\text{template})})$$

It is observed that the parasitic delay of a gate does not change with the size of the gate.

$$t_d = \kappa \cdot \left(\frac{R_{\text{template}}}{\alpha}\right) \cdot C_{\text{out}} + \kappa \cdot R_{\text{template}} \cdot C_{p(\text{template})}$$

However, the delay associated with a constant load changes inversely with the sizing factor α. Through substitution, delay can be expressed in terms of C_{in} and C_{out} of the gate instead of using α.

$$t_d = \kappa \cdot \left(R_{\text{template}} \cdot C_{\text{template}} \cdot \left(\frac{C_{\text{out}}}{C_{\text{in}}}\right) + R_{\text{template}} \cdot C_{p(\text{template})}\right)$$

2.3. Logical Effort Delay Model

In 1991 R. F. Sproull and I.E. Sutherland suggested that a technology independent delay could be obtained by normalizing the RC-delay model of a gate [4, 5]. They suggested that the delay of a gate be normalized to the per fanout delay of an inverter.

$$t_d = \kappa \cdot R_{\text{inv}} \cdot C_{\text{inv}} \left(\frac{R_{\text{template}} \cdot C_{\text{template}}}{R_{\text{inv}} \cdot C_{\text{inv}}} \cdot \left(\frac{C_{\text{out}}}{C_{\text{in}}}\right) + \frac{R_{\text{template}} \cdot C_{\text{parasitic}}}{R_{\text{inv}} \cdot C_{\text{inv}}}\right)$$

The technology dependent constant is referred to as τ.

$$\tau = \kappa \cdot R_{\text{inv}} \cdot C_{\text{inv}}$$

The logical effort (g), or relative drive capability, of each gate is given by:

$$g = \frac{R_{\text{template}} \cdot C_{\text{template}}}{R_{\text{inv}} \cdot C_{\text{inv}}}$$

The parasitic delay (p) of each gate is given by:

$$p = \frac{R_{\text{template}} \cdot C_{\text{parasitic}}}{R_{\text{inv}} \cdot C_{\text{inv}}}$$

The relationship of output load to input capacitance is referred to as the electrical effort (h) of the gate.

$$h = \frac{C_{\text{out}}}{C_{\text{in}}}$$

Using these terms, delay can be expressed as:

$$t_d = (gh + p) \cdot \tau$$

The logical effort of a gate can be determined by equalizing the resistance of the gate to the inverter and computing the ratio of input capacitances. The input capacitance is proportional to the sum of the gate widths attached to an input of the circuit. For example, input-a of the 2-input NOR gate in Figure 6

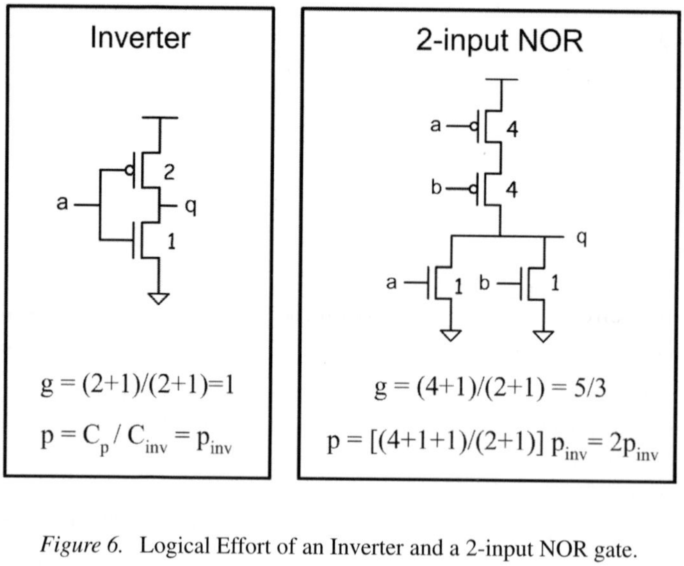

Figure 6. Logical Effort of an Inverter and a 2-input NOR gate.

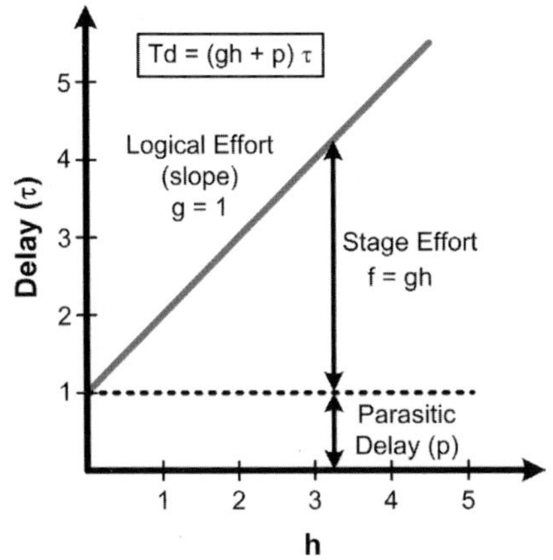

Figure 7. Logical Effort Delay Components.

has a total width of 5 which when normalized to the input capacitance of the inverter, yields a logical effort g of 5/3.

The parasitic delay can be determined from the ratio of transistor widths attached to the output node. For example, in the 2-input NOR gate the total transistor width attached to the output node is 6 which when normalized to the input capacitance of the template inverter, give a parasitic delay of $2p_{\text{inv}}$. To simplify analysis, it is often assumed that $C_{p(\text{inv})}$ equals C_{inv} which makes p_{inv} equal to 1.

A graphical representation of the LE terms is shown in Figure 7. The product of gh is referred to as the stage effort, f, and represents the delay associated with the output load of a gate. By plotting delay versus H the logical effort values can be obtained from simulation. The parasitic delay can be found from the delay intercept when h is 0, while the logical effort can be found from the slope of the delay versus h. To obtain delay in terms of τ, each delay target is normalized to the per fanout delay of the inverter.

3. Designing Circuits for Speed

Designing circuits for speed has been the focus of digital circuit designers since the inception of CMOS technology. To achieve better speed, designers initially focused on reducing the number of logic stages on the critical path. Designers soon realized that the fan-in and fan-out of circuits needed to be accounted for when analyzing circuits for speed [11]. As CMOS technology progressed, designers were also given the ability to modify transistor sizes to improve the performance of circuits. To address the issue of transistor sizing CAD tools, such as TILOS [12], were used to optimize the performance of

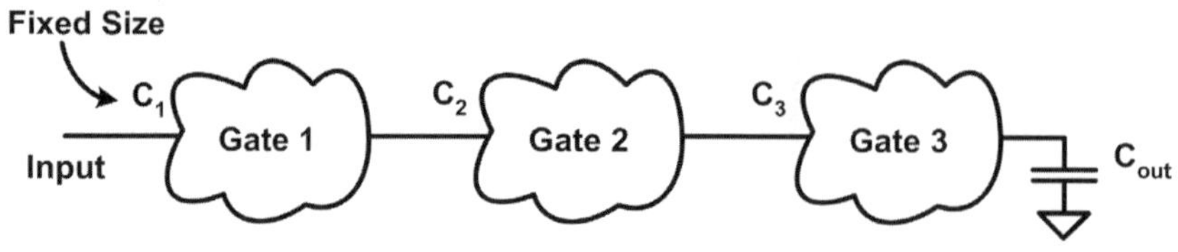

Figure 8. Chain of Gates with a Fixed Output Load and Fixed Input Size.

circuits. However, these tools offered designers little or no insight into why one design was faster than another or how gates should be sized for optimal delay. Logical Effort filled this void by providing designers with the ability to compare delay optimized digital circuits in an intuitive manner.

3.1. Delay Optimization of a Single Path Circuit

Logical Effort provides a method for optimizing the delay of a chain of gates driving a load. The constraints on the optimization are a fixed output load and a fixed input size. The derivation for delay optimal sizing of a chain of gates will be shown for the example in Figure 8.

The delay of the path can be expressed as:

$$T_{\text{path}} = \left[\left(g_1\frac{C_2}{C_1} + p_1\right) + \left(g_2\frac{C_3}{C_2} + p_2\right) + \left(g_3\frac{C_{\text{out}}}{C_3} + p_3\right)\right] \cdot \tau$$

The input capacitances of gates 1, 2 and 3 are referred to as C_1, C_2, and C_3 respectively. The minimum delay of the path with a fixed output load, C_{out}, and fixed input size, C_1, can be found by taking the derivative of the path delay with respect to C_2 and C_3.

$$\frac{\partial T_{\text{path}}}{\partial C_2} = \frac{g_1}{C_1} - g_2\frac{C_3}{C_2^2} = 0 \qquad \frac{\partial T_{\text{path}}}{\partial C_3} = \frac{g_2}{C_2} - g_3\frac{C_{\text{out}}}{C_3^2} = 0$$

Rearranging the expression yields:

$$g_1\frac{C_2}{C_1} = g_2\frac{C_3}{C_2} \qquad g_2\frac{C_3}{C_2} = g_3\frac{C_{\text{out}}}{C_3}$$

Expressed in terms of stage effort, $f_1 = f_2$ and $f_2 = f_3$. Thus the minimum delay of the path is achieved when the stage efforts of each gate. The optimal stage effort, f_{opt}, can be found from:

$$f_{\text{opt}} = \left(g_1\frac{C_2}{C_1} \cdot g_2\frac{C_3}{C_2} \cdot g_3\frac{C_{\text{out}}}{C_3}\right)^{1/3} = \left(\frac{C_{\text{out}}}{C_1} \cdot \prod_{i=1}^{3} g_i\right)^{1/3}$$

Generalized to an N-stage chain of gates:

$$f_{\mathrm{opt}} = \left(\frac{C_{\mathrm{out}}}{C_1} \cdot \prod_{i=1}^{N} g_i \right)^{1/N}$$

The following definitions are introduced to simplify discussion. The electrical effort or gain of a path, H, is defined as the ratio of output to input capacitance of the path.

$$H = \frac{C_{\mathrm{out}}}{C_{\mathrm{in}}}$$

The Logical Effort of the path, G, is defined as the product of the logical effort of the gates along the path.

$$G = \prod g_i$$

Using these simplifications, the optimal stage effort for a path is:

$$f_{\mathrm{opt}} = (GH)^{1/N}$$

The optimal delay for a chain of gates is given by:

$$T_{\mathrm{path}} = \left(N \cdot (GH)^{1/N} + \sum_{i=1}^{N} p_i \right) \cdot \tau$$

3.1.1. *Example of delay optimized sizing*

This example demonstrates how the sizes of the gates in Figure 9 are optimized such that the delay of the path with a fixed input size and fixed output load is minimal. The input capacitances of each gate on the path are referred to as $C_{\mathrm{in}}, C_2, C_3$, and C_4, respectively.

The optimal sizing is obtained from f_{opt}:

$$f_{\mathrm{opt}} = (GH)^{1/4} = \left(\left(\frac{5}{3} \cdot \frac{4}{3} \cdot \frac{5}{3} \cdot 1 \right) \cdot 21.87 \right)^{1/4} = 3$$

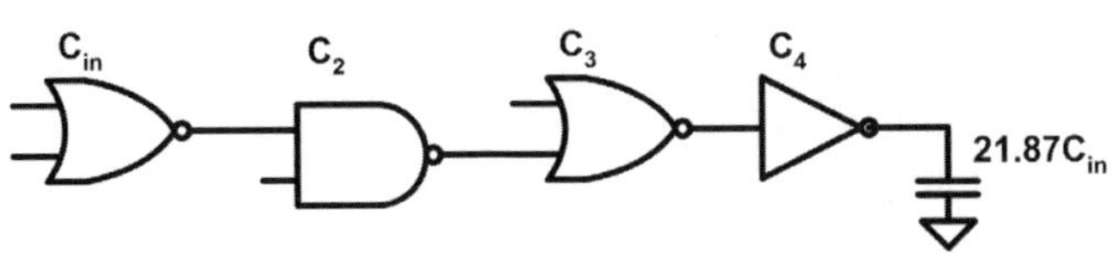

Figure 9. Example Chain of Gates.

The resulting optimal delay of the path is $T_d = (12 + 7p_{\text{inv}})\tau$. Using f_{opt}, the input capacitance of each gate can be computed as:

$$C_i = \frac{g_i \cdot C_{i+1}}{f_{\text{opt}}}$$

3.2. Delay Optimization of Circuits with Branching

Although the solution to the previous problem is useful for a simple chain of gates, it does not account for circuits with multiple paths. LE introduces branching (b) to allow for the analysis of multi-path circuits. Branching relates the off-path capacitance, $C_{\text{off-path}}$, to the on-path capacitance, $C_{\text{on-path}}$.

$$b = \frac{C_{\text{on-path}} + C_{\text{off-path}}}{C_{\text{on-path}}}$$

This often leads to confusion as the definition for electrical effort, h, includes the branching factor:

$$h = \frac{C_{\text{on-path}} + C_{\text{off-path}}}{C_{\text{in}}} = \left(\frac{C_{\text{on-path}} + C_{\text{off-path}}}{C_{\text{on-path}}}\right)\left(\frac{C_{\text{on-path}}}{C_{\text{in}}}\right)$$
$$= b \cdot \frac{C_{\text{on-path}}}{C_{\text{in}}}$$

When applying to a path it can be seen that

$$\prod_{i=1}^{N} h_i = H \cdot \prod_{i=1}^{N} b_i = HB \quad \text{where, } B = \prod_{i=1}^{N} b_i$$

Resulting in the following expression for f_{opt}:

$$f_{\text{opt}} = (GBH)^{1/N}$$

3.2.1. *Multi-Path circuit optimization example*

To achieve minimum delay in the multi-path circuit shown in Figure 10, the delay through Path A and Path B should be equal [13, 14].

The delay for Path A and B can be expressed as:

$$T_{\text{Path-A}} = [(g_1h_1 + p_1) + (g_2h_2 + p_2) + (g_3h_3 + p_3)] \cdot \tau$$
$$T_{\text{Path-B}} = [(g_1h_1 + p_1) + (g_4h_4 + p_4) + (g_5h_5 + p_5)] \cdot \tau$$

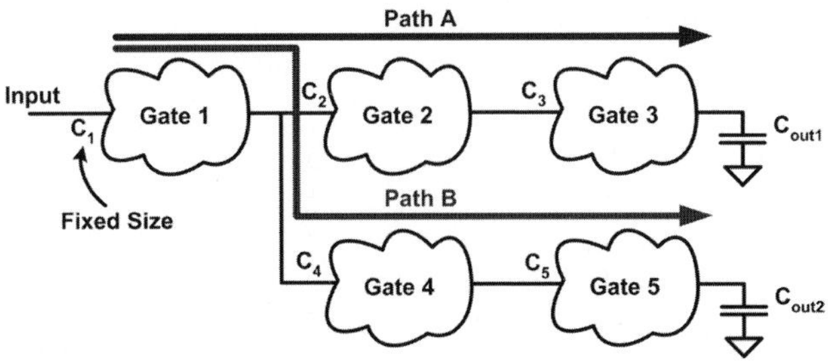

Figure 10. Example Multi-Path Circuit.

The branching at the output of Gate 1 for Path A and B can be determined as follows:

$$b_{\text{Path-A}} = \frac{C_2 + C_4}{C_2} \qquad b_{\text{Path-B}} = \frac{C_4 + C_2}{C_4}$$

Solving for C_2 and C_4:

$$C_2 = \frac{g_2 g_3 \cdot C_{\text{out1}}}{f_2 f_3} \qquad C_4 = \frac{g_4 g_5 \cdot C_{\text{out2}}}{f_4 f_5}$$

Substituting C_2 and C_4 into $b_{\text{Path-A}}$ and $b_{\text{Path-B}}$:

$$b_{\text{Path-A}} = \frac{\frac{g_2 g_3 \cdot C_{\text{out1}}}{f_2 f_3} + \frac{g_4 g_5 \cdot C_{\text{out2}}}{f_4 f_5}}{\frac{g_2 g_3 \cdot C_{\text{out1}}}{f_2 f_3}} \qquad b_{\text{Path-B}} = \frac{\frac{g_4 g_5 \cdot C_{\text{out2}}}{f_4 f_5} + \frac{g_2 g_3 \cdot C_{\text{out1}}}{f_2 f_3}}{\frac{g_4 g_5 \cdot C_{\text{out2}}}{f_4 f_5}}$$

Previously it was demonstrated that the optimal delay of a path without branching occurs when each stage has the same stage effort. Simplifying the delay to only include stage effort (by ignoring the parasitic delay difference between the Path A and B) the delay of each branch is equal when $f_2 = f_3 = f_4 = f_5$. Allowing for $b_{\text{path-A}}$ and $b_{\text{path-B}}$ to be expressed as:

$$b_{\text{Path-A}} = \frac{g_2 g_3 \cdot C_{\text{out1}} + g_4 g_5 \cdot C_{\text{out2}}}{g_2 g_3 \cdot C_{\text{out1}}} \qquad b_{\text{Path-B}} = \frac{g_4 g_5 \cdot C_{\text{out2}} + g_2 g_3 \cdot C_{\text{out1}}}{g_4 g_5 \cdot C_{\text{out2}}}$$

A special case for branching occurs when $g_2 g_3 = g_4 g_5$ and $C_{\text{out1}} = C_{\text{out2}}$. In this case $b_{\text{Path-A}} = b_{\text{Path-B}} = 2$.

While branching allows for off-path gate load to be included in LE, constant off-path loads of minimum sized gates and interconnect are not accounted for as they introduce nonlinearity into the branching computation. Further complicating branching are paths with different number of stages. Accurate accounting for these factors when optimizing for delay requires the use of numerical optimization.

Table 1. Delay Comparison of two circuits X and Y

Parasitic Delay (P)	Logic Complexity (GB)	Logic Stages (S)	Best Design for all H
$P_X = P_Y$	$G_X B_X = G_Y B_Y$	$S_X = S_Y$	Equal delay
$P_X = P_Y$	$G_X B_X < G_Y B_Y$	$S_X = S_Y$	X is faster
$P_X < P_Y$	$G_X B_X = G_Y B_Y$	$S_X = S_Y$	X is faster
$P_X < P_Y$	$G_X B_X < G_Y B_Y$	$S_X = S_Y$	X is faster
$P_X < P_Y$	$G_X B_X > G_Y B_Y$	$S_X = S_Y$	Depends on H
-	-	$S_X != S_Y$	Depends on H

3.3. Designing High-Performance Circuits

The delay optimal solution for a path has two components. A constant parasitic delay and a variable delay dependent on the gain of the path, H. As H decreases, the delay of the path approaches the parasitic delay.

$$T_{\text{path}} = \left(N \sqrt[N]{GBH} + \sum_{i=1}^{N} p_i \right)$$

The Logical Effort, G, of a path is constant, regardless of H. While branching, B, is approximately constant depending on the impact of nonlinearities such as wire and minimum sized gates with respect to H. These parameters define the inherent complexity of a circuit. We refer to the product of GB as the logic complexity of a circuit. By analyzing the logic complexity of a circuit in conjunction with its parasitic delay it is possible to compare circuits over a range of H to gain insight into designing high-performance circuits (Table 1).

From the table, it is seen that two circuits X and Y, which have the same number of stages and implement the same function, will always have the same delay if they have the same parasitic delay and logic complexity. Circuit X will always be the same speed or faster than Y if its parasitic delay is less than or equal to that of Y and its logic complexity is less than or equal to that of Y. However, if the circuit has less parasitic delay yet more logic complexity than the other circuit, the faster design will depend on the value of H. For implementations which use a different number of stages the best design depends on H.

4. Design in the Energy-Delay Space

CMOS technology scaling no longer has the favorable characteristics of constant power-density. As a result it is no longer possible to design solely for delay. Instead, both the energy and delay of a circuit must be accounted for. In this section we present a basic energy model which can be combined with RC-delay modeling to provide an energy estimate for LE delay optimized points. From these points the energy-delay space of digital circuits can be explored to identify the efficient region of operation and to identify energy-efficient characteristics of circuits.

4.1. Energy Model

An energy model which yields reasonable results that can be computed directly from gate size and output load is desirable (due to its compatibility with the transistor sizing described in section 3). For hand estimation, the dynamic energy of a circuit can be computed directly from the output load of the circuit, as:

$$E = C_{\text{load}} \cdot V_{\text{dd}}^2 = (C_p + C_{\text{out}}) \cdot V_{\text{dd}}^2$$

This model neglects the energy associated with short-circuit current and leakage. The model can be improved through simulation to include the energy associated with short-circuit current and leakage. The energy of a 2-input NAND gate obtained from simulation is shown in Figure 11. A linear dependence of energy on input size and output load is observed [15].

An offset can occur at zero size due to internal wire capacitance estimation, which can be accounted for by $E_{\text{internal-wire}}$. The dynamic energy associated

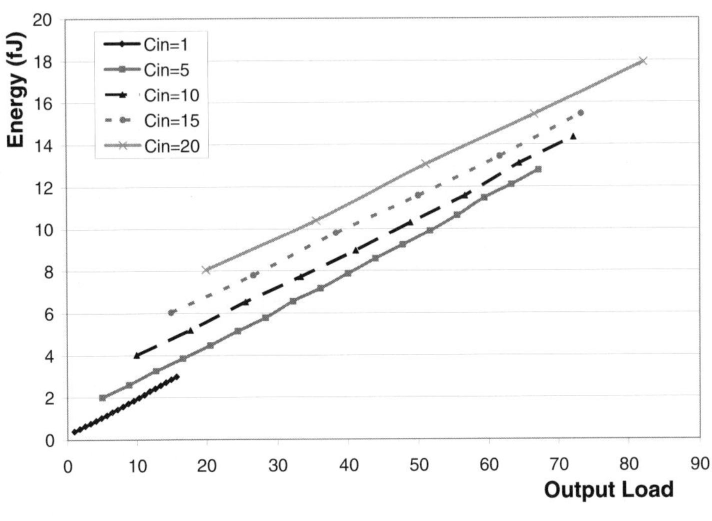

Figure 11. Energy Dependence on Input Size and Output Load for a 2-input NAND gate.

with the output of a gate can be expressed as:

$$E = E_p \cdot \text{gate size} + E_g \cdot C_L + E_{\text{internal-wire}}$$

E_p represents the energy per size and E_g represents the energy per output load. These terms can be obtained from simulation and directly account for the energy associated with output load and parasitic capacitance while providing a best fit for short-circuit and leakage current. The static energy of a gate per unit time, E_{leakage}, can be estimated by hand or obtained from simulation, from which the total static energy of the gate to be computed as $E = E_{\text{leakage}} \cdot \text{gate size} \cdot$ period. The switching activity of each gate is incorporated when estimating the energy of an entire circuit.

4.2. Minimal Energy Circuit Sizing for a Fixed Output Load and Fixed Input Size

To optimize a circuit with a fixed output load and a fixed input size it is first necessary to understand where the Logical Effort design point lies in the energy-delay space. The energy-delay space obtained through changing the sizes of the second and third inverter in a chain of three inverters with a fixed output load and fixed input size is shown in Figure 12. As can be seen, the solution space is vast even for such a simple circuit. In this solution space the

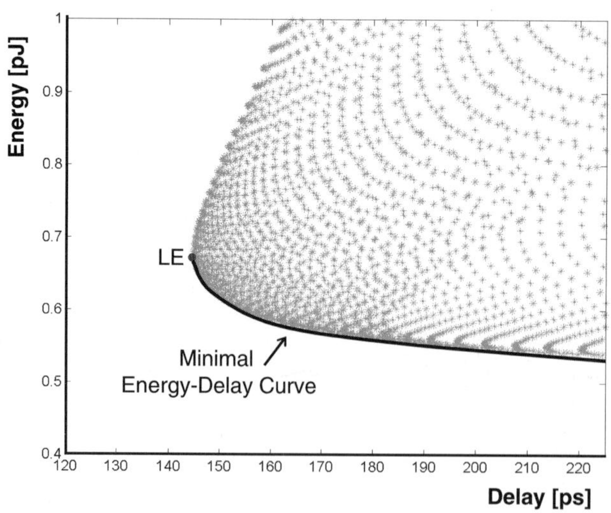

Figure 12. Energy-Delay Solution Space for a Chain of 3 inverters with a Fixed Output Load and Fixed Input Size.

delay optimized sizing of LE sets the performance limit for the circuit. Efficient design points in this solution space are those that achieve minimal energy for each delay. These points are obtained by relaxing the delay target from the LE point and resizing the circuit to reduce energy. The combined result of these optimizations yields the minimal Energy-Delay curve of a circuit for a fixed output load and a fixed input size.

It has been suggested that a tangent to this curve can be used to select an efficient design point [16–20]. For high-performance, some commonly used tangents are Energy. Delay^2(ED^2), Energy-Delay Product (EDP), and other ED^X metrics. The difficulty with designing for these metrics is that they can not be directly computed and can not be used to achieve a desired delay target or energy target. A minimal energy-delay curve for a fixed output load and a fixed input size obtained through transistor sizing along with various design metrics is shown in Figure 13.

The transistor sizings corresponding to each metric in Figure 13 are shown in Figure 14.

Energy decreases dramatically from the LE point at only a slight increase in delay. The rapid decent is due to the rippling affect of reducing the size of a gate that occurs later in the path. This weighting of gates along a path can be seen in the computation of the input capacitance of the *k-th* gate:

$$C_k = C_{\text{load}} \cdot \prod_{i=k}^{N} \frac{g_i}{f_i}$$

By changing the size of the *N-th* gate of the path by a factor α (equivalent to changing f_N by $1/\alpha$), the size of each preceding gate along the path also

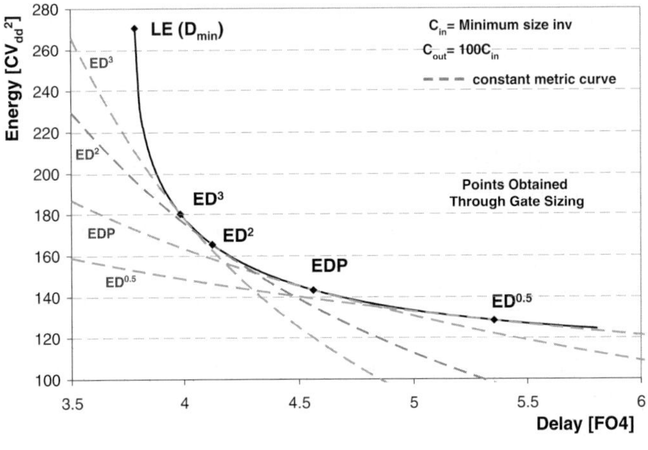

Figure 13. Minimal Energy-Delay Curve for a Chain of 6 inverters with Fixed output load and fixed input size.

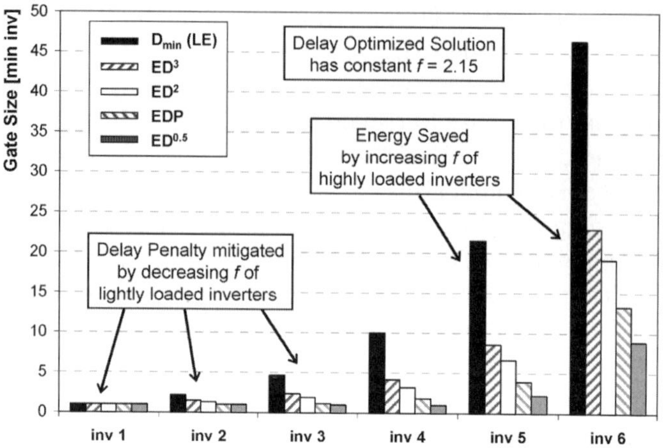

Figure 14. Corresponding Gate Sizing for Design Metrics on the Energy-Delay Curve.

changes by $1/\alpha$. It is this weighting of gates that causes the sizing to differ dramatically from the LE solution of equal f. By allowing the total delay of the path to be relaxed, the excess delay can be redistributed amongst the gates which contribute the most energy to the path (by changing f_i of these gates) to reduce the total energy [21, 22].

4.3. Circuit Sizing for Minimal Energy with a Fixed Output Load and Variable Input Size

In practice, circuit designers usually do not have the flexibility of degrading the performance of a circuit as it is often tied to the performance of the entire system. As a result, metrics which relate delay variation to energy variation are inapplicable at the circuit level. Instead the circuit should be designed for minimal energy at the desired performance target. As shown previously for a fixed delay, fixed input size and fixed output load there exists only one solution with minimal energy. However, if the input size is allowed to change, multiple energy solutions can be obtained at a fixed delay [21, 22]. The solution space obtained by varying the input size of a static 64-bit Kogge-Stone Adder [23] is shown in Figure 15.

The upper bound of the solution space consists of the delay optimized points obtained for each input size. The lower bound of the energy-delay space is constructed from the minimal energy points for each delay. Increasing the input size causes H to be reduced, allowing for performance to improve at the cost of increased energy. The minimum efficient input size for each delay is associated with the delay optimized point, while the maximum efficient input

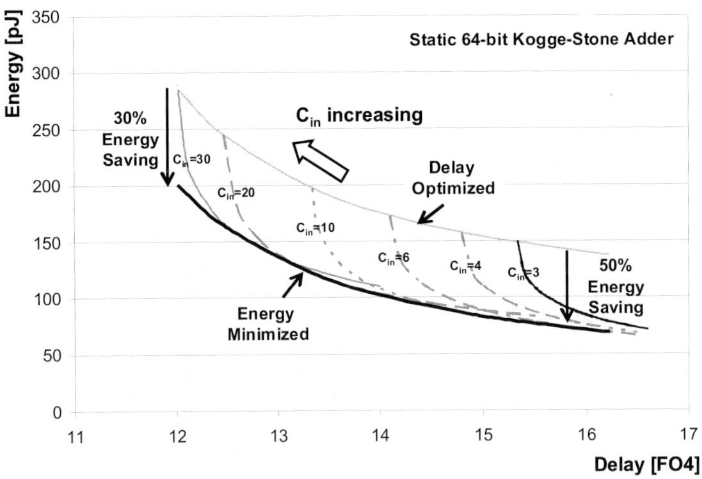

Figure 15. Energy-Delay Space for a Static 64-bit KS Adder with Fixed Output Load and Variable Input Size.

size for each delay is associated with the energy minimized point. By analyzing the complete energy-delay space of a circuit for a fixed output load, a potential 30–50% energy savings is observed in the adder example, depending on delay target.

4.3.1. Example: Energy minimization of an inverter chain

A chain of 6-inverters will be used to demonstrate how gate sizes change to achieve the same delay for different input sizes. The minimal energy sizings are shown in Figure 16 for several input sizes, with C_{out} equal to $100C_{min\text{-}inv}$ and a delay target of 18.9τ. As the input size is increased from minimal, a smaller delay can be achieved due to reduced H. The excess delay is redistributed amongst the gates to reduce total energy by reducing the sizes of the gates that impact energy the most and by increasing the sizes of the gates that have a smaller impact on energy to achieve the same delay. An increase in input size by 20% allows for a 22.3% reduction in energy. Further increases in input size yield savings at a diminishing rate.

4.3.2. Energy minimization of multi-path circuits

When optimizing circuits, the optimal solution occurs when the delay of each path from input to output is equalized [13, 14]. When analyzing the energy

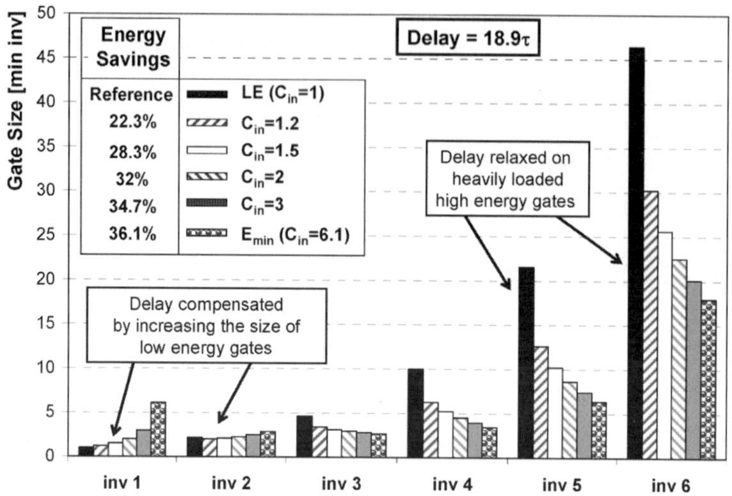

Figure 16. Gate Sizing of an Inverter Chain for Energy Reduction at a fixed Delay.

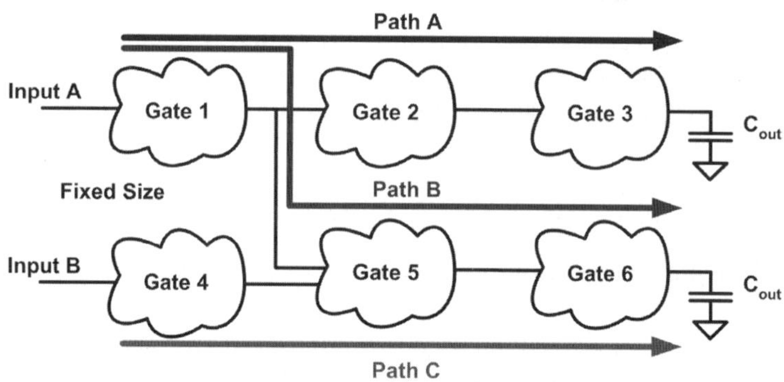

Figure 17. Multi-Path Circuit.

of a circuit, it is necessary to know the sizes of each gate in the circuit (not just those on the critical path). This further complicates the optimization process, as seen in the example of Figure 17. Paths A and B must now be optimized to include the constraint of having equal delay to that of Path C.

The exact solution to this problem requires a numerical approach such as convex optimization [24], from which we can obtain little to no intuition. In [21] we presented a simplified approach to analyze the energy-delay characteristics of an entire circuit. In this approach each gate is assigned to a logic stage, with every gate in the logic stage sized to have the same delay. Gates are assigned to stages starting from the input and moving towards the output. If a path has fewer stages than another path, the last gate of the path is sized to include the combined delay of the additional stages of the longest path. Using this approach the delay of each path in the circuit is always equal, allowing for optimization to be performed at the stage level. The optimization has only a few variables (equal to the number of stages) and can be performed using widely available optimizers such as Matlab and Microsoft® Excel's Solver.

5. Designing Energy-Efficient Digital Circuits

Designing energy-efficient digital circuits requires different guidelines than those developed from Logical Effort. In LE a chain of inverters optimized for delay is used to demonstrate the relative insensitivity of design implementation to number of stages (Figure 18). While delay is relatively insensitive around the optimal number of stages, energy is very sensitive to the number of stages. Inverter chains which contain more stages than delay optimal are always sub-optimal in terms of energy. This result is contrary to LE, and requires that the number of stages be carefully analyzed to obtain an energy-efficient design.

In Figure 19 the minimal energy-delay curves of several inverter chains are shown, each with the same output load and input size. It is observed that the five and six stage designs are never energy-efficient, while the two, three and four stage designs have regions of energy-efficiency depending on the desired operating target.

Contrary to delay based optimization, the location of gates within a chain impacts the energy characteristics of a circuit. For example, the two chains

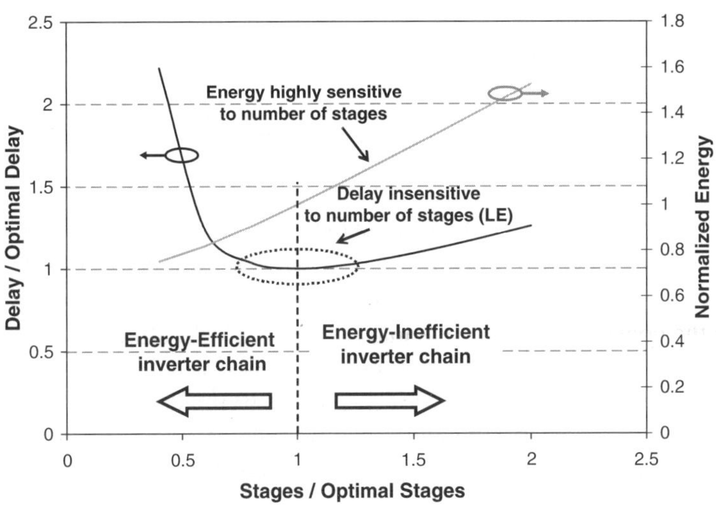

Figure 18. Optimal Number of Stages for an Inverter Chain.

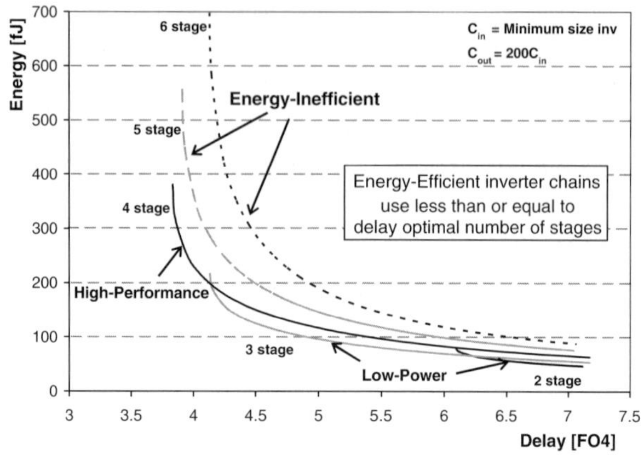

Figure 19. Optimal Number of Inverters for Fixed Output Load and Fixed Input Size with Varying Delay Target.

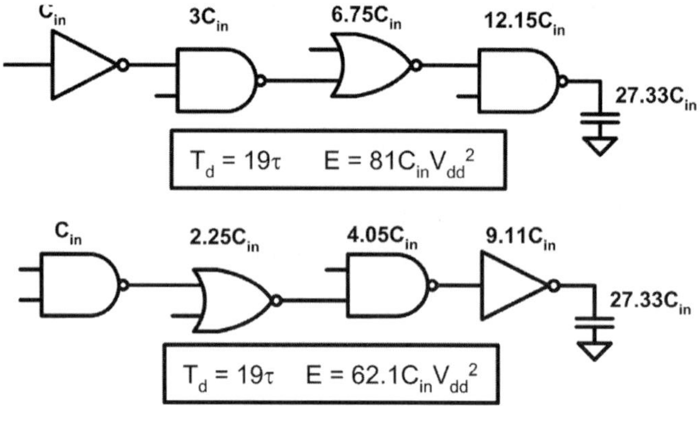

Figure 20. Energy Impact of Gate Placement in a Chain of Gates.

of gates in Figure 20 consist of the same gates, output load and input size, which results in the two paths having the same delay. However, the relative energy of each chain differs, from $81C_{\text{in}}V_{\text{dd}}^2$ to $62.1C_{\text{in}}V_{\text{dd}}^2$. Simpler gates, i.e. those with smaller g and p such as the inverter in the example, require less energy to drive a load than more complex gates. This is because for the same delay they present a smaller input capacitance to the previous gate and have less parasitic capacitance compared to more complex gates. Thus, simple gates should be placed in the most energy sensitive logic stage of a circuit whenever possible.

The arrangement of circuits also has implications on the optimal number of stages. For example, if we examine the impact of adding inverters to the

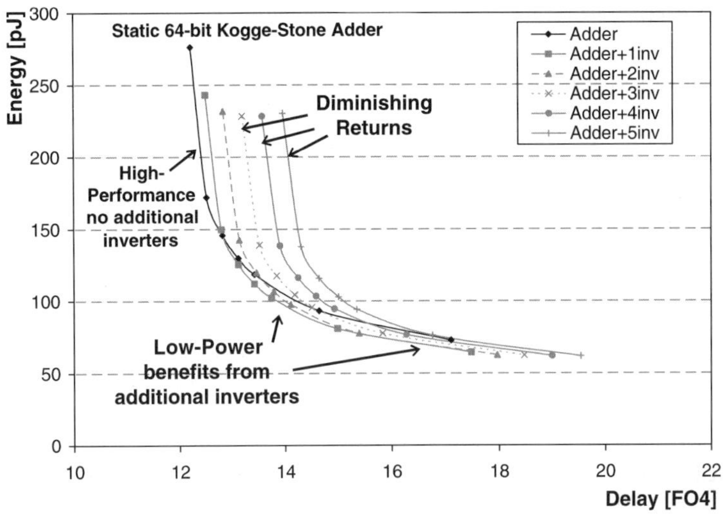

Figure 21. Impact of Buffers Insertion at the Output of a 64-bit KS Adder.

output of a 64-bit static KS adder as in Figure 21. Energy savings are obtained if up to 3 inverters are added at the output, although each occurs at a degraded performance target. Despite having more stages than delay optimal, the energy still decreases. This is because by adding simpler gates to the output, the size of the complex gates in the adder can decrease dramatically (similar to the example in Figure 20). Therefore, despite paying a slight delay penalty due to an extra logic stage, the energy of the design is decreased.

6. Conclusion

The design of digital circuits in current and future technologies requires an understanding of the energy-delay space. Design principles developed for optimizing delay, such as Logical Effort, no longer guarantee efficient designs when energy is considered. We have demonstrated that an energy model can be used in conjunction with standard RC-models to evaluate the energy-delay characteristics of a circuit. The analysis leads to the realization that ED^x metrics can not be used when designing a circuit for a fixed delay or energy. Instead circuits should be optimized for minimal energy at a fixed delay for a variety of system constraints. Using this approach a potential 30–50% energy savings can be achieved for circuits with no performance penalty compared to delay optimized results.

Acknowledgements

The authors would like to thank Hoang Dao and Milena Vratonjic for their comments and suggestions.

References

[1] Taur, Y. "CMOS design near the limit of scaling," *IBM Journal of Research and Development*, **2002**, *46*(2/3).
[2] International Technology Roadmap for Semiconductors, public.itrs.net.
[3] Meyerson, B. "How does one define "Technology" Now That Classical Scaling is Dead?", Keynote presentation, 42nd annual Design Automation Conference, Anaheim, CA, June **2005**.
[4] Sutherland, I.E.; Sproull, R. F. "Logical Effort: Designing for Speed on the Back of an Envelope," Advanced Research in VLSI, Proceedings of the 1991 University of California, Santa Cruz, Conference, Sequin, C.H. ed., MIT Press, **1991**, 1–16.
[5] Sutherland, I.E.; Sproull, R.F.; Harris, D. *Logical Effort Designing Fast CMOS Circuits*, Morgan Kaufmann Pub., **1999**.
[6] Horowitz, M. "Timing Models for MOS Circuits," PhD Thesis, Stanford University, December **1983**.
[7] Rubenstein, J.; Penfield, P.; Horowitz, M. A. "Signal Delay in RC Networks," *IEEE Transactions on Computer Aided Design*, **1983**, *Cad-2*(3), 202–211.
[8] Hodges, D.; Jackson, H. *Analysis and Design of Digital Integrated Circuits*, McGraw Hill, **1988**.
[9] Sakurai, T.; Newton, A. R. "Alpha-Power Law MOSFET Model and Its Application to CMOS Inverter Delay and Other Formulas," *IEEE Journal of Solid-State Circuits*, **1990**, *25*(2), 584–594.
[10] Weste, N.; Eshraghian, K. *Principles of CMOS VLSI Design A Systems Perspective*, Addison Wesley, **1992**.
[11] Oklobdzija, V. G.; Barnes, E. R. "On Implementing Addition in VLSI Technology," *IEEE Journal of Parallel and Distributed Computing*, **1988**, *5*, 716–728.
[12] Fishburn, P.; Dunlop, A.E. "TILOS: A Posynomial Programming Approach to Transistor Sizing," *International Conference on Computer Aided Design*, November **1985**, 326–328.
[13] Sundararajan, V.; Sapatnekar, S. S.; Parhi, K. K. "Fast and Exact Transistor Sizing Based on Iterative Relaxation," *IEEE Transactions on Computer Aided Design of Circuits and Systems*, **2002**, *21*(5), 568–581.
[14] Sapatnekar, S. *Timing*, Kluwer Academic Publishers, Boston, MA, **2004**.
[15] Oklobdzija, V. G.; Zeydel, B. R.; Dao, H. Q.; Mathew, S.; Krishnamurthy, R. "Comparison of High-Performance VLSI Adders in Energy-Delay Space", *IEEE Transaction on VLSI Systems*, **2005**, *13*(6), 754–758.
[16] Zyuban, V.; Strenski, P. "Unified Methodology for Resolving Power-Performance Tradeoffs at the Micro-achitectural and Circuit Levels", *IEEE Symposium on Low Power Electronics and Design*, **2002**.
[17] Zyuban V.; Strenski, P. "Balancing Hardware Intensity in Microprocessor Pipelines," *IBM Journal of Research and Development*, **2003**, *47*(5/6).
[18] Stojanovic, V.; Markovic, D.; Nikolic, B.; Horowitz, M.A.; Brodersen, R.W. "Energy-Delay Tradeoffs in Combinational Logic using Gate Sizing and Supply Voltage

Optimization," Proceedings of the 28th European Solid-State Circuits Conference, ESSCIRC'2002, Florence, Italy, September 24–26, **2002**, 211–214.

[19] Markovic, D.; Stojanovic, V.; Nikolic, B.; Horowitz, M.A.; Brodersen, R.W. "Methods for True Energy-Performance Optimization," *IEEE J. Solid-State Circuits*, **2004**, *39*(8), 1282–1293.

[20] Hofstee, H. P. "Power-constrained microprocessor design," in Proc. Int. Conf. Computer Design, **2002**, 14–16

[21] Dao, H.Q.; Zeydel, B.R.; Oklobdzija, V.G. "Energy Minimization Method for Optimal Energy-Delay Extraction", European Solid-State Circuits Conference, Estoril, Portugal, September 16–18, **2003**.

[22] Dao, H. Q.; Zeydel, B. R.; Oklobdzija, V. G. "Energy Optimization of Pipelined Digital Systems Using Circuit Sizing and Supply Scaling," *IEEE Transaction on VLSI Systems*, **2006**, *14*(2), 122–134.

[23] Kogge, P.M.; Stone, H.S. "A parallel algorithm for the efficient solution of a general class of recurrence equations", *IEEE Trans. Computers*, August **1973**, *C-22*(8), 786–793.

[24] Boyd, S.; Vandenberghe, L. *Convex Optimization*, Cambridge University Press, **2004**.

Chapter 3

CLOCKED STORAGE ELEMENTS IN DIGITAL SYSTEMS

Nikola Nedovic[1] and Vojin G. Oklobdzija[2]
[1]*Fujitsu Laboratories of America*
[2]*Integration Corp., Berkeley, California*

Abstract: Clocking is one of the most critical parts of each processor, often determining its performance and largely impacting its power consumption. The clocking subsystem and clocked storage elements in particular are responsible for an increasingly substantial portion of the circuit design improvements needed to accommodate the continuing scaling trends with each processor generation. This chapter describes the conventional clocking strategies and circuit techniques, and reviews the state-of-the art clocked storage elements used in modern microprocessors. In addition, it addresses some emerging methods aimed at handling incoming challenges in the microprocessor design.

Key words: clocked storage elements; latch; flip-flop; clock; clock jitter; clock skew; pipeline; clock frequency; power.

1. Introduction

Scaling of high-end microprocessors in the past two decades has delivered exponentially increasing performance at the price of exponentially increasing power consumption [1–8] (Figure 1). The design of the clocking subsystem and the *clocked storage elements* (CSE) is at the heart of this tradeoff. As the performance and the circuit complexity scale, the clocking subsystem is put to an increasingly substantial effort to synchronize the operation of the wide pipelines over ever-increasing die size. With the constant increase in the clock frequency and pipeline depth, and the reduction of the number of logic gates per pipeline stage, the clocked storage elements become a major

Vojin G. Oklobdzija and Ram K. Krishnamurthy (eds.),
High-Performance Energy-Efficient Microprocessor Design, 57–88.

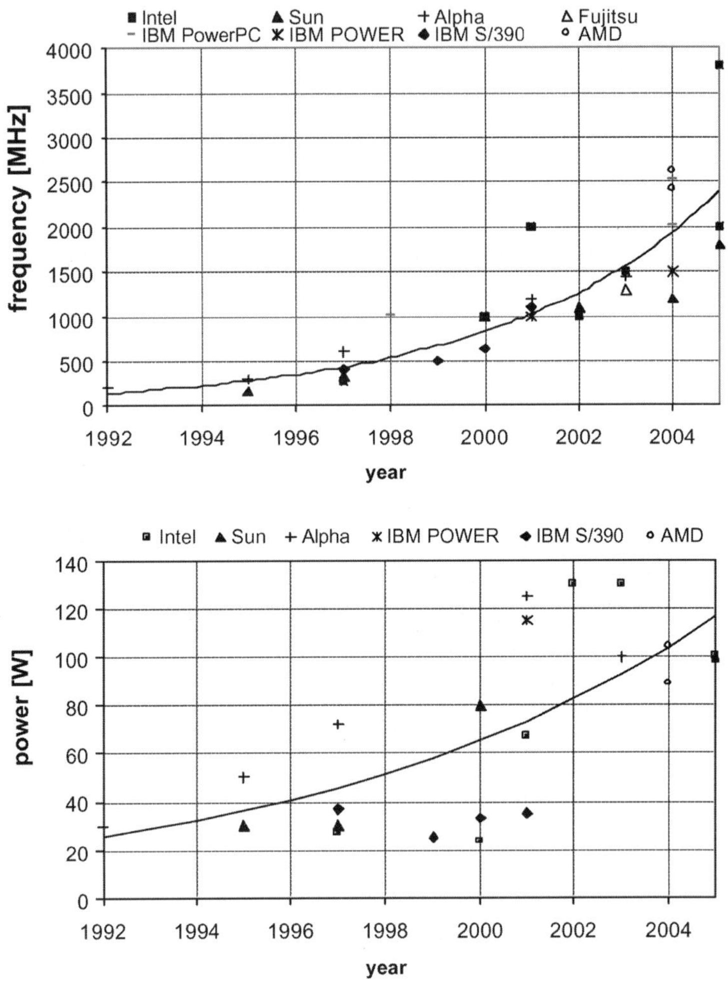

Figure 1. Microprocessor frequency and power trends.

contributor to the microprocessor performance and power consumption. In addition, the *clock uncertainties* do not scale with frequency due to the load mismatches, wire capacitances-dominated delays, and increased effect of the various sources of noise to the clock distribution system. With the high-end microprocessors operating frequencies well in the gigahertz range, clocking is today routinely responsible for 30–50% of their power consumption [1, 9, 10], and it is estimated to occupy up to about 30% of the cycle time. It is thus clear that the design of the clocked storage elements is critical to the overall performance of the entire microprocessor.

This chapter discusses the basic types and properties of the clocked storage elements and clocking strategies. It then reviews clocked storage elements used in the state-of-the-art microprocessors and points out the trends in this area. A number of circuit-level solutions that, combined with the choice of clocking strategy, may be used to improve the performance and power consumption tradeoffs are addressed. Finally, the chapter gives some practical issues in designing the clocked storage elements in microprocessors.

2. Clocked Storage Elements – Basics

The correct operation of a sequential system requires synchronization, i.e. a method of ensuring that the data processing occurs in an orderly manner. Even though this requirement is possible to meet using handshaking (asynchronous) signaling, the use of the global *clock* signal that provides the timing reference for the start and end of the computation is widely accepted as the most efficient synchronization method in complex systems such as microprocessors. In this synchronous framework a sequential system can be viewed using a finite state machine (FSM) model, as shown in Figure 2(a). While the asynchronous signaling promises an efficient way for synchronization of the computation in the increasingly challenging environment of the future microprocessors [11, 12],

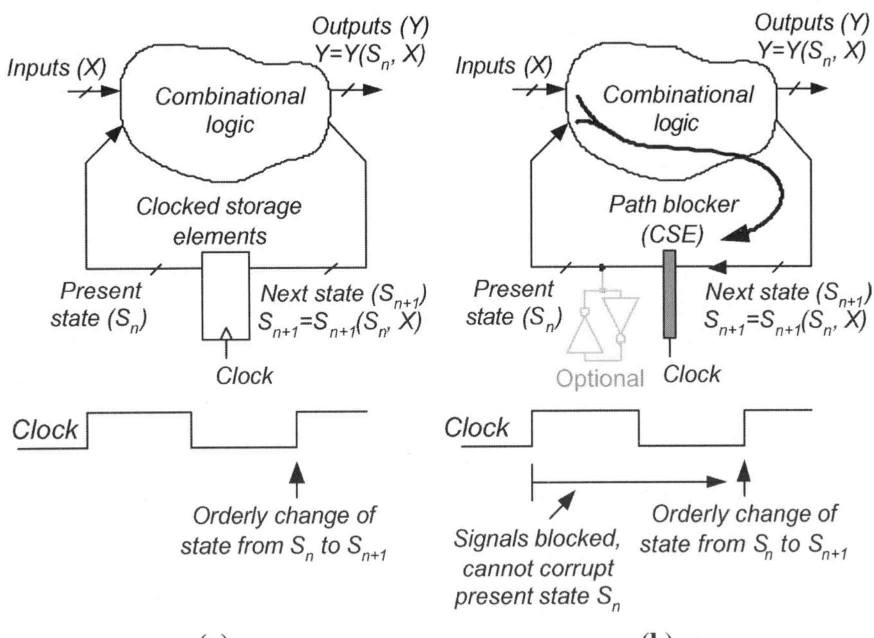

Figure 2. Finite state machine representation of a sequential system: (**a**) conventional representation, (**b**) clocked storage elements viewed as fast path blockers.

it has thus far failed to deliver a significant improvement over clocked (synchronous) methodology. The discussion in this chapter is limited to the synchronous environment.

Contrary to common belief, the main purpose of the clocked storage elements in pipelines, and generally in sequential circuits, is not to store the data computed in the previous cycle so that it can be used in the subsequent stage. Instead, their purpose is to prevent the signals carrying the *next state* (S_{n+1}) from disturbing the *present state* (S_n) signals, according to the timing imposed by the control *clock* signal. Thus, *CSE* blocks the fast paths during the cycle belonging to S_n. Even though they usually store the previously evaluated data, data storage is not their primary purpose. Therefore, any structure that blocks the propagation of the input according to the timing of the control clock signal can be considered a clocked storage element. This view of the finite state machine (FSM) [13], shown in Figure 2(b), encompasses *wave-pipelining* [14] and *opportunistic time borrowing* concept [15]. The most common clocked storage elements are *latches* and *flip-flops*, and most of the attention in this chapter will be dedicated to these structures, but other circuits not commonly associated with clocking, such as conventional domino gates, also belong to this category.

2.1. Non-idealities of the Clock Signal

Ideally, the clock distribution system provides the control clock signal with constant frequency to all termination points, so that the timing references for synchronization arrive to all clocked storage elements at the same moment in the cycle. In reality, however, timing uncertainties exist and must be taken into account in the design of a pipeline stage. These timing uncertainties manifest themselves in two ways: *clock jitter* and *clock skew*.

Clock jitter is the temporal fluctuation of clock arrival time. It can be characterized in two ways. *Cycle-to-cycle jitter* is the variability of the delay between the two consecutive clock edges. *Long-time jitter* is the absolute deviation of the clock edge arrival time over a long period of operation. Depending on the method of clock generation, the long-term jitter can be an order of magnitude larger than cycle-to-cycle jitter. Fortunately, cycle-to-cycle jitter is the most important parameter for the circuit design as it affects the timing constraints in the pipeline stage. In addition, it may be useful to define *short-term* clock jitter (*cycle-to-N-th-cycle* jitter) as the jitter between two clock edges within several (N) adjacent cycles. This definition allows us to account for the accumulating property of clock jitter in the systems where interdependence between data arrivals from several clock cycles exists. As the sources of cycle-to-cycle and short-term clock jitter are usually statistical in nature, they are typically quantified using standard deviation of their instantaneous value.

Clock skew is the spatial fluctuation of the CSE position in the system with respect to the temporally concurrent edges of the clock signal. It is measured at the different termination points of the clock distribution network. Clock skew is mainly caused by the capacitive load and driver strength mismatches in the different paths of the clock distribution system, resulting from the process variations.

In the early days of microprocessor design, clock skew was the dominant component of clock uncertainty, and it occupied a negligible portion of the clock cycle. With recent trends in clock frequency clock jitter introduced by the high-frequency clock generators (phase-locked loops) has become more pronounced [16]. In addition, as circuit complexity increases, the delay mismatches and substrate and power supply noise cause clock distribution clock uncertainties to increase relative to cycle time. As a result, overall clock uncertainty does not scale with clock cycle time, making clock timing non-idealities a major performance factor. For this reason, understanding the effects of clock uncertainties to the timing of the pipeline becomes crucial in high-performance microprocessor design.

2.2. Latch

A latch is a level-sensitive clocked storage element with the following functional behavior: when the control signal (clock) is at the active level the latch is transparent, and the output follows any transition at the input. When the clock is at the inactive level the latch is non-transparent, i.e. it holds the output state (Figure 3a,b). The transition of the clock signal from the active level to the inactive level is referred to as its *latching edge*, and the transition of the clock from the inactive level to the active level is the *releasing edge* of the clock.

The basic timing parameters of a latch are as follows:

- *Set-up time* – maximum allowed data arrival time with respect to the latching clock edge in order to correctly capture it.
- *Hold time* – minimum allowed data arrival time after the latching clock edge in order to correctly capture the previous value.
- *Clock-to-output delay* – delay between the releasing clock edge and the latch output switching to the new value.
- *Data-to-output delay* – delay between the data arrival and the output switching to the new value, assuming that the latch is in the transparent mode.

The most common configurations of transparent latches are pulsed latches and master–slave latches. A *pulsed latch* (PL) is a latch clocked locally with a clock signal of a very short active level, whose duration is made independent of the clock duty cycle – a *clock pulse* (CP) (Figure 3c). In this configuration

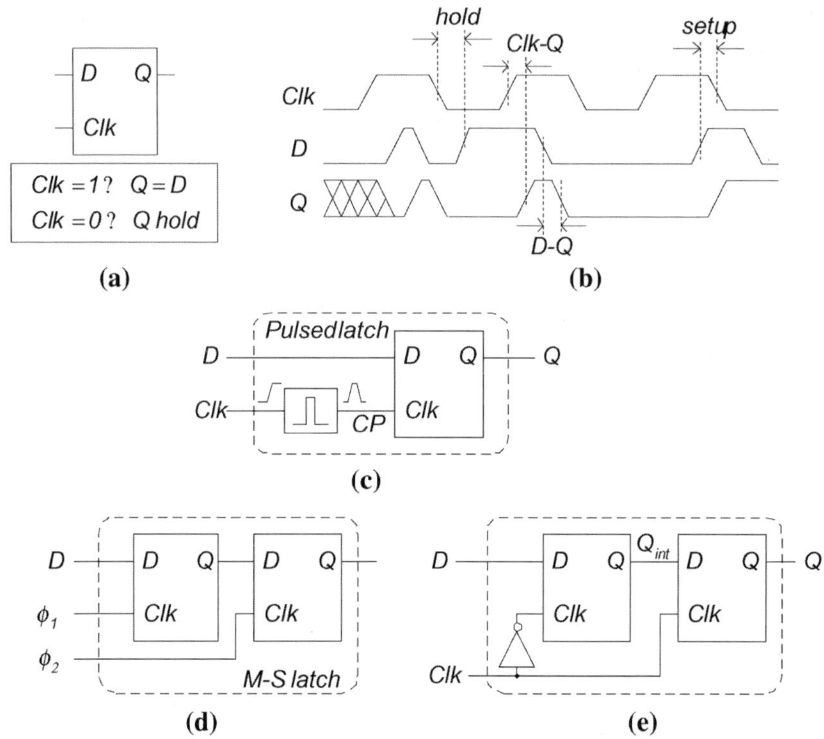

Figure 3. Latch definitions: (**a**) function, (**b**) timing diagram examples, (**c**) pulsed latch configuration, (**d**) master–slave configuration, (**e**) master–slave configuration emulating edge-triggered storage element.

the latch is transparent only during the duration of the clock pulse. If the clock pulse is narrow compared to the clock period the pulsed latch samples the data at the edge of the clock, i.e. it behaves as an edge-triggered storage element. However, note that this simplification generally does not hold in practice, as the worst-case pulse width over all process, voltage, and temperature (PVT) variations is usually not negligible and may even stretch to a quarter of a cycle or more in a very high-speed design. Thus, *pulsed latch* is a mythical creation. In reality we are dealing with the timing parameters that apply to a single latch system as described in Unger and Tan [17].

A master–slave latch (MS latch) is a configuration of transparent latches in which two latches are connected in series and clocked with two independent clock signals (Figure 3d). The input data is captured at the latching edge of the clock ϕ_1 to the front-end (master) latch, and released to output at the releasing edge of the clock ϕ_2 to the back-end (slave) latch. This achieves non-transparency in the system, if the two clocks, ϕ_1 and ϕ_2, are non-overlapping. The timing parameters and the behavior of the master–slave latch depend on the timing between master and slave clock phases, and they are described in detail in Unger and Tan [17].

The most common implementation of master–slave latch pair uses complementary clock phases so that at any moment only one of the latches is transparent (Figure 3e). In this configuration a master–slave latch behaves as an edge-triggered storage element. The MS latch shown in Figure 3e, in which the master latch is transparent on the low level of the clock and the slave latch is transparent on the high level of the clock, behaves as "rising-edge triggered" flip-flop. This is the most common source of confusion and the reason why the M-S latch combination is often referred to as a "flip-flop", which is incorrect.

2.3. Flip-flop

A *flip-flop* is an edge-sensitive clocked storage element that captures the value of the input during the active transition (edge) of the clock; otherwise the flip-flop is non-transparent, i.e. it holds the most recently captured value at the output. A flip-flop consists of two functional stages [13, 18] (Figure 4b):

- Pulse generator that produces a conditional set or reset pulse synchronous to the active edge of the clock, depending on the input value to be captured.

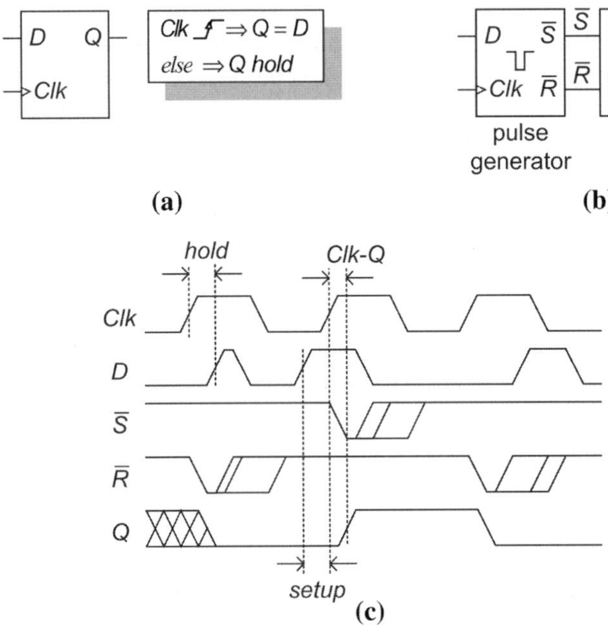

Figure 4. Flip-flop (**a**) function, (**b**) structure, (**c**) timing diagram examples.

- Capturing latch that captures the pulse created by the pulse generator and keeps the captured value.

As opposed to the operation of a pulsed latch, the pulse created by the pulse generator of a flip-flop contains information about the new data to be captured, and the capturing latch is used only to create a static output. Such a pulse need not have the short width as in pulsed latches, and it can be chosen to facilitate the circuit design.

A flip-flop is characterized by the following parameters (Figure 4c):

- *Set-up time* – maximum allowed data arrival time with respect to the capturing clock edge in order to correctly capture it.
- *Hold time* – minimum allowed data arrival time after the capturing clock edge in order to correctly capture the previous value.
- *Clock-to-output delay* – delay between the capturing clock edge and the flip-flop output switching to the new value.

A flip-flop designed to capture the data on the *low-to-high* transition of the clock signal is referred to as *rising-edge-triggered flip-flop*. If the capturing is achieved by the *high-to-low* transition of the clock signal, such a flip-flop is referred to as a *falling-edge-triggered flip-flop*.

3. Clocking Strategies

The design decisions that define the means of achieving data synchronization determine the *clocking strategy*. A pivotal decision in clocking strategy is the choice of the number and mutual timing relationship between clock phases. The most feasible options usually used in modern microprocessors are the *single-phase clock* and the *two-phase clock* as the simplest multiple-phase clocking strategy. While the multiple-phase clocking strategy provides the multiple timing references per cycle that allow better immunity to fast path hazards, the single-phase clock is usually the strategy of choice in modern microprocessor design as it offers the needed simplicity of clock generation and distribution. The most common clocked storage elements used with the single-phase clock are flip-flops, while the transparent latches are usually associated with the multiple-phase clock. The timing analysis in the system with a single-phase clock and flip-flops can also be applied to the master–slave latch with complementary clock phases and the pulsed latch [17].

3.1. Single-phase Clocking Strategy with Flip-flops

The system that uses a single-phase clock, flip-flops, and static logic is shown in Figure 5. This system must satisfy two timing constraints. First, the

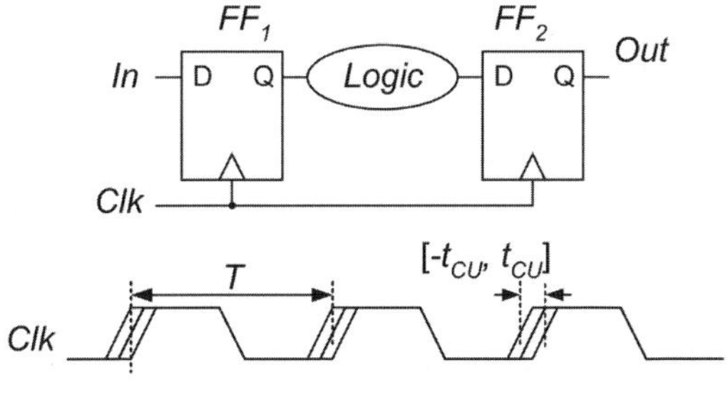

Figure 5. Pipeline stage in a system with flip-flops.

data must arrive to the receiving flip-flop FF_2 at the earliest for hold time t_{H2} after the clock, assuming the earliest possible arrival of the clock to the releasing flip-flop FF_1, the latest possible arrival of the clock to the receiving flip-flop FF_2, the shortest clock-to-output delay of $FF_1(t_{cq1})$, and the shortest logic delay t_l [17]. This constraint, known as *fast path requirement*, is described by Equation (1).

$$t_{cq1} + t_l \geq t_{cu} + t_{H2} \tag{1}$$

Second, the data released by the flip-flop FF_1 must arrive to the receiving flip-flop in time to be captured properly. This condition must be met even with the assumption of the latest possible clock edge at the releasing flip-flop FF_1, the earliest possible clock edge at the receiving flip-flop FF_2, the worst-case clock-to-output delay t_{CQ1} and data traveling through the slowest logic path with delay t_L [17]. This *slow path requirement* is defined by Equation (2).

$$t_{CQ1} + t_L + t_{SU2} \leq T - t_{CU} \tag{2}$$

In Equations (1) and (2), T represents the clock period, and t_{cu} and t_{CU} represent the worst-case time difference between the clock arrivals to the releasing and receiving flip-flop due to the clock uncertainties (jitter and skew). Equations (1) and (2) illustrate one of the key advantages of the single-phase clock scheme that uses flip-flops – the simplicity of the timing analysis that eliminates interdependencies between multiple cycles and multiple pipeline stages. As will be shown later, this is not the case with clocking strategies based on transparent latches.

Note that the clock uncertainty t_{cu} in Equation (1) is not equal to the clock uncertainty t_{CU} in Equation (2). Since fast path requirement in Equation (1) applies to the concurrent clock edges at the releasing and receiving flip-flops, t_{cu} equals the delay mismatch of the two paths from the same source to the different destinations transferring the *same* clock edge. Therefore, only the part of the clock distribution system from the point where the paths to the releasing and receiving flip-flop diverge accounts for t_{cu}. In contrast, the releasing and receiving clock edge in slow paths originate from the two consecutive edges of the reference clock. Therefore, t_{CU} in Equation (2) consists of the jitter from the clock generator and the clock distribution system, and the clock skew between the releasing and receiving flip-flops. Although simple, this observation allows us to make two important points: (1) the clock jitter from the clock generator does not contribute to the clock uncertainty in the fast paths, and (2) placing the releasing and receiving flip-flops close to each other (ideally, in the same local clock domain) helps meeting both fast path and slow path timing requirements.

In addition to defining the specific timing requirement, Equation (2) also points out the timing metrics of the clocking subsystem by indicating the amount of time that the clocking strategy takes from the clock cycle for the purpose of synchronization. This time is the sum of the clock-to-output delay of the releasing flip-flop, setup time of the receiving flip-flop, and the clock uncertainty between the two flip-flops. This synchronization overhead can be divided into the sum of the setup time and the clock-to-output delay, called *data-to-output delay*, as the measure of the performance of the flip-flop, and the clock uncertainty, as the performance metrics of the clock distribution system.

The data-to-output delay characteristic of a flip-flop provides a variety of useful information about its timing properties [13, 18]. It is typically a convex function of its data-to-clock delay, due to the change in clock-to-output delay from the nominal constant value when data arrives much before the clock, to abruptly rising near the failure region (Figure 6). The data-to-clock delay for which the data-to-output delay characteristic reaches its minimum is referred to as the *optimal setup time*. Note that this setup time definition is somewhat different from the conventional *ad-hoc* methods that define setup time as the data-to-clock delay for which the clock-to-output delay increases for a certain

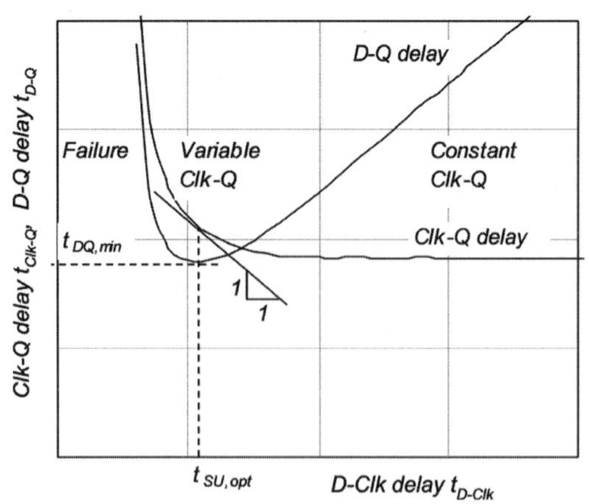

Figure 6. Data-to-output delay of a flip-flop (© 2004 IEEE).

percentage of its nominal value, or for which certain critical nodes experience a (unwanted) glitch of some predefined magnitude. Furthermore, recalling that the data-to-output delay represents the overall timing overhead of the flip-flop, it becomes obvious that a flip-flop whose characteristic is flat around the minimum has a capability to reduce the variation of the output arrival time as the data-to-clock delay changes around optimal setup time. This property, called *soft clock edge*, can be achieved using flip-flops based on transparency windows, pulsed latches or master–slave latches with overlapping clocks. The soft clock edge property has two important implications. First, it enables the flip-flop to absorb the clock uncertainty by means of delivering its output approximately at the same moment regardless of the clock arrival time, when the data arrival time is constant. This property of absorbing clock uncertainties can be quantified using the clock uncertainty absorption coefficient that represents the portion of the clock uncertainty not reflected at the output [18, 19]. Second, the flatness of the data-to-output characteristic indicates a limited transparency of the flip-flop to the input data when the clock arrival time is constant. This property is normally associated with level-sensitive clocking strategies. It allows for increasing the maximum operating frequency by using the technique known as *time borrowing*, which consists of assigning more time for computation to the slow logic in some pipeline stages at the expense of the time assigned to faster logic blocks in other pipeline stages. As the clock uncertainty absorption and time borrowing exploit the same soft clock edge property, these techniques are essentially equivalent.

3.2. Two-phase Clocking Strategy with Transparent Latches

A system with two-phase clocking, transparent latches, and static logic, together with the timing relationship between the clock phases, is shown in Figure 7. In this configuration, also known as a *split-latch system*, the consecutive latches are controlled by the alternating clock phases and separated by the logic blocks. In this way the pipeline stage is divided into two sub-stages.

3.2.1. Timing in a split-latch system

A split-latch system must satisfy one setup time and one hold time requirement for each of the two latches, totaling four independent constraints. Assuming that the earliest data arrives to each latch before the releasing edge of its clock, it can be shown [19] that the hold time requirements yield the following constraints:

$$t_{l1} \geq V_{12} + t_{H2} - t_{cq1} + t_{CU,R1L2} \quad (3)$$

$$t_{l2} \geq V_{21} + t_{H1} - t_{cq2} + t_{CU,L1R2} \quad (4)$$

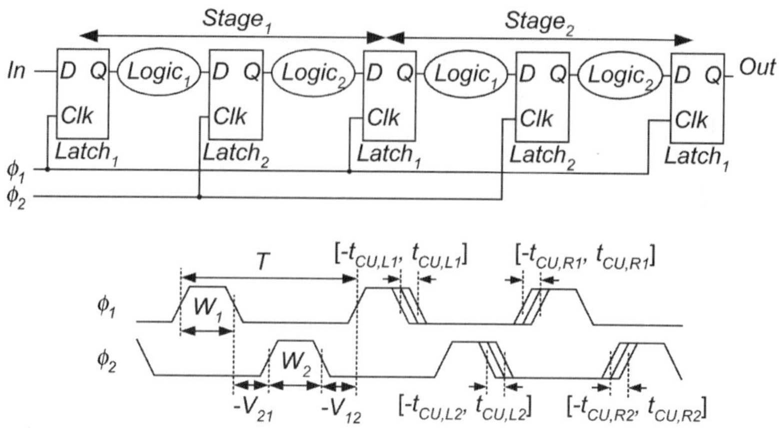

Figure 7. A system that uses split-latch configuration and non-overlapping clocks.

In Equations (3) and (4) t_{l1} and t_{l2} represent the minimum logic delays of the logic blocks $Logic_1$ and $Logic_2$ that satisfy the hold time requirements, V_{12} is the overlap between rising edge of ϕ_1 and falling edge of ϕ_2, V_{21} is the overlap between rising edge of ϕ_2 and falling edge of ϕ_1, t_{H1} and t_{H2} are the hold times of $Latch_1$ and $Latch_2$, and t_{cq1} and t_{cq2} are the minimum clock-to-output delays of $Latch_1$ and $Latch_2$. The clock uncertainty between the releasing edge of ϕ_1 and latching edge of ϕ_2 is denoted $t_{CU,R1L2}$, and the clock uncertainty between the latching edge of ϕ_1 and releasing edge of ϕ_2 is denoted $t_{CU,L1R2}$. The significance of the Equations (3) and (4) is that they show that, with a sufficient amount of clock separation, where clock separation is indicated by negative overlaps V_{12} and V_{21}, the split-latch system guarantees immunity to fast-path hazard by design. This property is very appealing for complex microprocessor design as it reduces or completely eliminates the need for time-consuming and error-prone process of fixing the fast path violations in the design flow.

Since the latches are transparent for a portion of the clock cycle, the timing depends not only on the mutual relationship between the clock phases but also on the moment when the data enters the pipeline stage. We can contain the timing analysis within one stage if we assume that the time arrival of the latest input occurs at the same moment in the cycle in all pipeline stages. In this case, it can be shown [19] that the setup time requirements yield the following constraints:

$$t_{L2} \leq T + V_{21} - t_{SU1} - t_{CQ2} - t_{CU,L1} - t_{CU,R2} \quad (5)$$

$$t_{L1} + t_{L2} \leq T + W_1 - t_{CQ1} - t_{DQ2} - t_{SU1} - t_{CU,L1} - t_{CU,R1} \quad (6)$$

$$t_{L1} \leq T + V_{12} - t_{CQ1} - t_{SU2} - t_{CU,R1} - t_{CU,L2} \quad (7)$$

$$t_{L1} + t_{L2} \leq T + W_2 - t_{CQ2} - t_{DQ1} - t_{SU2} - t_{CU,R2} - t_{CU,L2} \quad (8)$$

$$t_{L1} + t_{L2} \leq T - t_{DQ1} - t_{DQ2} \quad (9)$$

Equations (5) and (6) correspond to the setup time requirement of $Latch_1$ when the data is released by the rising edge of the clock ϕ_2 from the preceding $Latch_2$ (Equation 5), or the rising edge of the clock ϕ_1 from the preceding $Latch_1$ (Equation 6). Similarly, Equations (7) and (8) correspond to the setup time requirement of $Latch_2$ when the data is released by the rising edge of the clock ϕ_1 from the preceding $Latch_1$ (Equation 7), or the rising edge of the clock ϕ_2 from the preceding $Latch_2$ (Equation 8).

In contrast to Equations (5)–(8), Equation (9) defines the timing constraints when the data arrives to all latches at the time they are transparent. It is interesting to note that in this case the minimum logic delay in the stage does not depend on setup times or clock uncertainty. This observation leads us to the two important properties of the level-sensitive systems. First, as long as the latest data arrives to each latch at the time when it is transparent, the level-sensitive system is immune to clock uncertainty. Second, if we give up on our requirement that the input data must arrive to all pipeline stages at the same moment in the cycle, we can tolerate stages with propagation delay larger than nominal, thus violating Equation (9), provided that this time is compensated for by the faster logic in other stages. This property allows another form of time borrowing, and it is summarized in Equation (10):

$$T = \frac{1}{N}\sum_{i=1}^{2N}(t_{DQ,i} + t_{L,i}) \tag{10}$$

Equation (10) states that the minimum clock period T in the pipeline is not constrained by the delay of its slowest stage. Instead, it depends on the average delay of logic and latches in all stages. Again, note that the essential requirement for time borrowing is that the data arrives to all latches during the time they are transparent [18].

The beneficial properties of immunity to fast path hazards and clock uncertainty, as well as the time borrowing, make split-latch system very appealing choice of clocking strategy, and it has been used in the design of some high-end processors [20]. However, it has two serious drawbacks that undermine its overall performance. First, its critical path consists of two transparent latches, which presents a large timing overhead compared to the single-edge system that uses flip-flops [19, 21]. Second, as can be seen from Equations (3–10), the timing analysis of a split-latch system is much more complicated than that based on flip-flops due to larger number of clock phases and interdependence between signal arrivals from multiple cycles. For these reasons most of the microprocessors designed today are based on single-edge clocking strategy with flip-flops. The choice of this strategy is usually accompanied by the sophisticated CAD tools used to identify and fix the hold time violations, and in some cases by the use of fast soft clock edge flip-flops in the critical paths to help reducing effects of clock uncertainty and exploit limited time borrowing.

3.2.2. Other latch-based strategies

Once we have determined the timing constraints for a split-latch system it is easy to reduce them to other latch-based strategies. For example, if the logic block $Logic_1$ is eliminated in Figure 7, the split-latch system reduces to a master–slave-based configuration, where $Latch_1$ becomes master latch, and $Latch_2$ becomes slave latch. Similarly, if the delay of $Logic_1$ is forced to zero, the timing analysis of a split-latch system is directly applicable to a master–slave-based strategy. Note that if the master and slave clocks are non-overlapping, Equations (5–9) collapse into Equation (2) with effective clock-to-output delay equal to the sum of the clock-to-output delay of the slave latch (t_{cq2}) and the clock separation between the master and slave clocks ($-V_{21}$). In other words the master–slave configuration with non-overlapping clocks trades fast path immunity for delay.

Similarly, we can obtain the timing requirements of the single clock phase strategy that uses a single latch if we start from the master–slave configuration, force positive overlap between clocks ($V_{21} > 0$), and combine the master and slave latches into a single latch clocked with the logical AND of the master and slave clocks. The setup and hold times of this single latch are equivalent to the setup and hold time of the original master latch, its clock-to-output delay is equal to the clock-to-output delay of the original slave latch, and its data-to-output delay is equal to the sum of the data-to-output delays of the master and slave latches. With this modification the timing requirements for a master–slave-based system reduce to those for a single-latch-based system [17].

$$t_l \geq W + t_H - t_{cq} + t_{CU,RL} \tag{11}$$

$$t_L \leq T + W - t_{SU} - t_{CQ} - t_{CU,LR} \tag{12}$$

$$t_L \leq T - t_{DQ1} \tag{13}$$

Note that the single-latch-based system achieves very low timing penalty in the slow paths (only one data-to-output delay of a latch), and good time borrowing and clock uncertainty absorption capabilities. However, as shown in Equation (11), these beneficial properties are traded for the need for large minimum logic delays in order to prevent hold time violations. For this reason the single-latch systems are almost exclusively used with the locally controlled narrow clock pulse widths (pulsed latches) in order to minimize fast path hazards.

4. Clocked Storage Elements in Commercial Microprocessors

This section reviews the clocked storage elements used in modern high-end commercial microprocessors. Through a chronological review we will see how frequency scaling drove a dominant trend in high-performance microprocessors from the slower latch-based strategy to faster flip-flops, to even faster pulsed latches with the capability of clock uncertainty absorption. In each step of this evolution the designers traded the reliability of the clocking strategy, i.e. immunity to fast path hazards, for its speed and simplicity. Although an important performance parameter at all times, the clocking energy consumption has only recently emerged as a limiting factor.

4.1. Alpha 21064 (1992)

The first microprocessor in the Alpha family, Alpha 21064 [20], made a radical change from the traditional Digital's four-phase clocking strategy by employing the level-sensitive split-latch scheme with two complementary clock phases. This strategy used modified true single phase clock (TSPC) transparent latches [22], shown in Figure 8, which enabled level-sensitive clocking with a single-wire clock. The main issue with this type of clocking strategy was the introduction of the fast path hazards that were eliminated by using careful design methodology.

The clocking strategy of Alpha 21064 allows for the simple single-wire clock distribution with shallow clock hierarchy. The delay penalty for clocking is the delay of two transparent latches. This penalty is further reduced by embedding various logic gates within the latch.

4.2. Power PC 603 (1994)

The Power PC 603 processor adheres to the clocking methodology of IBM that relies on LSSD methodology, which demands the use of transparent latches and supports full scanability [23]. It employs a conventional master–slave

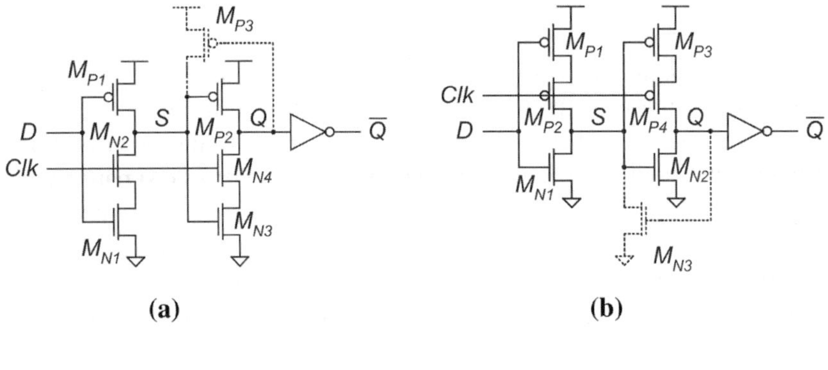

Figure 8. Alpha 21064 latches (© 1992 IEEE): (**a**) transparent on high level of clock, (**b**) transparent on low level of clock.

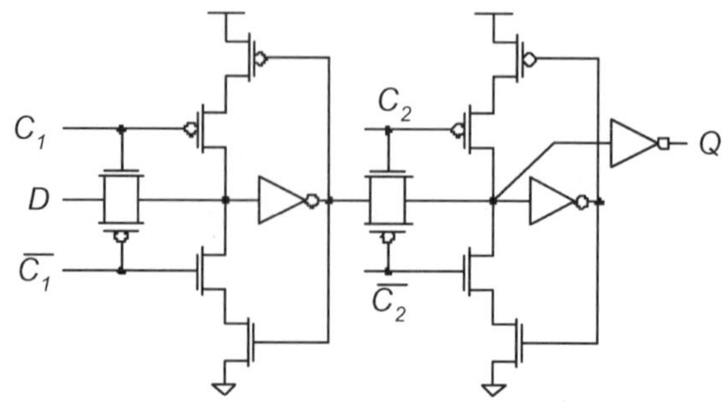

Figure 9. Transmission gate master–slave latch (© 1994 IEEE).

structure, with latches implemented using transmission gates (TGMS latch [24]). The simplified version of transmission gate master–slave latch used in the Power PC 603 is shown in Figure 9. The master and slave clocks supplied to the TGMS are generated in the local clock regenerator, which also controls the timing of the clock phases and the scan clock, and allows for shutting off (gating) the local clock.

The internal switching activity of the TGMS latch is mainly driven by the input data, which naturally reduces the energy of the storage element when the input activity is low. The delay of the TGMS is moderately large compared to clocked storage elements used in later microprocessors, as it consists of the propagation delays through two latches. The input must be protected from the noise by either incorporating an additional inverter driver in front of the transmission gate as the part of the TGMS, or placing a logic gate driver physically close to the latch.

4.3. AMD K6 (1996)

The hybrid latch flip-flop [25], (Figure 10), used in the AMD K6 processor, is a soft-edge flip-flop whose operation is based on generating a short transparency period synchronous to the edge of the clock. This transparency period, or the transparency window, is the portion of the clock period where both clock Clk and delayed reverse-polarity clock Ckd, are high. During the transparency window the first stage of the flip-flop conditionally generates a pulse based on the level of the input D. This pulse generator in the first stage is implemented using the static three-input CMOS NAND gate. The second stage captures the pulse generated by the first stage and extends it to the duration of the clock cycle, thus acting as a latch. The static output is then buffered to ensure signal integrity and sufficient driving capability.

Due to its simple structure, the HLFF is one of the fastest clocked storage elements used in the industry. Furthermore, it has a compact layout, as it requires only 14 transistors in addition to the three inverters for generating

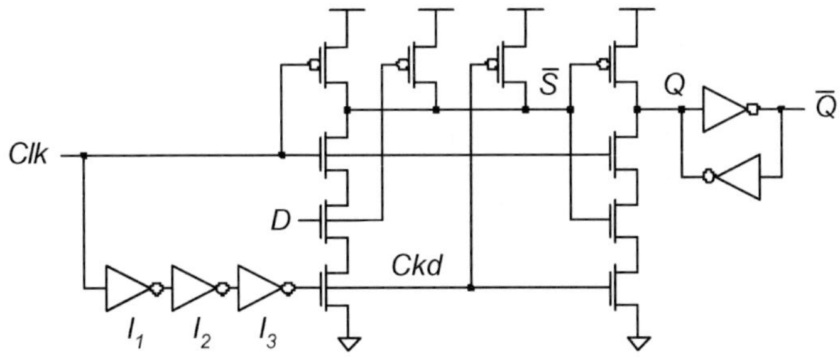

Figure 10. Hybrid latch flip-flop (© 1996 IEEE).

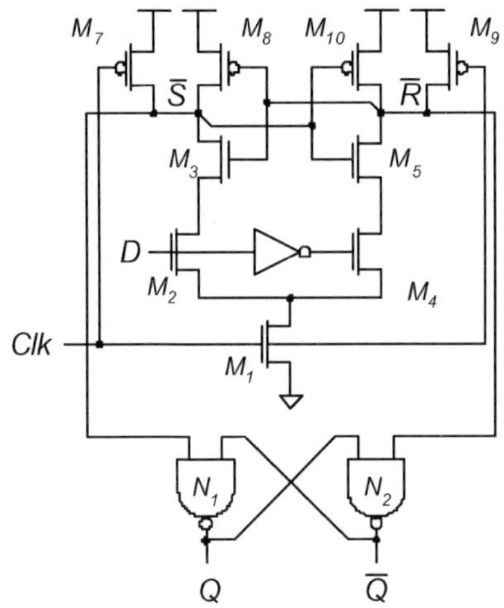

Figure 11. Alpha 21264 flip-flop (© 1998 IEEE).

signal *Ckd* that can be shared among multiple flip-flops. Finally, since HLFF is transparent for a short time after the rising edge of the clock, it has good clock uncertainty absorption property. The most serious drawbacks of this flip-flop are its large energy consumption, which is a consequence of the large internal switching activity, the hazard of the glitch at the output, and somewhat complex implementation of the second stage latch that presents a large capacitive load to the clock.

4.4. Alpha 21264 (1997)

With the increasing demand for performance the third Alpha microprocessor, Alpha 21264, abandoned the slower split-latch scheme used in Alpha 21064 and Alpha 21164 in favor of a faster edge-triggered strategy [26, 27]. The flip-flops used in this clocking strategy were based on the dynamic sense-amplifier flip-flop proposed in Matsui *et al.* [28] shown in Figure 11 [9]. This flip-flop features the differential first-stage pulse generator that compares the voltage levels on inputs D and $\overline{D}$ at the rising edge of the clock. Once the voltage difference between the $\overline{S}$ and $\overline{R}$ outputs of the pulse generators develops, it is further amplified using the positive feedback that disables the discharge of the opposite output. The second stage of the flip-flop is the simple S-R capturing latch, implemented using the back-to-back NAND gates. The NAND implementation of the S-R latch causes unnecessary delay degradation, as the longest signal path must propagate through the two heavily loaded outputs [44, 45]. In critical paths this S-R latch was replaced by the push–pull S-R latch that decouples the two output propagation paths.

4.5. UltraSPARC-III (2000)

The Sun UltraSPARC-III uses the fast edge-triggered flip-flop family whose operation is based on generating the transparency window. The basic flip-flop in this family is the semi-dynamic flip-flop SDFF [29, 30] (Figure 12a). The

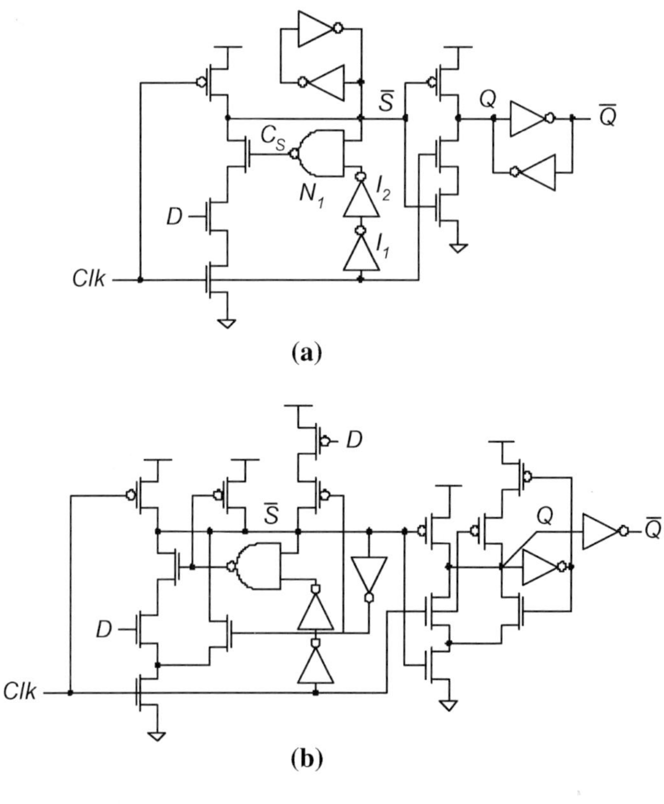

Figure 12. Sun UltraSPARC-III flip-flops: (**a**) basic semi-dynamic flip-flop (© 1999 IEEE), (**b**) final version (© 2000 IEEE)

SDFF is transparent during the time window determined by the delay through the two inverters I_1 and I_2, and the NAND gate N_1 (Figure 12a), which defines the transparency window after the rising edge of the clock Clk. Depending on the logic value of the input D in the transparency window, the first stage of the flip-flop conditionally generates the pulse valid for the duration of the clock width. The second stage of the flip-flop converts the pulse to the static output, which is then buffered for improving the signal integrity and driving capability.

The main feature of the SDFF, compared to the HLFF, is the domino-like implementation of the first-stage pulse generator. This implementation allows for straightforward logic embedding, and eliminates the need for the transparency window in the second stage, reducing its implementation to the simple TSPC-based dynamic-to-static latch [22]. In addition, the SDFF is fast due to a short n-MOS-dominated critical path. It also employs the conditional shut-off mechanism in the first stage that uses positive feedback to improve low-to-high setup time and further improve performance. However, similar to the HLFF, due to the precharge/evaluation operation of the pulse generator, the internal consumption of the SDFF is large even if input switching activity is low. In addition, like the HLFF, the SDFF suffers from a glitch at the output in cycles when both input and previous output are high. This glitch further increases power consumption and reduces the noise robustness of the subsequent logic. The occurrence of the glitch can be eliminated without a significant performance penalty, as shown in Nedovic [31].

The final version of the flip-flop used in the UltraSPARC-III is shown in Figure 12b. It is modified to use the conditional keepers instead of back-to-back inverters used in SDFF in order to achieve sufficient robustness to the soft errors.

4.6. Fujitsu SPARC64 (2003)

With the continuing frequency-scaling trend, microprocessor designers resort to even faster clocking strategies to accommodate growing performance demand. The Fujitsu SPARC64 microprocessor that employs the pulsed latch-based strategy with the capability of clock tuning and clock uncertainty absorption [7] is an example of this tendency. The pulsed latch, shown in Figure 13, is implemented as a simple transmission gate-based design with added functionality for scanability and asynchronous reset. The clock is distributed using a tree structure for the purpose of reducing energy consumption. The local clock buffer has the capability to shut off the clock for further energy saving, and to fine-tune the clock delay to the latches in steps of 20 ps. In addition, the SPARC64 clocked storage element comprises another latch connected in series to the main pulsed latch. The second latch acts as a slave in a master–slave configuration used in non-critical paths with the goal of reducing fast path hazards, and as the scan output [23].

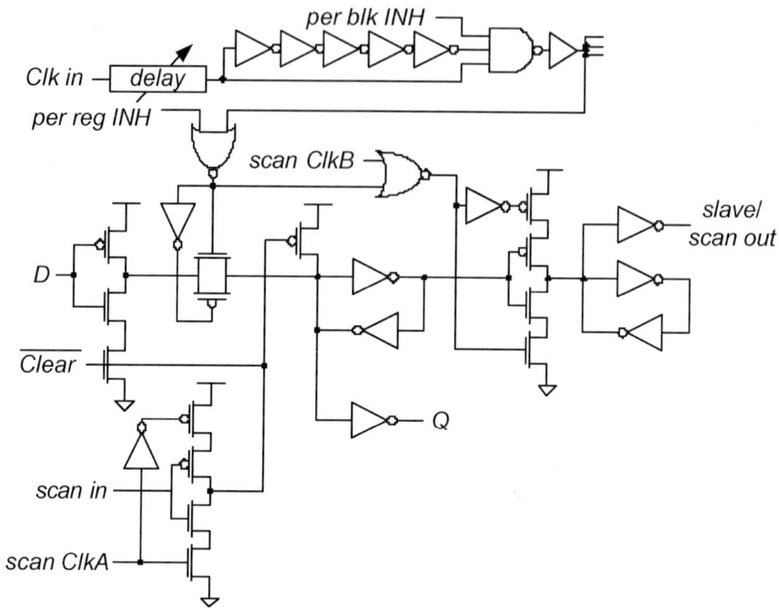

Figure 13. SPARC64 pulsed latch (© 2003 IEEE).

The tunable clock delay allows for optimizing clock arrivals to the pipeline stages with uneven delays. In addition, the clock pulse of the SPARC64 latch is relatively wide (about 15% of the cycle time [7]), which provides the soft clock edge and allows for absorbing clock uncertainties. However, such pulse width causes a large hold time, which requires the use of sophisticated CAD tools for identifying and fixing the fast path hazards.

5. High-performance and Low-power Circuit Techniques

As technology scaling itself is not sufficient to accommodate the increasing demand for microprocessor performance, circuit design is required to provide speed improvements in each product generation. Furthermore, as high-end microprocessors are reaching power consumption limits imposed by heat removal simultaneously to the severe obstacles in technology scaling due to process variability and sub-threshold leakage, circuit design must focus on reducing overall energy consumption while still delivering a sufficient performance. As clocking and clocked storage elements significantly affect microprocessor performance and energy consumption, the innovations in this area receive particularly wide attention.

This section describes several circuit-level techniques aimed at improving performance or reducing energy consumption of clocked storage elements in microprocessor pipelines. These techniques are divided into performance improvement techniques, where we address the design of soft clock edge flip-flops; logic embedding and latchless pipelines, and energy-saving techniques, which include local clock gating and dual edge triggering.

5.1. Design of Clocked Storage Elements with Soft Clock Edge

As shown in Section 3.1, a soft clock edge is the property of the clocked storage element to exhibit a limited transparency around the capturing clock edge, which is manifested by the flatness of its data-to-output characteristic. This property is beneficial for reducing the effect of clock uncertainty and for compensation for unbalanced pipeline stage delays by means of allowing for limited time borrowing between the stages [13, 18]. In order to design a storage element with such characteristics one must define a narrow timing window in which the storage element is transparent to the input. This property can be achieved in three ways. The first option is to introduce delay in the path that locks the state of the pulse generator of the flip-flops. This method is illustrated in Figure 14a that shows the differential skew tolerant flip-flop [32]. In this design the pulse generator is made transparent to the input by delaying the transition of the pulse generator outputs $\overline{S}$ and $\overline{R}$ for the propagation delay of the inverter and the NOR gate before using them to shut off the subsequent transitions of the complement pulse generator output $\overline{R}$ or $\overline{S}$. In addition, this approach achieves significant speedup compared to the original sense-amplifier flip-flop by relocating the computation of some functionality of the pulse generator outside of the critical path.

The second option for achieving the soft clock edge property is to clock a latch with a narrow pulse after each clock edge, as shown in Figure 14b. In this configuration, described in Section 2.2 as the pulsed latch, the storage element is transparent to the input during the duration of the clock pulse. The clock pulse is usually generated locally to the latch in order to minimize the variation in the pulse width and reduce the effects of noise [33].

The third method for achieving the soft clock edge in the clocked storage elements is overlapping master and slave clocks in the master–slave latch configuration, Figure 14c. If the overlap between the releasing edge of the slave clock and the latching edge of the master clock is larger than the delay through the master latch, the latest data arrives to the slave latch only after it becomes transparent. As a result, the master–slave pair is transparent to the input during the short time after the releasing edge of the slave clock. Similar to the pulsed latches, the overlap between master and slave clocks is generated locally.

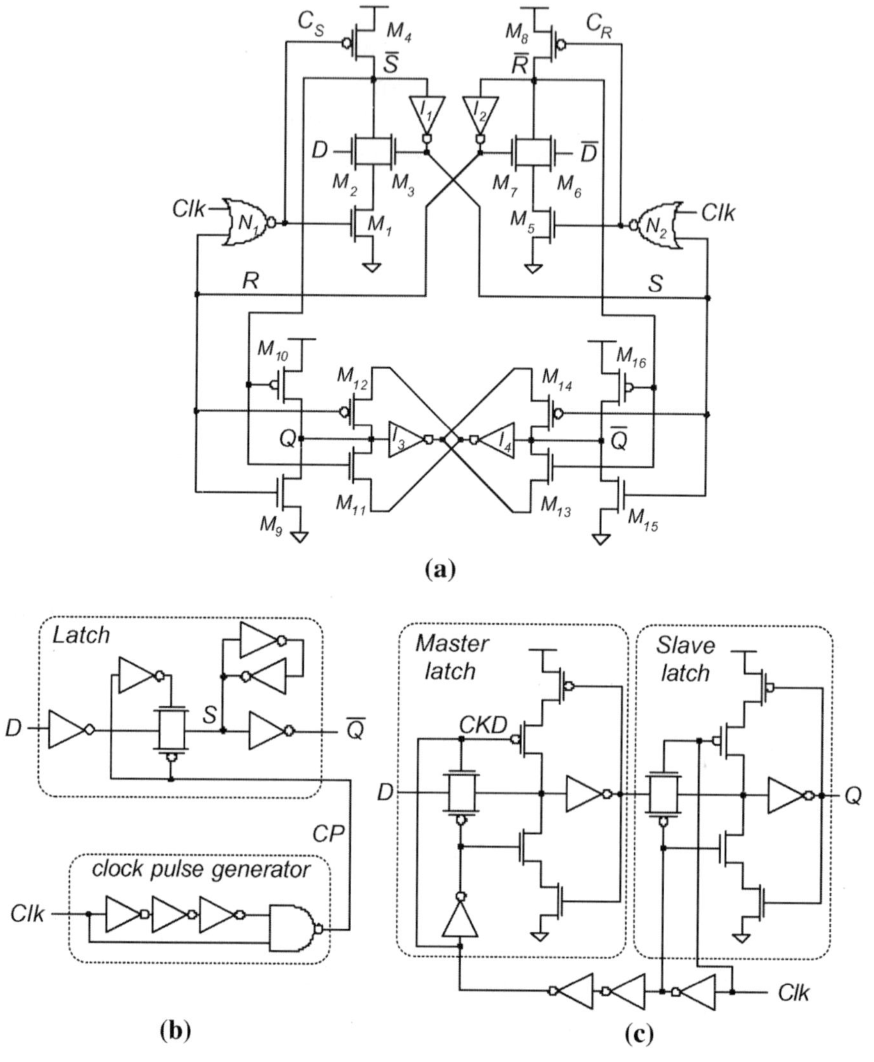

Figure 14. Methods for designing clocked storage elements with soft clock edge: (**a**) flip-flop (© 2003 IEEE), (**b**) pulsed latch (© 2001 IEEE), (**c**) master–slave latch with overlapped clocks.

5.2. Logic Embedding

In order to reduce the timing overhead for clocking in critical paths or for testing purposes the clocked storage elements routinely incorporate a limited logic function. The penalty of this logic embedding depends on the topology of the clocked storage element. Historically, logic embedding was first proposed in a static latch known as Earl's latch in 1965 [34]. As this implementation

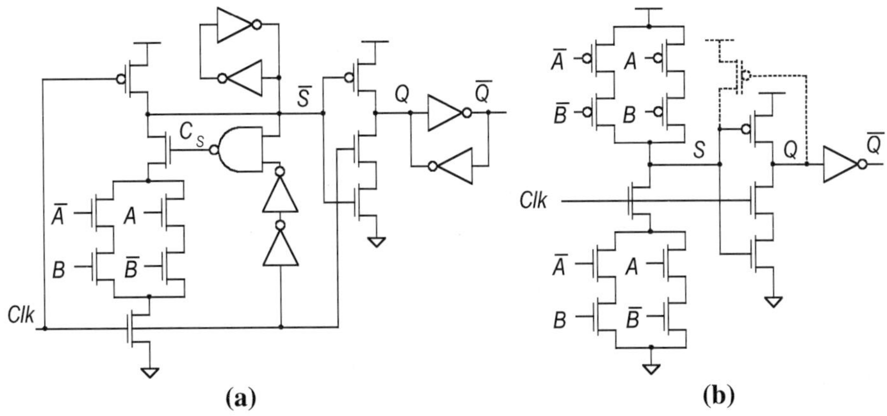

Figure 15. Embedding of XOR logic function: (**a**) NMOS (SDFF), (**b**) CMOS (TSPC).

allows for embedding logic in both stages of the latch, Earl's latch can be used to nearly eliminate the clocking overhead.

The clocked storage element topologies most suitable for logic embedding employ the dynamic logic-like first stage, thus allowing fast and simple implementation of the domino-style logic gate within the storage element. This logic embedding capability is illustrated in Figure 15a that shows the semi-dynamic flip-flop (SDFF) with the XOR logic gate incorporated in its first stage [29]. Note that the logic embedding may increase the transistor stack height, which in practice diminishes the overall delay improvement obtained from grouping the logic with the storage element.

Another implementation of logic embedding is demonstrated in Dobberpuhl *et al.* [19], in which the logic function is incorporated in the first stage of the TSPC latches used in the split-latch configuration (Figure 15b). Due to the latch topology the logic must be implemented in both pull-up and pull-down paths, which makes this implementation less efficient than the one shown in Figure 15a.

5.3. Latchless Pipelines

In modern microprocessors, depending on the design methodology, time-critical pipeline stages are sometimes implemented using dynamic logic [5]. Analyzing the domino logic gate as a typical choice of the dynamic logic family [35, 36] (Figure 16a), we can see that it possesses two features that allow it to control the timing of the signal propagation through the pipeline. First, the fast evaluation data cannot propagate through the domino gate before the releasing (rising) edge of the clock. Second, since the signaling is monotonous by convention, once evaluated data does not change until the pre-charging

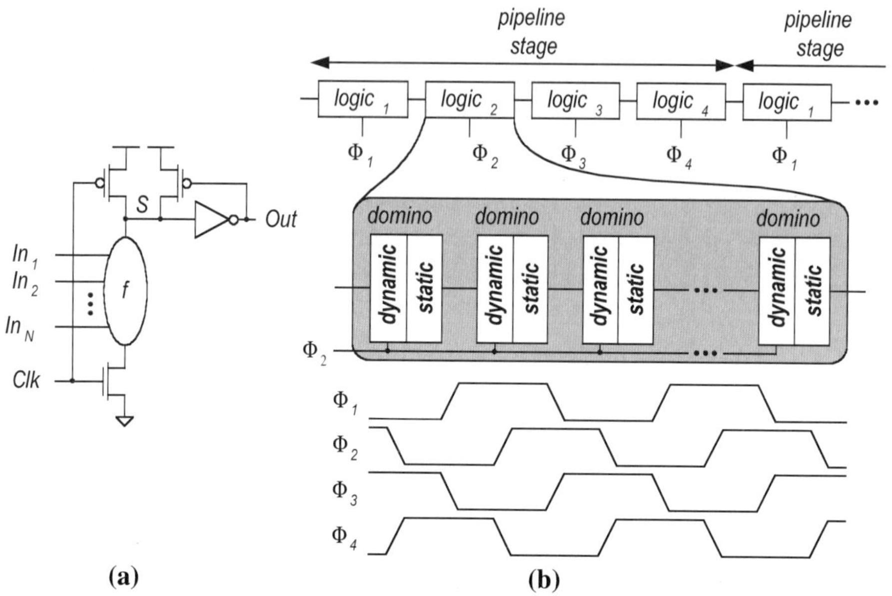

Figure 16. Four-phase skew tolerant domino strategy: (**a**) single-ended footed domino gate (© 1984 IEEE), (**b**) structure of the pipeline stage (© 1997 IEEE).

(falling) edge of the clock, even if the input is invalidated. A suitable clocking strategy can be applied that exploits these properties of domino logic gates to ensure proper data synchronization without explicit latches or flip-flops. The key of such a clocking strategy is to divide the pipeline stage into several substages, each clocked with a separate clock phase, and enforce the timing of the clock so that the data in a substage gets invalidated (pre-charged) only after it has been used in the subsequent substage.

An example of the latchless strategy is shown in Figure 16b [15, 37] together with clock phase timing relationship. In this configuration, known as four-phase *skew-tolerant domino* or *opportunistic time borrowing*, each of the four substages consists of conventional footed domino gates, and is controlled by the delayed clock with respect to that used to clock the previous substage. The strategy ensures that the data enters each pipeline stage one cycle later than it enters the previous stage by delaying the earliest data evaluation between the substages at each of the four edges. The overlap between adjacent clock phases is the essential feature of the skew tolerant domino strategy, as it enables the data evaluated in one substage to be used by the subsequent substage before it is pre-charged.

The skew-tolerant domino strategy has several beneficial properties. First, it uses entire cycle time for useful logic computation. Second, as the latest data can always be set to arrive to each individual domino gate during the time when its clock is high, this strategy is tolerant to the clock uncertainty approximately up to the amount of the overlap between the adjacent clock stages. Hence it allows for clocking with zero performance penalty and for the straightforward implementation of the time-borrowing technique. In addition, for any data propagation path, the receiving clock edge always occurs for $T/4$ before the releasing clock edge for the subsequent data. This property is a direct consequence of using multiple (four) clock reference edges in one cycle, and it practically eliminates the fast path hazard by design.

The disadvantages of the skew-tolerant domino strategy are the need for generating and distributing multiple clock phases, and the increased complexity of the design and timing analysis, as there are no hard boundaries between the stages. In addition, as the relative impact of the clock uncertainty continues to increase with the frequency scaling and process variations, it may become impossible or inefficient to guarantee the overlap between the adjacent clock phases, which may compromise the functionality of the skew-tolerant domino strategy.

A variation of latchless strategy commonly used in the high-performance microprocessors uses the footless domino gates with multiple locally delayed clock phases. Another implementation of the latchless strategy based on domino logic style is presented in Itanium 2 [5]. This strategy uses two global clock phases and specialized dynamic gates with the capability of delaying the pre-charge until the evaluated data is consumed by the subsequent pipeline stage. The use of the pre-charge delay makes the input to all pipeline substages valid during the entire time of its evaluation.

5.4. Local Clock Gating

Both in high-performance and low-power microprocessors there is an increasing demand for saving clocking energy. The clock gating comes as one of the most straightforward methods for energy savings by reducing the switching activity of clock or internal nodes when it is detected that such activity does not perform any useful work. The clock gating can be aimed at the clock distribution (global clock gating), when it usually consists of shutting off the clock to the individual functional blocks during idle periods, or at the clocked storage elements (local clock gating), when it statistically reduces the switching activity of the internal nodes.

The basic principle of local clock gating is to shut off an internal energy-costly computation if it is detected that this computation would not result in changing the state of the storage element. One method of implementing this functionality is used in data look-ahead latch [38], shown in Figure 17a. In this pulsed latch the clock activity is statistically reduced by keeping the latch opaque if the monitoring circuit that consists of an XOR gate detects that the latch output Q is equal to the new latch input D. However, this implementation

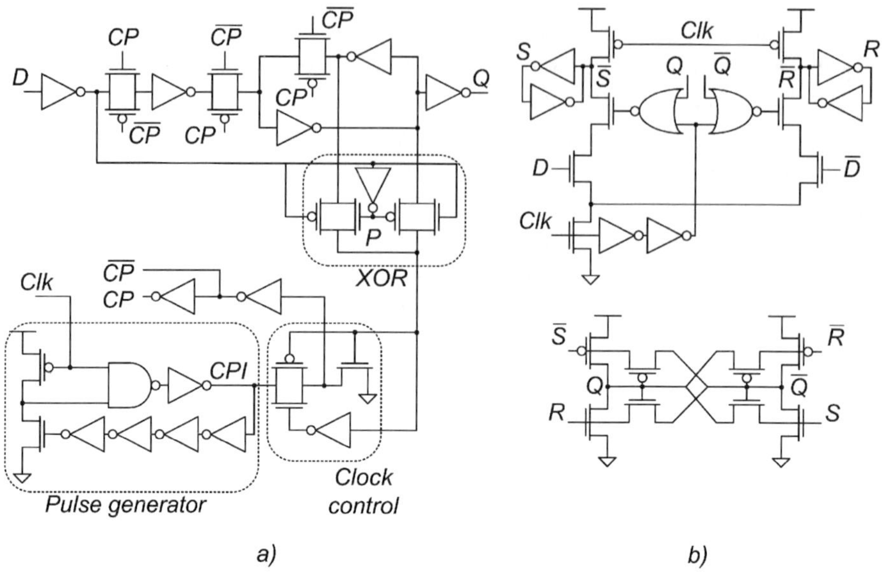

Figure 17. Local clock gating: (**a**) data transition look-ahead latch (© 1998 IEEE), (**b**) conditional capture flip-flop (© 2000 IEEE).

has a severe drawback manifested as the increase of the setup time in order to ensure the minimum width of the clock pulse. This delay increase significantly diminishes the effectiveness of the data look-ahead latch.

Another way of implementing a local clock gating technique is illustrated in Figure 17b that shows the differential conditional capture flip-flop (CCFF) [39]. The CCFF avoids the timing penalty of the activity monitoring circuit by conditioning the generation of the transparency window of the flip-flop with the state of the output. If the output is high, the assertion of node C_S and discharge of node $\overline{S}$ are disabled, even if input D is high. Similarly, if output is low, node $\overline{R}$ cannot be discharged even if input D is low. This scheme reduces the switching activity of nodes $\overline{S}$ and $\overline{R}$, and statistically saves energy consumption. At the same time, the delay of the CCFF is not compromised compared to the basic sense amplifier flip-flop since the output monitoring is implemented in the non-critical circuit for generating the transparency window. Similar to the CCFF, Nedovic and Oklobdjiza [40] propose statistical reduction of the internal switching activity by conditionally disabling the pre-charge of the pulse generator depending on the current state of the flip-flop, thus simplifying circuit topology.

5.5. Dual-edge Triggering

Dual-edge triggering is a clocking strategy that reduces energy consumption of the clock distribution system by using both edges of the clock for data synchronization. This strategy allows for achieving the same data throughput with halved clock frequency compared to the reference single-edge-triggered clocking strategy. In order to fully exploit the potential for energy savings, the dual-edge clocking strategy must meet two criteria [41]. First, the delay, energy consumption, and clock load of the dual-edge-triggered clocked storage elements it uses must be comparable to those of single-edge-triggered clocked storage elements. This may be a challenging requirement, as dual-edge triggering implies considerable increase of circuit complexity with respect to single-edge-triggered clocked storage elements in order to provide additional functionality. Second, due to the use of both clock edges, the clocking strategy must be able to control the clock duty cycle with uncertainty comparable to the uncertainty of the clock cycle time.

Two examples of dual-edge-triggered clocked storage elements are shown in Figure 18. The transmission gate latch-mux (TGLM) [42] (Figure 18a) is

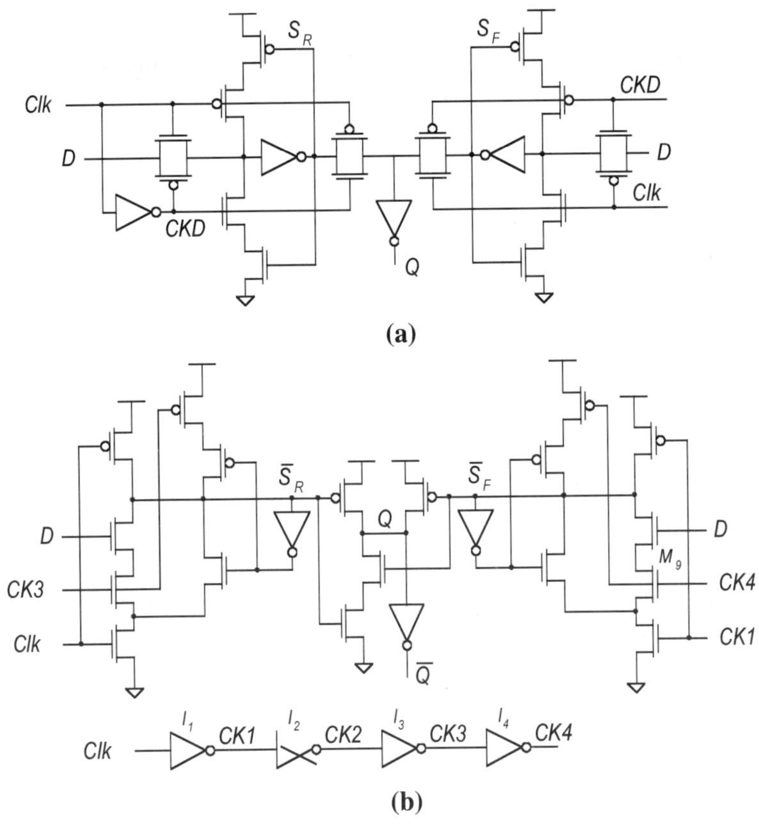

Figure 18. Dual-edge triggered clocked storage elements: (**a**) transmission gate latch mux (© 1996 IEEE), (**b**) symmetric pulse generator flip-flop (© 2002 IEEE).

the representative of latch-mux structure and the dual-edge counterpart of the transmission-gate master–slave latch [24] clocked with the complementary clock phases. It consists of two latches transparent at opposite levels of the clock, and a multiplexor that selects the output of non-transparent latch. The TGLM offers minimal increase in circuit complexity over its single-edge-triggered counterpart. However, the considerable increase in clock load may offset energy savings due to lower clock frequency by the increase of switching capacitance in the clock distribution network.

The symmetric pulse generator flip-flop (SPGFF) [43], shown in Figure 18b, consists of two pulse generators active at the opposite edges of the clock, and the NAND gate that combines outputs of pulse generators into the static output. After each clock edge the transparency window is generated, during which the corresponding pulse generator can evaluate. At any moment one of the nodes S_X and S_Y holds the value of the input sampled at the most recent edge of the clock, and the other one is pre-charged high. The critical path of the circuit consists of a fast dynamic gate and a simple static CMOS NAND gate. Thus the static output is available after two fast logic stages, making the delay of the SPGFF comparable to the fastest single-edge-triggered clocked storage elements. In addition, it presents a very small load to the clock distribution network, thus enabling the full energy savings potential of the dual-edge strategy.

6. Practical Issues in Design of Clocked Storage Elements

The clocked storage elements in microprocessors are presented to application-specific requirements that may significantly influence their design and choice of clocking strategy. Typically, clocked storage elements are categorized as either standard or time-critical. Microprocessor designers commonly use standard clocked storage elements in large volumes throughout the microprocessor, and resort to the use of time-critical clocked storage elements only in critical paths.

Clearly, design requirements for standard and time-critical clocked storage elements are very different. As time-critical clocked storage elements are used in slowest paths, the main concern in their design is speed in order to meet the target cycle time. A desirable property of such time-critical clocked storage elements is a soft clock edge for clock uncertainty absorption and time borrowing. Furthermore, the relatively small number of time-critical clocked storage elements in a microprocessor allows for an aggressively sized design with little attention to their power consumption, layout compactness, and clock load. Similarly, since time-critical clocked storage elements are normally placed in slow paths that are less probable to experience fast path hazards, the hold time is usually of no major concern in their design. The typical critical clocked storage elements in modern microprocessors are pulsed latches and fast flip-flops with a soft clock edge property.

Given the exponentially increasing trend in clocking power and the wide use of standard clocked storage elements in microprocessors, their choice and circuit design is usually more power- and size-aware. This choice is further supported by the observation that the large delay is in fact beneficial, as it helps reduce fast path hazards, while it usually bears little relevance to meeting the cycle time of non-critical paths. Standard clocked storage elements are typically implemented using conservatively sized slow and robust master–slave latches. Note that the improving capability of CAD tools to identify fast path hazards and to optimize a larger number of paths to reduce overall power consumption results in replacing many non-critical clocked storage elements with time-critical ones [7]. This trend alters the design requirements and somewhat blurs the boundaries between the two groups.

Finally, due to the great complexity of modern microprocessors, most if not all clocked storage elements are required to incorporate a scan function in order to facilitate testing. The scanability of clocked storage elements allows for loading arbitrary test vectors, thus bringing the microprocessor in a desired state, and observing the result of the operation of each pipeline stage. During the scan operation the system clock is shut off. The practical methods for embedding the scan mechanism into clocked storage elements depends on its topology and performance requirements. Typically, fast flip-flops use the logic embedding property and combine the scan clock with the local clock in the local clock buffer [30]. The latch-based systems usually provide a parallel input path for scan in the master latch and reuse the slave latch [7, 23, 24].

7. Conclusion

Regardless of whether the technology scaling trend continues or not, the clocking strategy, and specifically the design of clocked storage elements, will remain a key factor for further performance scaling in microprocessors. As the number of logic gates per pipeline stage decreases and the importance of clock uncertainty and clock-related power consumption continues to grow, microprocessor designers must devote increasing attention to the choice and design of clocked storage elements. Accordingly, the past two decades saw a considerable shift in view of the role of clocking in microprocessors. This shift is vividly illustrated by the change of the dominant clocking strategy from multiple-phase clock latch-based to a single-wire fast flip-flop- or pulsed latch-based strategy, often incorporating logic embedding and a soft clock edge property.

The further development of clocked storage element design within the synchronous framework will arguably attempt to satisfy two conflicting requirements. In high-performance microprocessors the trend is to further minimize the timing overhead and the effect of the process variations and mismatch to the clocking by using fast and simple clocked storage elements with logic embedding, and incorporating soft boundaries between the pipeline stages. At the same time the increase in microprocessor complexity, clock frequency, and comparative share of the clocking in overall microprocessor power mandate that power consumption must be a major design objective for a clocking subsystem. Accordingly, clocked storage elements must be designed to accommodate for the power budget, as they are a significant contributor to overall microprocessor power. The success of microprocessors in the near future will to a large extent depend on the choice of clocked storage elements capable of meeting the above contradicting criteria.

References

[1] Anderson, C.J. *et al.* "Physical design of a fourth-generation POWER GHz microprocessor", *IEEE Int. Solid-State Circuits Conf. Dig. Tech. Papers*, February **2001**, 232–233, 451.

[2] Heald, R. *et al.* "A third generation SPARC V9 microprocessor", *IEEE J. Solid-State Circuits*, **2000**, *35*(11), 1526–1538.

[3] Hofstee, P. *et al.* "A 1-GHz single-issue 64b powerPC processor", *IEEE Int. Solid-State Circuits Conf. Dig. Tech. Papers*, February **2000**, 92–93.

[4] Jain, A. *et al.* "A 1.2GHz Alpha microprocessor with 44.8GB/s chip pin bandwidth", *IEEE Int. Solid-State Circuits Conf. Dig. Tech. Papers*, February **2001**, 240–241.

[5] Naffziger, S.D.; Colon-Bonet, G.; Fischer, T.; Riedlinger, R.; Sullivan, T.J.; Grutkowski, T. "The implementation of the Itanium 2 microprocessor", *IEEE J. Solid-State Circuits*, **2002**, *37*(11), 1448–1460.

[6] Stinson, J.; Rusu, S. "A 1.5GHz third generation Itanium processor", *IEEE Int. Solid-State Circuits Conf. Dig. Tech. Papers*, February **2003**, 252–253.

[7] Ando, H. *et al.* "A 1.3GHz fifth generation SPARC64 microprocessor", *IEEE Int. Solid-State Circuits Conf. Dig. Tech. Papers*, February **2003**, 246–247, 491.

[8] Hart, J. *et al.* "Implementation of a 4th-generation 1.8GHz dual-core SPARC V9 microprocessor", *IEEE Int. Solid-State Circuits Conf. Dig. Tech. Papers*, February **2005**, 186–187.

[9] Gronowski, P.E.; Bowhill, W.J.; Preston, W.J.; Gowan, R.P.; M.K.; Allmon, R.L. "High-performance microprocessor design", *IEEE J. Solid-State Circuits*, **1998**, *33*(5), 676–686.

[10] Naffziger, S.; Stackhouse, B.; Grutkowski, T. "The implementation of a 2-core multithreaded Itanium family processor", *IEEE Int. Solid-State Circuits Conf. Dig. Tech. Papers*, February **2005**, 182–183, 592.

[11] Sutherland, I.E. "Micropipelines", *Comm. ACM*, *32*, **1989**, *32*, 720–738.

[12] Furber, S.B.; Garside, J.D.; Gilbert, D.A. "AMULET3: a high-performance self-timed ARM microprocessor", *Proc. Int. Conf. on Computer Design*, October **1998**, 247–252.

[13] Oklobdzija, V.G. "Clocking and clocked storage elements in a multi-gigahertz environment", *IBM J. Res. Dev.*, **2003**, *47*(5/6), 567–584.

[14] Burleson, W.P.; Ciesielski, M.; Klass, F.; Liu, W. "Wave-pipelining: A tutorial and research survey", *IEEE Trans. Very Large Scale Integration (VLSI) Systems*, **1998**, *6*(3), 464–467.

[15] Harris, D.; Huang, S.C.; Nadir, J.; Chu, C-H.; Stinson, J.C.; Ilkbahar, A. "Opportunistic time-borrowing domino logic", U.S. Patent No. 5,517,136, May **1996**.

[16] Mule', A.V.; Glytsis, E.N.; Gaylord, T.K.; Meindl, J.D. "Electrical and Optical Clock Distribution Networks for Gigascale Microprocessors", *IEEE Trans. Very Large Scale Integration Systems*, **2002**, *10*(5).

[17] Unger, S.H.; Tan, C.J. "Clocking schemes for high-speed digital systems", *IEEE Trans. Computers*, **1986**, *C-35*(10), 880–895.

[18] Oklobdzija, V.G.; Stojanovic, V.M.; Markovic, D.M.; Nedovic, N.M. Digital system clocking: high-performance and low-power aspects, New York: John Wiley, **2003**.

[19] Nedovic, N. "Clocked storage elements for high-performance applications", Ph.D. thesis, University of California Davis, June **2003**.

[20] Dobberpuhl, D.W. *et al.* "A 200-MHz 64-b dual-issue CMOS microprocessor", *IEEE J. Solid-State Circuits*, **1992**, *27*(11), 1555–1567.

[21] Stojanovic, V.; Oklobdzija, V.G. "Comparative analysis of master–slave latches and flip-flops for high-performance and low-power systems", *IEEE J. Solid-State Circuits*, April **1999**, *34*(4), 536–548.

[22] Yuan, J.; Svensson, C. "High-speed CMOS circuit technique", *IEEE J. Solid-State Circuits*, **1989**, *24*(1), 62–70.

[23] LSSD Rules and Applications, *Manual 3531*, Release 59.0, IBM Corporation, 29 March **1985**.

[24] Gerosa, G. *et al.* "A 2.2W, 80MHz superscalar RISC microprocessor", *IEEE J. Solid State Circuits*, **1994**, *29*(12), 1440–1452.

[25] Partovi, H.; Burd, R.; Salim, U.; Weber, F.; DiGregorio, L.; Draper, D. "Flow-through latch and edge-triggered flip-flop hybrid elements", *IEEE Int. Solid-State Circuits Conf. Dig. Tech. Papers*, February **1996**, 138–139.

[26] Gieseke, B.A. *et al.* "A 600MHz superscalar RISC microprocessor with out-of-order execution", *IEEE Int. Solid-State Circuits Conf.*, February **1997**, 176–177.

[27] Bailey, D.W.; Benschneider, B.J. "Clocking design and analysis for a 600-MHz Alpha microprocessor", *IEEE J. Solid-State Circuits*, **1998**, *33*(11).

[28] Matsui, M.; Hara, H.; Uetani, Y. *et al.* "A 200 MHz 13 mm^2 2-D DCT macrocell using sense-amplifying pipeline flip-flop scheme", *IEEE J. Solid-State Circuits*, **1994**, *29*(12), 1482–1490.

[29] Klass, F. "Semi-dynamic and dynamic flip-flops with embedded logic", *Symp. on VLSI Circuits, Dig. Tech. Papers*, June **1998**, 108–109.

[30] Klass, F.; Amir, C.; Das, A. *et al.* "A new family of semidynamic and dynamic flip-flops with embedded logic for high-performance processors", *IEEE J. Solid-State Circuits*, **1999**, *34*(5), 712–716.

[31] Nedovic, N.; Oklobdzija, V.G. "Dynamic flip-flop with improved power", *Proc. Int. Conf. on Computer Design*, September **2000**, 323–326.

[32] Nedovic, N.; Oklobdzija, V.G.; Walker, W.W. "A clock skew absorbing flip-flop", *IEEE Int. Solid-State Circuits Conf. Dig. Tech. Papers*, February **2003**, 342–343.

[33] Tschanz, J.; Narendra, S.; Chen, Z.; Borkar, S.; Sachdev, M.; De, V. "Comparative delay and energy of single edge-triggered and dual edge-triggered pulsed flip-flops for high-performance microprocessors", *Int. Symp. Low Power Electronics and Design, Dig. Tech. Papers*, August **2001**, 147–152.

[34] Earl, J. "Latched carry-save adder", *IBM Tech. Disclosure Bull.*, **1965**, *7*(10), 909–910.

[35] Krambeck, R.H.; Lee, C.M.; Law, H.-F.S. "High-speed compact circuits with CMOS", *IEEE J. Solid-State Circuits*, **1982**, *SC-17*(3), 614–619.

[36] Oklobdzija, V.G.; Kovijanic, P.G. "On testability of CMOS-domino logic", *Proc. FTCS-14: 14th IEEE Int. Conf. on Fault-Tolerant Computing*, **1984**, 50–55.

[37] Harris, D.; Horowitz, M. "Skew-tolerant domino circuits", *IEEE J. Solid-State Circuits*, November **1997**, *32*(11), 1702–1711.

[38] Nogawa, M.; Ohtomo, Y. "A data-transition look-ahead DFF circuit for statistical reduction in power consumption," *IEEE J. Solid-State Circuits*, **1998**, *33*(5), 702–706.

[39] Kong, B.-S.; Kim, S.-S.; Jun, J.-H. "Conditional capture flip-flop technique for statistical power reduction", *IEEE Int. Solid-State Circuits Conf. Dig. Tech. Papers*, February **2000**, 290–291, 465.

[40] Nedovic, N.; Oklobdzija, V.G. "Improved hybrid latch flip-flop design", *Proc. 13th Symp. Integrated Circuits and Systems Design*, September **2000**, 211–215.

[41] Nedovic, N.; Oklobdzija, V.G. "Dual-edge triggered storage elements and clocking strategy for low-power systems", *IEEE Trans. on Very Large Scale Integration Systems*, **2005**, *13*(5), 577–590.

[42] Llopis, R.P.; Sachdev, M. "Low power, testable dual edge triggered flip-flops", *Int. Simp. Low Power Electronics and Design, Dig. Tech. Papers*, **1996**, 341–345.

[43] Nedovic, N.; Oklobdzija, V.G.; Aleksic, M.; Walker, W.W. "A low power symmetrically pulsed dual edge-triggered flip-flop", *Proc. 28th European Solid-State Circuits Conf.*, September **2002**, 399–402.

[44] Nikolic, B.; Stojanovic, V.; Oklobdzija, V.G.; Wenyan Jia, Chiu, J.; Leung, M. "Sense amplifier-based flip-flop", *IEEE Int. Solid-State Circuits Conf. Dig. Tech. Papers*, February **1999**, 282–283.

[45] Stojanovic, V.; Oklobdzija, V.G. FLIP-FLOP, US Patent No. 6,232,810, 15 May **2001**.

Chapter 4

STATIC MEMORY DESIGN

Nestoras Tzartzanis
Fujitsu Laboratories of America
E-mail: nestoras@us.fujitsu.com

Abstract: In this chapter, we explore the design of static memory structures. We start with the description of the operation of the single-port six-transistor SRAM cell. Subsequently, we discuss issues related to the design of memory peripherals such as voltage-mode and current-mode differential reads, single-ended reads, and control logic. We present design techniques to reduce SRAM dynamic and static power. We also cover issues pertaining to SRAM reliability, testing, and yield. Subsequently, we show how to implement efficient multi-port register file storage cells and peripherals and we present an example of replica-based self-timed control logic.

Key words: Static memory; SRAM; register file; cache; six-transistor SRAM cell; multiport memory; differential memory reads; voltage-mode sense amplifiers; current-mode amplifiers; single-ended memory reads; low-power SRAM; lowvoltage SRAM.

1. Introduction

During the microprocessor evolution, memories became an integral part of microprocessor design. The first integrated microprocessors contained only register files as storage for temporary data, while their memory system was entirely located off chip. The first microprocessors with on-chip caches were reported in 1987 [1, 2]. Current microprocessor chips include up to three levels of cache memory [3–5]. Furthermore, the total on-chip memory capacity increased from a few kilobytes to several megabytes. Consequently, there is

Vojin G. Oklobdzija and Ram K. Krishnamurthy (eds.),
High-Performance Energy-Efficient Microprocessor Design, 89–119.

a strong demand for dense, fast, and energy-efficient memories. In addition, there are some trade-offs between density, speed, and energy dissipation that can be made, depending on memory design specifications. For instance, the primary concern of multi-port register files (RF) is their delay time and clock frequency. At the other end of the spectrum, density and energy efficiency are more important than speed for single-port SRAM intended for second- or third-level caches. Although advances in CMOS technology have facilitated the integration of large SRAM inside microprocessor chips, they also have created new design challenges such as reduced noise immunity and excessive dynamic and static power dissipation.

In this chapter, we provide some basic single-port SRAM and multi-port RF design guidelines. Although SRAM and RF share many design similarities, they also differ in many aspects. Single-port SRAM use the same bit lines for both writes and reads. Typically, multi-port RF have dedicated write and read ports, and hence, bit lines are used either for writes or reads. Furthermore, SRAM have larger bit capacity than RF and they commonly require both row (i.e. word) and column (i.e. bit) decoding.

Next, we describe the single-port six-transistor memory cell including static noise margin considerations. We continue with some basic SRAM array organization and we focus on different bit-line architectures. We study both differential and single-ended read-out circuits. Subsequently, we present techniques to reduce both dynamic and static power in SRAM. Memories contain many high-capacitance nodes such as address, word, bit, and data lines that are major contributors of dynamic power. Their power dissipation can be reduced by lowering their voltage swing, decreasing their switching capacitance, or recovering some of their energy. Since SRAM are the densest microprocessor blocks in terms of transistor width per area unit, leakage power has emerged as a primary design issue. Proposed techniques to reduce leakage power include supply voltage control, body bias control, and sleep transistors. We finish the SRAM subsection with a discussion on reliability and testing. Then, we shift our focus on RF design starting with various RF cell topologies and continuing with RF bit-line architectures and control logic. Finally, we conclude with a summary.

2. Single-port SRAM Design

In this section, we analyze various design configurations of SRAM key components such as the six-transistor memory cell, typical memory array arrangements, bit-line architectures, and control logic. We also cover techniques to reduce SRAM dynamic and static power and we address SRAM reliability and testing issues.

2.1. Six-transistor Storage Cell

The six-transistor memory cell (Figure 1) consists of two pull-down or drive nFET (MN_1 and MN_2), two pull-up or load pFET (MP_1 and MP_2) and two access or transfer nFET (MA_1 and MA_2). The differential bit lines (BL and $\overline{BL}$) serve as inputs for writes and outputs for reads. The word line (WL) enables the cell. During writes, one of the bit lines is driven to the supply voltage V_{dd} while the other is clamped to ground, forcing the new data into the cell. During reads, both bit lines are precharged before WL is enabled. When WL is enabled, depending on the data stored in the cell, one of BL or $\overline{BL}$ is discharged.

Sizing the six transistors is a delicate procedure since there are several conflicting criteria that must be considered. First, the cell must be as small as possible to improve SRAM density. Second, the access and pull-down nFET must be strong enough to discharge the highly-capacitive bit line within a specified time. Third, making the access nFET strong increases the bit-line capacitance which affects its discharge time. Fourth, increasing the strength of either the access or the pull-down nFET increases the bit-line leakage current which increases static power and reduces the bit-line noise immunity. Fifth, the cell static noise margin (SNM) that determines its stability, i.e. its immunity to process, temperature, and supply voltage variations must be maximized given other specifications.

Some of the design criteria listed above depend on the system specifications. For instance, fast access time is more important than high density for a level-1 cache. In contrast, high density and low static power are more important than fast access time for a level-3 cache. Thus, for example, three levels of cache designed for the same microprocessor were optimized for different specifications: the first level was optimized for latency [6], the second level for bandwidth [7], and the third level for area efficiency [3]. Regardless of other specifications, ensuring correct operation by improving noise immunity is key to the design of any SRAM. In this subsection we study the cell static noise

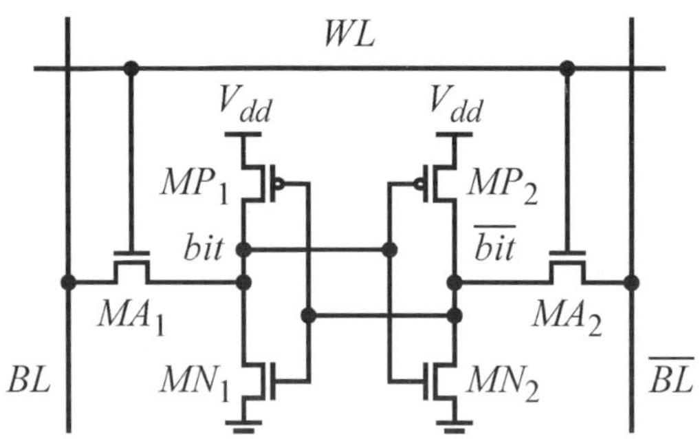

Figure 1. Six-transistor SRAM cell.

margin. Noise issues related to the bit lines due to their leakage and coupling are addressed in later subsections.

The SRAM cell SNM is defined as the minimum dc noise voltage that when applied to its internal nodes *bit* and $\overline{bit}$, causes the data stored in the cell to flip [8, 9]. In other words, SNM is the maximum dc noise voltage that the cell can tolerate. When the cell is idle, i.e. during data retention, *WL* is 0 V and the cell flip point depends on the voltage transfer characteristics of the cross-coupled inverters. The cell however is more sensitive to noise during reads when *WL* is V_{dd} and both bit lines *BL* and $\overline{BL}$ are precharged to V_{dd}. Then the voltage of the internal node (*bit* or $\overline{bit}$) that is "0" raises to above 0 V. The voltage level that this node raises to depends on the ratio of the transconductances between the pull-down nFET and the access nFET, which is called the cell ratio *r* [9, 10]. Assuming that both nFET have the same channel length, *r* is determined by the width ratio of the pull-down nFET to the access nFET. Furthermore, during reads the access nFET is connected in parallel to the pull-up pFET changing the voltage transfer characteristics of the cross-coupled inverters [9, 10].

Figure 2 shows the voltage transfer characteristics for an SRAM cell designed for a 90 nm CMOS process during data retention ($WL = 0$ V) and during read ($WL = V_{dd}$) for different cell ratios *r*. The cell SNM is graphically represented by the side of the maximum square that fits within the voltage transfer characteristics of the two cross-coupled inverters. During data retention the cell SNM is approximately the same, independent of *r*. However, during read, the cell SNM significantly improves for increasing *r* since the voltage of the "0" node reduces from 196 mV ($r = 1$), to 110 mV ($r = 2$), and

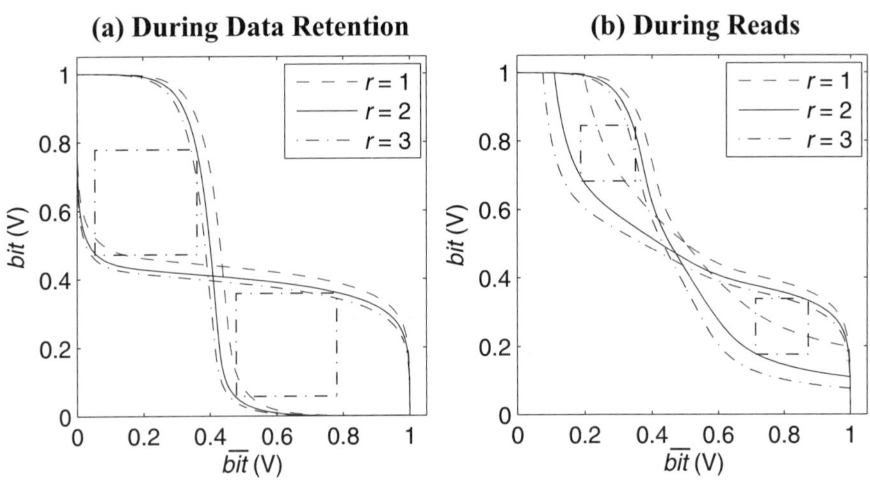

Figure 2. SRAM cell noise margins for different cell ratios *r* (**a**) during data retention and (**b**) during reads; squares are drawn for $r = 2$.

finally to 76 mV ($r = 3$). As CMOS technology scales, SRAM cell SNM tends to decrease because of supply voltage and threshold voltage scaling [9]. To mitigate the problem, in some deep sub-micron CMOS technologies, high-threshold voltage FET are available for SRAM cells [4].

Other issues to consider in sizing the cell FET are that reads must be non-destructive and writes must cause the cell to change state. The former can be accomplished by meeting the cell SNM requirements so that the voltage rise of the "0" node is lower than the cell flip voltage. Since the access devices are nFET, they are more effective in passing a "0". Assume that the cell stores a "1" (*bit* is V_{dd}) and a "0" must be written (*BL* is 0 V). For the write to be successful, the access nFET MA_1 must over-power the pull-up MP_1. Once the voltage of *bit* is lower than the threshold voltage of the inverter formed by MP_2 and MN_2, $\overline{bit}$ switches to V_{dd} and the cell flips. If both the access nFET and pull-up pFET are minimum-size, the access FET would be stronger than the pull-up pFET due to the higher mobility of electrons compared to holes. Increasing the strength of access nFET improves write speed at the expense of reducing the cell SNM and increasing the cell size, the bit-line capacitance, and the bit-line leakage current.

2.2. Organization

Typically, SRAM are organized in a single array which is divided into blocks. For instance, in ref. 11 the SRAM array has three levels of hierarchical word decoding: global, subglobal, and local. The advantage of this architecture compared to a flat design is that it reduces the decoding time since local word lines that connect to the cells are relatively short and contain small capacitance. Furthermore, it also reduces dynamic power since only the bit lines of the active blocks are enabled. Another approach applicable to large caches is to divide the cache into subarrays or macros [3–5]. Individual SRAM macros can be used as "tiles" to fill irregularly shaped areas achieving high area efficiency. Using SRAM macros to build the cache also results in low dynamic power if the idle macros are disabled [12].

A typical SRAM macro (Figure 3) consists of the cell array, a row decoder (including word line drivers), a predecoder, a column decoder (including bit-line precharge circuitry), sense amplifiers for reads, and bit-line drivers for writes. SRAM macros usually are small enough that it is not necessary to divide them into blocks. The cell array contains n rows and m columns. The column decoder operates as a demultiplexer for writes and a multiplexer for reads selecting k out of m bits that are accessed. Usually, column decoding combines 8 or 4 adjacent row bits to 1 input (writes) or output (reads) bit. Although the row decoder is shown on the side of the array, usually it is placed in the middle of the array to reduce the word-line RC delay. Signals *SE* and

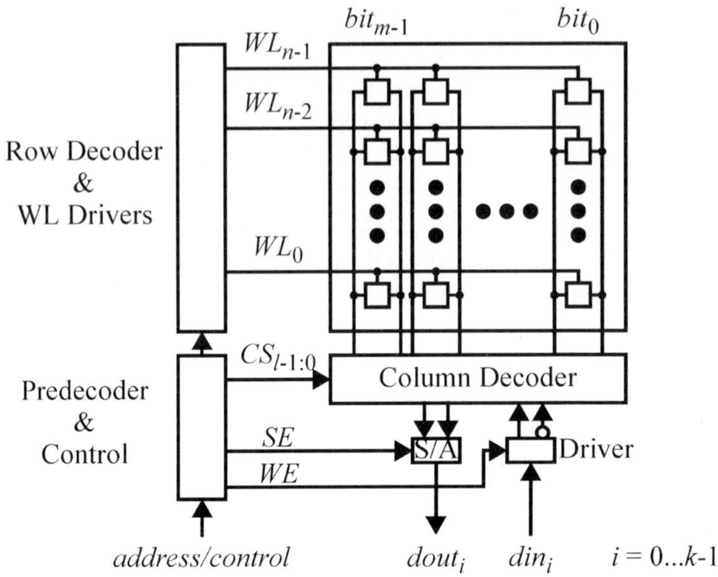

Figure 3. Typical SRAM macro organization.

WE enable the sense amplifiers (for reads) and the write drivers (for writes), respectively. Signals $CS_{l-1:0}$ control the column decoder with $l = 8$ or $l = 4$. Although not shown in Figure 3, address, control, and input data are generally latched at the boundary of the SRAM macro. Output data could be latched depending on the timing specifications.

Predecoders and decoders can be designed using static NAND or NOR gates followed by inverter buffers. They can be sized similarly to standard CMOS gates based on their output load [13]. It is also common for their devices to be skewed for fast rise time to reduce access time. Furthermore, they can be self-reset to produce pulsed word lines.

2.3. Bit-line Architecture

Next to the storage cell, the bit-line architecture is the most important design aspect in an SRAM. As discussed earlier, the cell FET are relatively weak in order to keep the cell small. Therefore, discharging the highly-capacitive and highly-resistive bit line through the cell is a slow process. Moreover, the bit-line architecture determines the design of the rest of the peripheral logic as well as the control logic. Traditionally, the inherently differential structure of the SRAM bit lines has been used to speed-up the read operation through signal amplification.

A typical bit-line structure containing two adjacent columns ($BL_a/\overline{BL_a}$ and $BL_{a-1}/\overline{BL_{a-1}}$) is shown in Figure 4. MP_5, MP_6, MP_7, and MP_8 precharge the bit lines while MP_9 and MP_{10} equalize them to ensure both bit lines within a pair are at the same potential before the cell is read. FET pairs MN_1/MP_1, MN_2/MP_2, MN_3/MP_3, and MN_4/MP_4 form transmission gates that perform the column decoding task, i.e. they connect one out of the two bit-line pairs to $d_i/\bar{d}_i$. During reads, first the bit lines are precharged. Subsequently, the word line is enabled along with the column decoder. Finally, when enough signal difference has been built between d_i and $\bar{d}_i$, the sense amplifier is enabled. During writes, one of the precharged bit lines is discharged through MN_5 or MN_6 depending on din_i, while the column decoder and the word line are enabled. During writes, cells that are located in the non-selected bit lines maintain their data since both of their bit lines are left precharged similar to a (non-destructive) read operation. Although for simplicity two columns are shown in Figure 4, typically four or eight bit-line pairs are connected to $d_i/\bar{d}_i$ through the column decoder transmission gates. In the extreme case that there is only one column per output bit, these transmission gates are not required since their purpose is to multiplex the bit lines.

Applying differential bit-line sensing during reads not only decreases the read delay, but also improves noise immunity due to common-mode noise

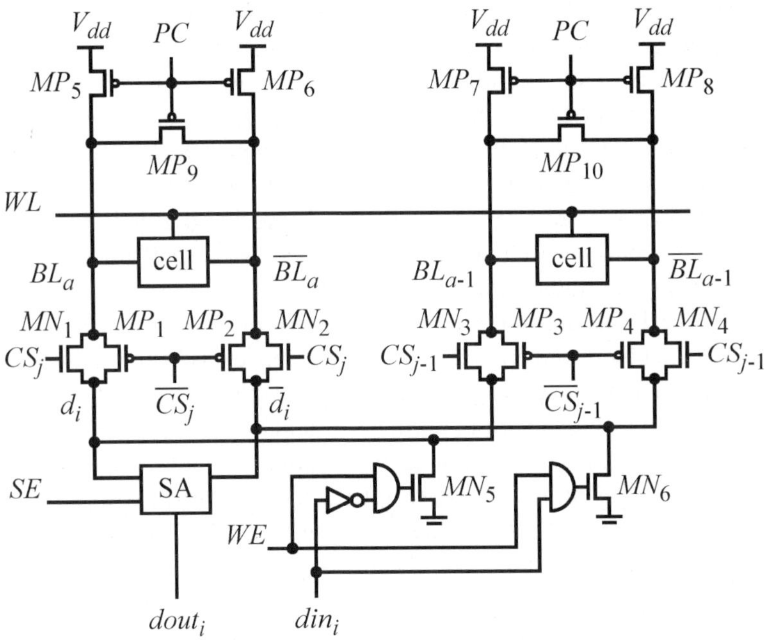

Figure 4. Typical SRAM bit-line structure.

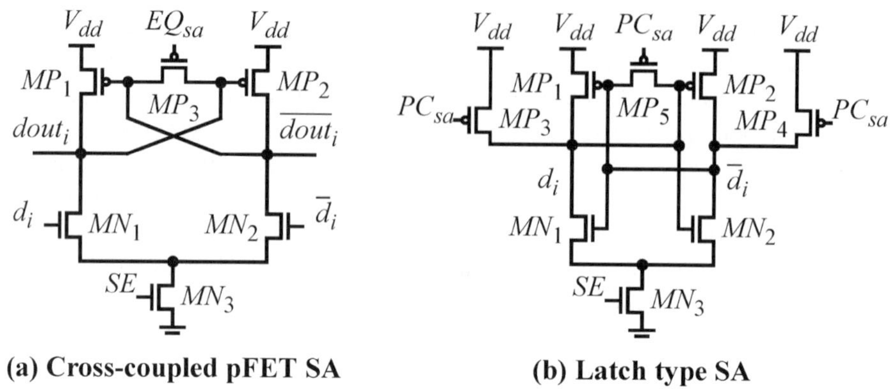

Figure 5. Sense amplifier circuit topologies.

rejection. Figure 5 shows two commonly used voltage-mode sense amplifiers. The cross-coupled pFET sense-amplifier (CCPSA) was introduced as a replacement to paired current-mirror sense amplifiers [14, 15]. CCPSA not only dissipates less power than current-mirror SA, but also it is faster due to the high gain of the pFET positive feedback. Its outputs must first be reset. One way to reset them is to equalize them. Although in Figure 5 this task is accomplished using a pFET (MP_3), it may require a transmission gate or an nFET depending on the supply and FET threshold voltages. The drawback of CCPSA is that it requires careful timing and it is prone to transistor mismatches.

The latch-type sense amplifier (LTSA) which is based on inverter crosscoupling is another approach. In this case the SA inputs are also its outputs. LTSA requires to be reset before the amplification occurs. LTSA is more effective when reset to a voltage close to the metastability point of the cross-coupled inverters. However, because the bit lines are precharged, it is typically initialized by precharging its internal nodes. This can be performed by explicitly precharging the internal nodes (as shown in Figure 5) or by the bit-line precharge pFET through the column decoder transmission gates. The sense amplifier is enabled when enough voltage difference has been built between d_i and $\bar{d}_i$. When enabled, one of these nodes is completely discharged while the other is charged to V_{dd}. Therefore, unless d_i and $\bar{d}_i$ are disconnected from the bit lines during sensing by disabling the column decoder transmission gates, LTSA discharges an entire bit line, resulting in unnecessary power dissipation and poor performance. Device mismatches in the LTSA can cause it to latch to the wrong state if the differential input signal is not strong enough when the LTSA is enabled. Excessive circuit simulations are required to determine the minimum voltage difference for correct operation. In addition to corner simulations, Monte Carlo simulations with statistical variations on the FET device parameters are necessary. Typically, increasing the FET length beyond the minimum channel length allowed by the CMOS process results in more robust operation.

Many different sense-amplifier circuit topologies have been used in the past. The exact topology depends on the CMOS process and the sense amplifier requirements. For instance in ref. 16, which uses a hierarchical bit-line structure, LTSA is used for the local bit lines, whereas a variant of CCPSA is used for the global bit lines. It has also been common to employ multi-stage amplification schemes [11, 14, 15, 17]. However, as the CMOS technology and the supply voltage scale, in most cases a single amplification stage is sufficient.

Current-mode sense amplifiers have been successfully used in SRAM [18, 19]. Instead of voltage difference they amplify a current difference in the bit lines. Their merit compared to voltage-mode sense amplifiers is their insensitivity to bit-line capacitance. However, as CMOS technologies scale, interconnect capacitance per unit length remains about constant or slightly decreases, whereas interconnect resistance per unit length increases [20]. Consequently, the resistance factor dominates in the bit-line RC delay.

Typically, in current-mode sensing schemes, the bit lines are clamped either to the supply voltage or to ground. When the cell is enabled, there is some current drawn in one of the differential bit lines depending on the data stored in the cell. Figure 6 shows a current-mode sense amplifier that is based on the topology presented in ref. 18 and was used in a register file [21]. MP_1, MP_2, MP_3, and MP_4 form a current sensing structure. MP_5 and MP_6 clamp the multiplexed bit lines d_i and $\bar{d}_i$ to the supply voltage V_{dd}. Before the sense

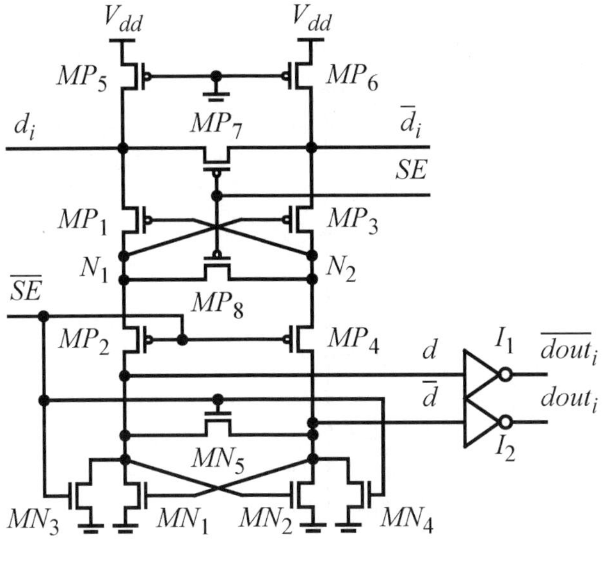

Figure 6. Current-mode sense amplifier (© 2004 IEEE).

amplifier is enabled, MP_7 equalizes d_i and $\bar{d}_i$ to V_{dd}. Likewise, MP_8 equalizes nodes N_1 and N_2 to $V_{dd} - V_{thp1}$, where V_{thp1} is the threshold voltage of MP_1 and MP_3. Also, MN_3, MN_4, and MN_5 pre-discharge nodes d and $\bar{d}$. When the sense amplifier is enabled the potential of N_1 and N_2 splits depending on the current difference between d_i and $\bar{d}_i$. One node discharges to $2 \cdot V_{thp2}$ (where V_{thp2} is the threshold voltage of MP_2 and MP_4) and the other charges to V_{dd}. Therefore, one of MP_1 or MP_3 turns off while the other is still conducting. One of d and $\bar{d}$ is charged to V_{dd} while the other remains at 0 V.

Recently, bit-line subthreshold leakage current has emerged as a serious noise problem in SRAM. The conventional approach of amplifying the bit-line differential signal is susceptible to subthreshold currents of the other non-accessed cells attached to the same bit line, reducing the differential voltage available for sensing. To overcome this problem, SRAM designs use only one of the bit lines for reads [6, 7] and the sense amplifier is replaced by a static gate. Since bit lines must be fully discharged, reads can be very slow. To speed up the operation, hierarchical bit-line architectures are adopted with local bit lines connecting to only eight or 16 cells.

Figure 7 shows a single-ended bit-line architecture similar to that in ref. 7. It is assumed that SRAM cells are arranged in blocks that consist of 16 rows

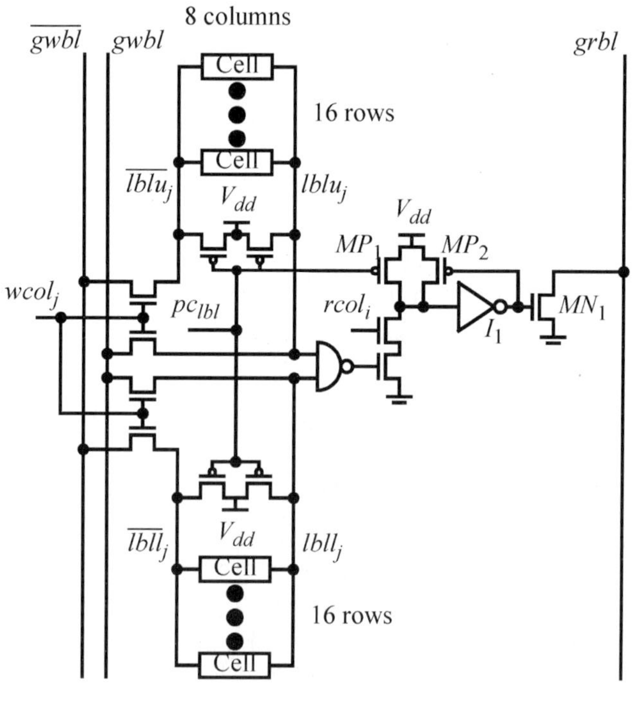

Figure 7. Single-ended bit-line structure.

and eight columns. The eight columns are multiplexed to produce one data bit. To reduce overhead two opposing blocks share the same column select and multiplexing circuit. All components in Figure 7 except MP_1, MP_2, MN_1, and I_1 are replicated for each column. Local bit-line pairs $lblu_j/\overline{lblu}_j$ and $lbll_j/\overline{lbll}_j (j = 0 \ldots 7)$ connect the 16 cells within each row. For writes, one of the eight write column select signals, $wcol_j (j = 0 \ldots 7)$ is activated. This signal connects one of the local bit-line pairs to the global write bit-line pair $gwbl/\overline{gwbl}$ through pass nFET. The global write bit-line pair connects to one local bit-line pair of each opposing block, instead of just one local bit-line pair. Although this leads to more energy dissipation since precharged local bit lines are unnecessarily discharged, it does not cause any malfunction since only one word line is activated. It could be possible to have two sets of write column select signals, i.e. a total of 16, at the expense of increased area overhead for the column select circuit. For reads, local bit lines from opposing blocks are multiplexed using a static NAND gate. The NAND gate drives a dynamic AND–OR gate. Read column select signals, $rcol_j (j = 0 \ldots 7)$ select one out of the eight columns to conditionally discharge the AND–OR gate. The output of this gate drives an nFET that pulls down a global read bit line *grbl*. Local bit lines as well as the AND–OR gate are precharged by signal pc_{lbl}. Global read bit lines are also precharged by a different precharge signal pc_{grbl} (not shown in Figure 7). Global read bit lines can also be split in two pieces which are multiplexed in the middle of the array.

More aggressive techniques include bit-line leakage compensation [22] and bit-line leakage tolerance [23]. In the former, the bit-line leakage current is detected and the same amount is then injected in the bit line to cancel out its leakage, allowing differential sensing. In the latter, reads are single-ended in a domino-like configuration but the local bit lines are pre-discharged instead of precharged. As shown in ref. 23, this configuration causes the access nFET of non-selected cells to be reverse biased, which reduces their leakage current.

2.4. Control Logic

Generating all the control signals required by an SRAM is an involved design task. Most of those signals must be delivered within very stringent time requirements to ensure proper operation. For instance, if the sense amplifiers are enabled before sufficient voltage or current difference has been built in their inputs, it is likely that their outputs would be incorrect. Furthermore, as caches occupy increasingly larger areas, their designers must consider more extreme process and voltage variations. Even if the cache is divided into small SRAM macros, all those macros must be synchronized to deliver the data following certain timing specifications.

Typically, the access time of second- or third-level caches is longer than the microprocessor clock cycle time. The conventional approach to improve cache bandwidth is pipelining [4, 7]. Not only the cache access operations, but the SRAM macros themselves can be pipelined to meet the bandwidth requirements [24]. Since it is difficult to break SRAM operations into pipeline stages, a wave pipeline approach has been used [24, 25]. In ref. 24 the micro-operations are input-triggered, self-resetting circuit blocks. In ref. 25 a PLL is employed to generate internal clock phases that control the SRAM operation. The design challenge of wave pipelined SRAM is that eventually their output must be synchronized with the system clock signal. Similar to wave pipelining, the 24 MB, multi-cycle cache reported in ref. 5 was designed to operate asynchronously eliminating delay associated with clock skew, latch delays, and cycle margins. Also, since it was not necessary for the asynchronous operation to be split within clock cycle boundaries, it was possible for accesses to complete in five cycles instead of the projected eight cycles that an equivalent synchronous design would require.

Control signals within an SRAM macro can be generated in a self-timed manner tracking the delay of the micro operations. For reads, those micro operations would be address predecoding, row decoding, memory access, column decoding, sense enable, and latch enable. The different ways that those control signals are generated fall into three categories: (a) using clock edges or phases [26], (b) using inverter delay chains [24, 27], and (c) using circuit replicas [28]. It has been shown [28] that the latter approach tolerates better process, voltage, and temperature variations, since the replicas match very well the actual circuits and they are located close to them. Also, gate, interconnect, and cell delays vary differently even if all of them are in the same chip. Therefore, using inverters to track interconnect and cell delays could be ineffective or it would require larger margin than necessary. In Section 3 a replica-based self-timed control logic for a register file will be discussed in detail.

2.5. Techniques for Reducing Dynamic Power

Power dissipation has been a key design consideration for microprocessors. Memories, being the largest blocks in microprocessor chips, are major contributors to their dynamic and static power dissipation. This subsection presents some techniques for reducing SRAM dynamic power. The prevalent approach for reducing dynamic power in CMOS digital circuits is to lower their operating voltage. Although this is simple with static CMOS, low-voltage memory operation causes the cell SNM to degrade. Also, the bit-line discharge time can substantially increase, especially coupled with the use of high-threshold voltage FET in the cell. Low-voltage SRAM have been reported [29, 30]. In ref. 29

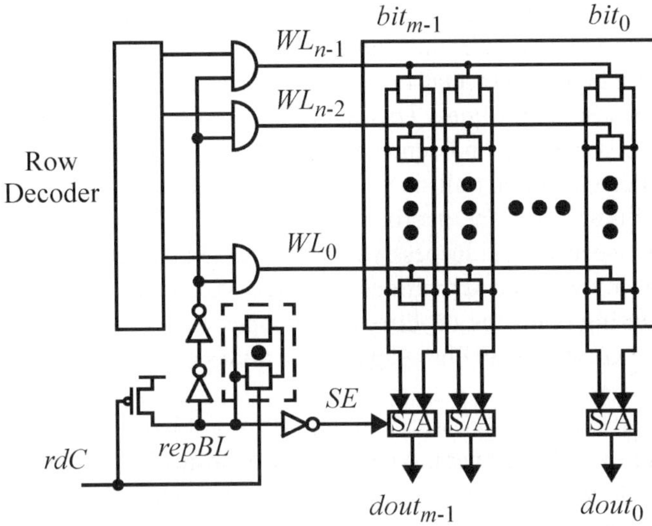

Figure 8. Using replica bit line to control the bit-line voltage swing during reads.

a replica circuit is used to track bit-line delay for a wide range of operating voltages (i.e., 1–5 V). In ref. 30 the word-line voltage is boosted to compensate for the low-voltage operation.

As mentioned earlier, a small voltage difference between the bit lines is sufficient for a voltage-mode sense amplifier to produce the output data. Therefore, it is not necessary to fully discharge the bit lines during reads. In ref. 28, replica-based techniques are exploited to reduce bit-line voltage swing (Figure 8). The replica bit line (*repBL*) and the actual bit lines start being discharged at the same time. Since *repBL* is discharged faster than the actual bit-lines, when it is fully discharged, the actual bit-lines are partially discharged. When discharged, *repBL* disables the word lines and enables the sense amplifiers. Two bit-line replica circuits are presented. One is based on capacitance ratio with the replica bit line being a fraction of the actual bit lines. The other is based on cell current ratio, with the dummy cell drawing a multiple of cell current, thus discharging the replica bit line much faster than the actual cells discharge the actual bit lines.

Although bit-line power can be straightforwardly reduced for reads, a full voltage swing is commonly required for writes. In ref. 31, bit-line swing is limited to $V_{dd}/2$ during writes by using $V_{dd}/2$ as the bit-line reference voltage and applying half-swing pulses. The same technique is also extended in reducing decoder power dissipation. In ref. 32, a low reference voltage (i.e. 20% of V_{dd}) is used for the bit lines which swing by 10% of V_{dd} during writes. Reads are performed using current-mode sense amplification while bit lines are clamped to the reference voltage. To improve cell stability, word lines swing to a reduced voltage during reads. In ref. 33, the sense amplification of the bit-line voltage difference during reads is extended for writes. The cell itself is a sense amplifier that is set to the new data based on a small voltage difference in the bit lines. This write scheme requires two control signals per row (i.e. word line and sense enable). Extra area overhead includes an nFET in series to the cell pull-down nFET to form a latch-type sense amplifier (similar to Figure 5b). The overhead of the extra nFET can be amortized by sharing it between several successive cells. In ref. 34 the bit lines are resonantly powered during writes and their energy is recovered and recycled through an LC tank. During reads the bit lines are clamped to 0 V. The energy recovery approach is extended to all high-capacitance nodes including address, data, and word lines.

The SRAM design proposed in ref. 35 combines bit-line low-voltage swing with low-voltage operation. Specifically, instead of ground, the sources of the pull-down nFET are connected to a variable supply called source line. Reads are sped up by driving the source line to a negative voltage which forward biases the substrate of the pull-down nFET. During writes the source line is left floating, i.e. in high impedance, which makes it possible to flip the cell with a small voltage difference in the bit lines since the pull-down nFET are inactive. When the write completes, the source line is set to ground which triggers the cell to operate as a latch-type sense amplifier. As shown in ref. 35, the speed-up obtained by using the negative-voltage source line can be traded for power reduction by lowering the supply voltage. An improved design based on the source-line concept was presented in ref. 36. The source line is driven to a positive voltage for non-selected cells and to 0 V for accessed cells eliminating thus the need of a negative supply voltage.

Reducing bit-line switching capacitance is another approach to decreasing dynamic dissipation orthogonal to low-voltage operation. Typically, all bit lines switch during reads or writes even if only a few of them are selected through the column decoder. Dividing the SRAM array into blocks and using a hierarchical word-line scheme minimizes the number of switched bit lines [37]. Likewise, hierarchical bit-line schemes inherently result in reduced bitline switching capacitance since the array is divided into horizontal segments and local bit lines connect to relatively few cells (Figure 7). With proper row decoding it is possible to limit the switching bit-line capacitance to local bit lines within one segment. Although global bit lines span the entire cell array, their capacitance is small since it includes mostly wiring capacitance.

Another target for power reduction is the data bus that connects multiple SRAM macros or the cache to the processor. As is the case with any microprocessor bus, low-voltage-swing [28, 37, 38] or energy-recovery [34] techniques can be employed for power reduction in the data bus.

2.6. Techniques for Reducing Static Power

The on-going scaling of CMOS process technology has made FET static current, which translates directly into static power, one of the most important issues in microprocessor design. Caches, being the densest and largest components in current microprocessors, are among the top contributors to static power. According to ref. 39, static current accounts for 30% of level-1 cache power and 80% of level-2 cache power in a representative 0.13 μm microprocessor.

There are four different current sources that contribute to static power in the CMOS process: (1) the subthreshold current between the drain and the source of an off FET due to the weak inversion in the channel region, (2) the gate tunneling current between the gate and the channel through the gate oxide, (3) the gate-induced drain leakage (GIDL) current between the drain and the channel when the FET is off and there is a voltage difference between the gate and the drain, and (4) the P-N junction reverse saturation and tunneling current.

The subthreshold leakage current depends mostly on the FET threshold voltage. Using SRAM-specific high-threshold voltage FET for the cells [4] not only improves the cell SNM but also reduces its subthreshold leakage dissipation. The gate-tunnel leakage depends exponentially on the gate oxide thickness and the gate-to-source voltage. It has been reported that a 2-Å decrease in the gate oxide thickness causes a 10-fold increase in the gate-tunnel leakage [40] and it is more pronounced when the FET is on [41]. The GIDL current depends on the gate-to-drain voltage difference V_{gd} and becomes noticeable when $|V_{gd}|$ is equal to the supply voltage V_{dd} [40]. However, for current CMOS processes GIDL is insignificant in carefully engineered devices compared to subthreshold leakage current [41]. Junction reverse saturation current is insignificant for current CMOS processes, but future highly-doped 50-nm CMOS processes are predicted to introduce a significant amount of junction tunneling current [42].

Figure 9 shows the effect of subthreshold and gate-tunnel leakage currents [40, 41], which dominate during nominal operation, on an SRAM cell. It is assumed that *bit* is at V_{dd}, the word line is at 0 V, and both bit lines are at V_{dd}. Static power reduction techniques seek to dynamically control the leakage current paths [39], the FET threshold voltages [43], or the cell supply voltages [36, 40, 41, 44–47]. These techniques can be applied either during standby mode or during active mode. The former requires that the entire SRAM array is disabled when the microprocessor is idle, whereas the latter requires that only unused sections of the SRAM are disabled while selected SRAM sections are made available for access. Active mode static power reduction is more difficult since it incurs control timing complexity and possible delay overhead. Either standby or active mode leakage control requires dynamic power and area overhead, especially if the cell size must be increased. For any viable technique, leakage power reduction must outweigh the dynamic power overhead. Finally,

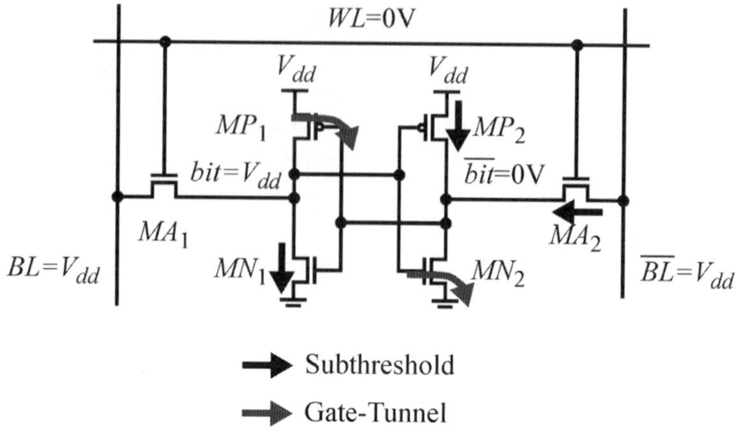

Figure 9. Dominant leakage components in SRAM cells.

these techniques degrade the cell SNM at least during data retention when the cells are disabled.

Controlling the leakage path (Figure 10a) was primarily proposed for reducing the subthreshold leakage current [39]. An nFET (MN_3) is inserted between the pull-down nFET and the ground creating an internal virtual ground (V_{gnd}). MN_3 is activated when the cell is accessed connecting the word line to its gate. One nFET can be shared between several cells in the same row, thus reducing the overhead. Depending on the supply and the FET threshold voltages, when MN_3 is off the cell SNM is reduced because one of the cell internal nodes (*bit* or $\overline{bit}$) that nominally should be grounded is floating and its voltage rises to above 0 V. Therefore, data retention could be problematic. In ref. 39, it is reported that data retention is possible with proper transistor sizing. It should be noted that this approach also reduces gate-tunnel leakage, since the gate-to-source voltage of the FET that would draw gate-tunnel leakage current is reduced.

An alternative approach to reduce subthreshold leakage current is to control the FET threshold voltage by dynamically adjusting their substrate bias [43] (Figure 10b). Voltages V_{nw} and V_{pw} connect to the n-well of the pFET and the p-well of the nFET, respectively. For unused cells, V_{nw} and V_{pw} reverse bias all cell FET. For selected cells they are driven to their nominal settings, V_{dd} and 0 V, respectively, before the word line is enabled. The area overhead includes the V_{nw} and V_{pw} wiring in addition to the supply and ground voltages. It is possible however for multiple rows to share the same wells to reduce the overhead. Also, the substrate voltage control must be synchronized to the word-line signal. More important, it is necessary to provide the supply voltages for reverse bias from off chip or generate them on chip using dc–dc converters. Unless using off-chip passive components, dc–dc converters have

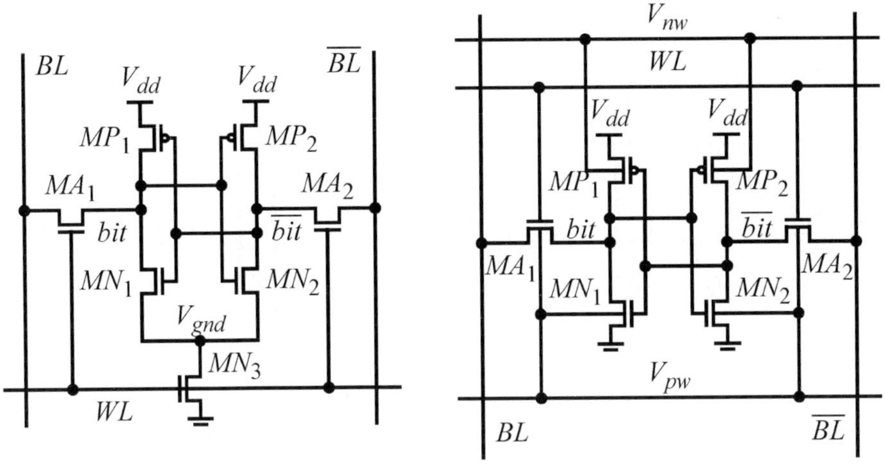

(a) Controlling the leakage paths **(b) Controlling the FET threshold voltages**

Figure 10. Cell static power reduction by controlling (**a**) the leakage paths and (**b**) the FET threshold voltages.

low efficiency, which may be acceptable depending on V_{nw} and V_{pw} current requirements. Other process-dependent problems afflicting body-bias techniques include GIDL current increase, especially for high-threshold FET [48], and negative bias temperature instability (NBTI) [49].

Controlling the cell ground and/or power supplies (Figure 11) is used primarily to suppress gate-tunnel leakage by reducing the internal gate-to-source voltages of the FET that draw gate-tunnel leakage [40, 41, 45, 46]. The ground supply V_{ssv} is raised above 0 V for unused cells and is set to 0 V for selected cells [40, 45, 46]. Likewise, the power supply V_{ddv} is lowered below V_{dd} for unused cells and is set to V_{dd} for selected cells [41]. Although the variable ground and power supplies (V_{ssv} and V_{ddv}) connect only to the sources of pull-down nFET and pull-up pFET respectively, their substrates connect to the nominal supply voltages, i.e. ground for nFET and V_{dd} for pFET. In addition to gate-tunnel leakage, this technique suppresses subthreshold leakage currents since some of the FET are reverse biased. Varying the ground supply reverse biases the pull-down and access nFET [40] and varying the power supply reverse biases the pull-up pFET [41]. The variable supply voltages must be restored to their nominal settings before the word line is enabled. This technique also requires extra supply voltages which can be provided off-chip or generated on-chip using diode-connected FET [41, 45, 46]. As expected, the cell SNM is degraded during data retention.

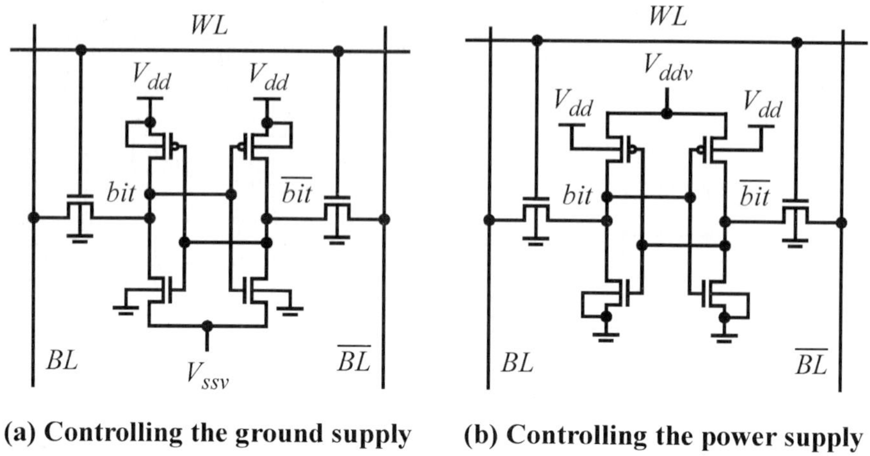

Figure 11. Cell static power reduction by controlling (**a**) the ground and (**b**) the power supply.

2.7. Reliability and Testing

In this subsection we address several issues related to SRAM reliability and testing. First, we present techniques to improve the cell SNM when low-voltage operation is necessary. Second, we discuss bit-line coupling issues which could cause the SRAM to malfunction. Third, we describe soft errors and error checking and correcting (ECC) mechanisms. Fourth, we address memory redundancy to improve yield. Finally, we touch on SRAM testing, specifically built-in-self-test (BIST) circuits.

Depending on the CMOS process it could be difficult to obtain adequate cell SNM for low-voltage applications, especially if the cell is implemented with low threshold voltage FET. In order to improve the cell SNM, its supply voltage is boosted when accessed similar to the approach used to reduce leakage dissipation (Figure 11b) [50–52]. In ref. 50, the array voltage is boosted when the SRAM is activated, i.e. the cells operate at higher voltage for both reads and writes. Typically, the cell SNM needs to be improved during reads. Also, the higher cell supply voltage reduces the bit-line discharge time. In ref. 51 the supply voltage of the accessed row is bootstrapped when its word line is activated. This is accomplished by running the supply lines next to word lines and using their capacitance coupling for the bootstrapping. In ref. 52 the cell power supply voltage is boosted only for reads. As discussed earlier, during writes, cells in non-selected columns perform dummy reads since their precharged bit lines are left floating. To ensure that the dummy reads are non-destructive, the supply voltage of the non-selected columns is boosted, whereas cells in the columns selected for writes are connected to the low supply voltage.

Minimizing bit-line coupling capacitance is important to ensure correct SRAM operation. Bit lines can couple either to word lines or to bit lines from neighboring columns. Bit-line coupling capacitance to word lines is relatively small since they run on different directions. Moreover, both bit lines are coupled to word lines and hence they are both affected the same way. Common-mode noise is rejected when differential sensing is used. Coupling between bit lines of neighboring columns could be detrimental. The conventional approach to reduce the bit-line to bit-line coupling capacitance is to use twisted bit-line routing (Figure 12) [53, 54]. Twisting the bit lines not only reduces the coupling capacitance, but also converts the coupling interference into common-mode interference which is rejected with differential sensing. More aggressive bit-line twisting routing methods span multiple pairs and can reduce bit-line coupling capacitance by up to 44% [55].

There are two causes of malfunction in memories: soft errors and permanent defects. Soft errors are non-recurring errors caused by alpha particles [54, 56] and cosmic ray neutrons [57] that hit the cell storage node, or by internal circuit noise due to lack of adequate design margins. Here we discuss only alpha and cosmic-ray-induced soft errors, although either type of error can be corrected with appropriate hardware or software. The occurrence of soft errors is statistical and their probability increases as the storage node charge decreases and the memory capacity increases. As technology scales, both storage node capacitance and supply voltage decrease and therefore the storage node charge

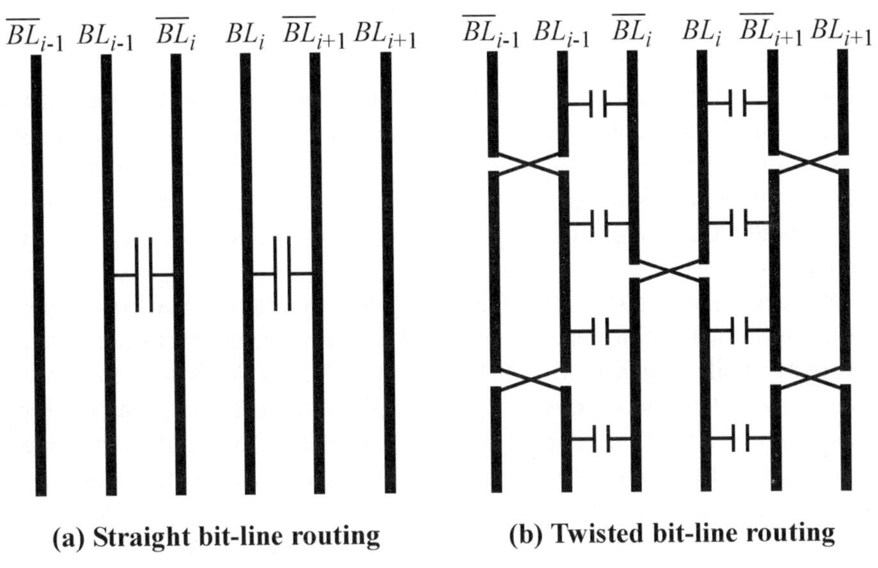

Figure 12. (**a**) Straight and (**b**) twisted bit-line routing.

decreases. In general the storage node charge should be kept larger than a critical charge Q_c which is technology-dependent and decreases as technology scales [54, 58]. Simple methods to estimate the cosmic-ray-induced soft errors rates (SER) have been proposed [58]. In ref. 59, SER immunity was improved 3.5 orders of magnitude by artificially increasing the storage node capacitance. Any type of soft errors can be handled with ECC mechanisms. Parity bits are accessed along with data bits. Depending on the number of parity bits used for ECC and their assignment to groups of data bits, it could be possible to correct single-bit errors. For m data bits and k parity bits, the following expression should be satisfied [54]: $2^k \geq m + k + 1$. Unlike alpha particles that affect single cells, soft errors caused by cosmic rays can disturb multiple neighboring cells [60], which would be costly to correct using ECC. However, the probability of multiple soft errors in a single word can be minimized by increasing the number of columns that are multiplexed to generate one bit of data [40, 60, 61]. To eliminate the ECC timing overhead during reads, a hidden-ECC mechanism is proposed in ref. 61. SRAM words are read and checked when the SRAM is idle. When a soft error is detected, the corrected data is written back to the SRAM.

Meeting yield specifications and further improving yield is very important for the success of any microprocessor. Yield problems arise from material defects and process variations. SRAM defects can either be scattered, affecting localized cells, or accumulated, affecting sections of SRAM [54]. The former can be treated similar to soft errors, i.e. using ECC mechanisms, but more often they are treated through row or column redundancy. The latter can be treated through block redundancy. In either redundancy scheme, defective cells or blocks are identified during initial testing. Fuses are then used to physically remap redundant circuitry to replace the defective circuitry. Accesses to defected columns or rows are redirected to redundant ones. For reliability, the fuses are located far from the SRAM and their settings are stored in soft registers during power up [26]. Depending on the cache organization, different redundancy strategies are possible [5, 26]. It has been shown that a combination of both ECC and redundancy is very effective to achieve high yield in memories [62].

Memories can be tested effectively using BIST circuits. BIST is especially useful for SRAM testing, since its area and performance overhead can be kept extremely small. When used with an on-chip PLL to generate a full-speed system clock, BIST can eliminate or greatly reduce the need for expensive high-speed, high-bandwidth wafer probing. Typically, BIST blocks are programmed through a serial scan chain to generate different address and data patterns and execute many types of test. At minimum, a BIST block must perform 100% stuck-at fault (nets shorted to power or ground), address uniqueness, and data retention tests. Various data patterns to detect crosstalk noise and worst-case leakage current operability are also essential. Following test, a BIST block reports the defective circuits for fuse repair, and retests the final reconfigured memories.

3. Multi-port Register File Design

Register files are multi-port static memories with dedicated read or write ports. They are smaller in terms of bit capacity but faster than SRAM. This section is an overview of RF design covering the following subjects: storage cell, organization, bit-line architecture, and control logic.

3.1. Storage Cell

Similar to SRAM, the RF storage cell is based on cross-coupled inverters. However, since write and read operations are not coupled, its design is typically simpler than the six-transistor SRAM cell. As we see next, reads can be assured to be non-destructive and not to degrade the cell static noise margin. Although RF have relatively small bit capacity, it is still necessary to keep their storage cell small. In addition to keeping the RF size small, this would also result in short delay time and low-power dissipation since internal word-line and bit-line wire length would be minimized. The cell size can be wire- or transistor-limited. Writes and/or reads can be differential or single-ended. Apparently, single-ended operations minimize the number of wires that run through the RF array. Another issue to be considered is bit-line to bit-line coupling which is different than in SRAM. In RF, bit lines can couple to other bit lines of the same cell that belong to a different port. To prevent bit-line coupling from causing the RF to malfunction, bit lines can be shielded with power lines or they can be spaced farther apart than the minimum metal space allowed by the CMOS process.

Different combinations of write and read port circuit topologies are possible. Figure 13 shows several differential and single-ended write ports. The first one is based on the SRAM cell. NFET M_1 and M_2 must be replicated for each write port. In a single-ended alternative (Figure 13b), one of the access nFET is replaced with a transmission gate, making it possible to transfer a strong "0" as well as a strong "1" using one bit line per port [63] at the expense of a differential word line. Inverter I_2 should be a weak feedback inverter so that it can be over-powered by the bit-line driver and the transmission gate. The write port shown in Figure 13c requires single-ended bit lines and word lines. The word line negative polarity is generated inside the cell and is used to disable the feedback path of the cross-coupled inverters [64]. When the cell contains more than one write port, all write word lines are fed into a NOR gate and the output of the NOR gate drives M_2 disabling the feedback when any one of

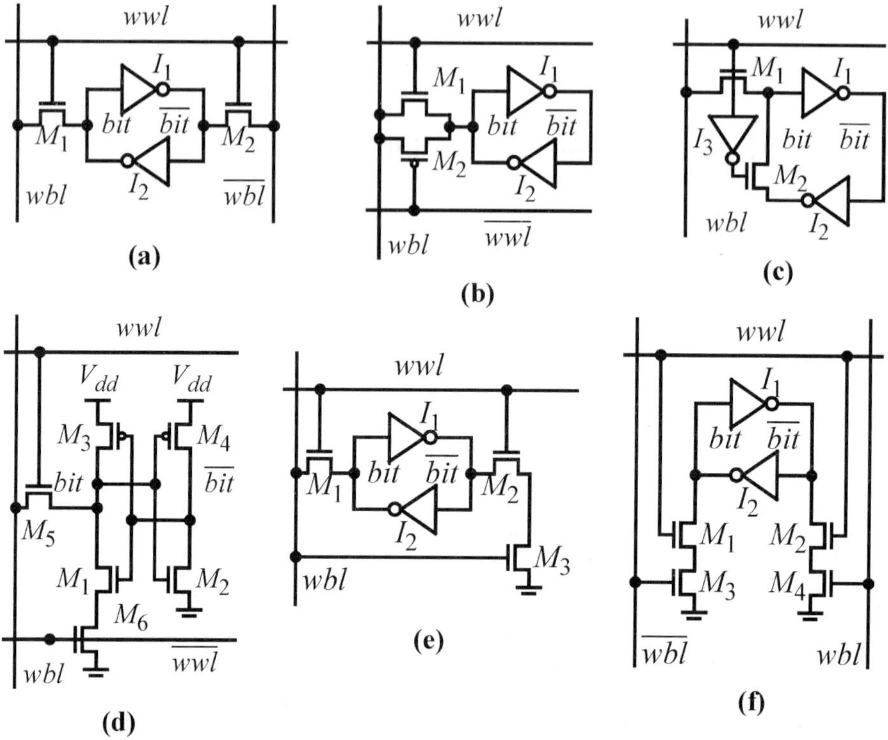

Figure 13. Register file write-port circuit configurations.

the write word lines is enabled. A similar approach is shown in Figure 13d, in which the feedback inverter is disabled through a stack nFET (M_6) in its pull-down stack [65]. The next configuration (Figure 13e) relies on nFET passing a strong "0" to accomplish write operations with single-ended write bit and write word lines [66, 67]. When *wbl* is "1", $\overline{bit}$ is pulled-down through M_2 and M_3. Finally, the circuit topology shown in Figure 13f resembles a set/reset latch. Depending on the input, one of the pull-down stacks sets the cell to "0" or "1". The advantage of this design is that when both *wbl* and $\overline{wbl}$ are "0" the cell maintains its old data. This feature can be useful when certain bits in a word should be masked out such as byte-wide write operations.

RF reads are typically single-ended based on precharged domino logic (Figure 14a). When the read word line (*rwl*) is enabled the read bit line (*rbl*) is conditionally discharged depending on the data stored in the cell. NFET M_1 and M_2 are replicated for each read port. To balance the internal capacitance, both *bit* and $\overline{bit}$ can be used to drive the read-port pull-down nFET with those driven by *bit* producing negative polarity outputs. Figure 14b shows an example of a complete RF storage cell [67]. M_1, M_2, and M_3 are replicated for each write

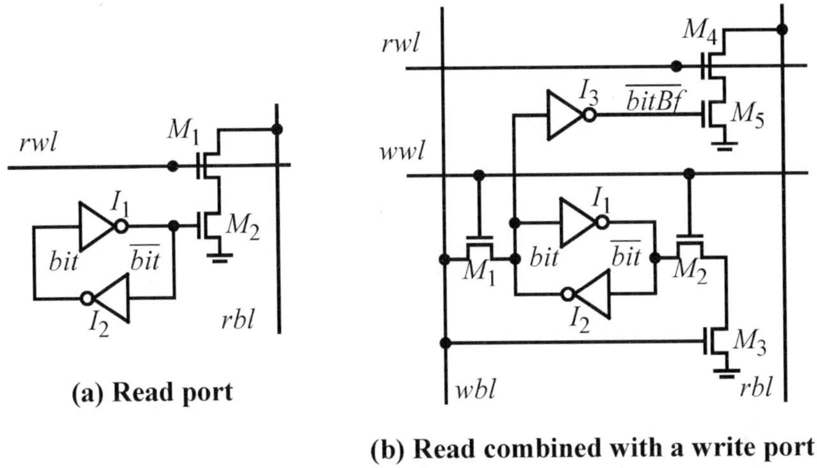

Figure 14. Commonly used read port (**a**) stand-alone and (**b**) combined with a write port (© 2002 IEEE).

port and M_4 and M_5 are replicated for each read port. Inverter I_3 is inserted to isolate a large 10-read-port load from the $bit/\overline{bit}$ nodes. Compared to the six-transistor SRAM cell, reads do not interfere with the data stored in the cell, allowing a wide freedom in sizing the RF cell FET. Furthermore, low-voltage operation can easily be achieved [68]. Other RF read-out methods have been proposed using single-ended sense-amplifiers instead of precharged bit lines. However, they result in excessive static dissipation [69, 70].

3.2. Organization

RF are simpler to organize than SRAM because usually they contain 128 or less words, which makes column decoding unnecessary. Also, RF cells are larger than SRAM cells, which simplifies the design of the peripheral circuits that must match the cell pitch. In a typical organization the port addresses are first predecoded and then decoded (Figure 15). The decoder also includes the word line drivers. Static CMOS gates or precharged domino logic [67] can be used for predecoding and decoding. It is possible to place the decoders in the middle of the array to reduce the word-line resistance and capacitance. This is desirable especially for RF that support half-word accesses since the array can be split in most significant (MS) and least significant (LS) bits. Read bit lines are fed to a read-out circuit and potential latches. When read bit lines are precharged, latching the output can result in dynamic power savings by preventing the propagation of the bit-line precharging and discharging sequence to the RF output. For write ports, input data is first latched and then inverter

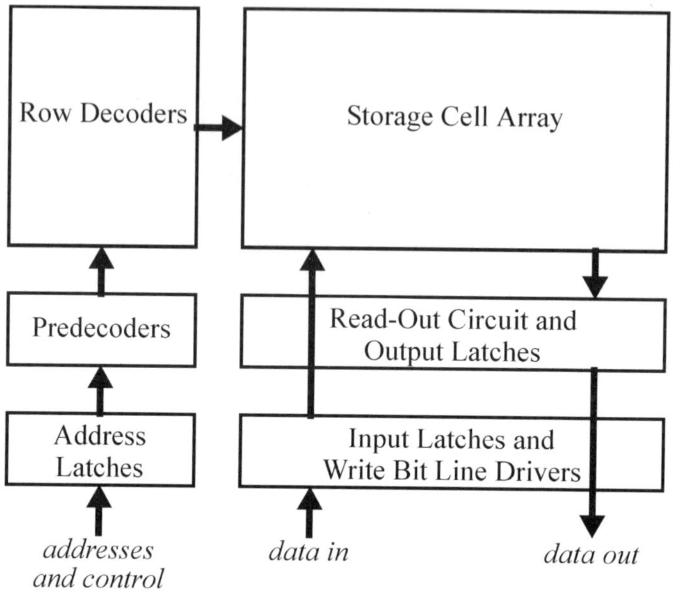

Figure 15. RF organization and floor plan.

buffers are used to drive the write bit lines. If the write bit lines are too long (i.e. the RF contains too many words), they can be split into sections and repeaters can be used to drive each section. The same applies to the outputs of the predecoder which span equal length.

3.3. Read Bit-line Architecture

As is the case with SRAM, bit-line leakage has also been a key problem in RF design. The difference is that RF read-out circuits are already mostly based on single-ended reads with precharged domino logic. Using a single bit line does not completely solve the leakage problem since the bit line may still be discharged below the threshold voltage of the receiving gate due to subthreshold current of the cells attached to it. To ensure proper circuit operation, a feedback pFET is inserted with required strength proportional to the number of cells on the bit line. The feedback pFET typically considerably increases the read access time. To overcome the problem bit lines are partitioned into smaller sections that include eight or 16 cells [67, 71, 72]. Global bit lines are used to drive the output (Figure 16). In case (a), each global bit line runs across the entire RF array and outputs are available at the bottom of the array. In case (b), global bit lines are also split into two pieces. Those half global bit lines are multiplexed in the middle of the array.

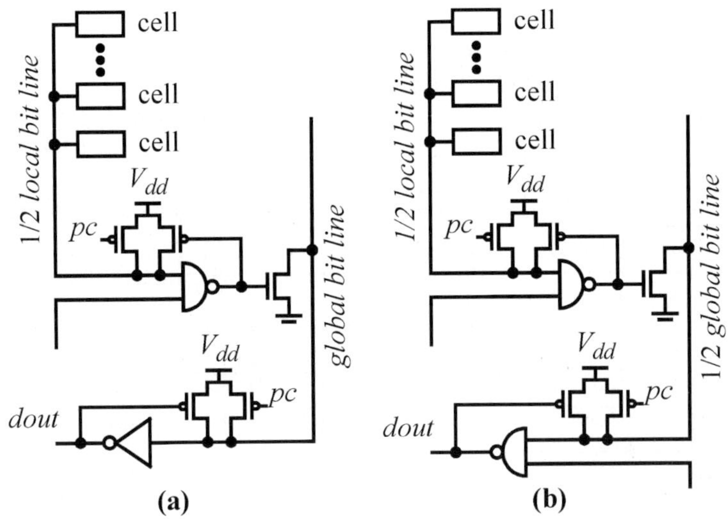

Figure 16. RF read bit-line structures (© 2004 IEEE).

Several techniques to improve bit-line noise immunity have been proposed. In ref. 72, the order of the read-port stack nFET (Figure 14a) is reversed; *rwl* drives the bottom nFET and also precharges the internal node between the two nFET when non-selected. Also, a gate is inserted to mask the gate voltage of the top nFET that is driven by the cross-coupled inverters to 0 V unless *rwl* is enabled. When the bit line is precharged both the drain and the source of the top nFET in the stack are at V_{dd} for the non-selected cells cutting off the path of the bit-line subthreshold leakage current. The overhead is five FET per port. In ref. 21, current-mode differential signaling is used to address the subthreshold leakage bit-line noise. The single-ended bit-lines are split into two pieces and a sense amplifier is placed in the middle. Differential sensing is achieved by generating a reference current in the nonselected half bit line. The nFET subthreshold leakage current is reduced since this technique makes it possible to increase their channel length by 80%.

3.4. Replica-based Self-timed Control Logic

In this subsection we present a replica-based self-timed control logic implemented in a 34 word × 64 bit, 10-read/6-write port RF [67]. This control logic can be modified for use in an SRAM. It should be noted that the RF supports write-through operations (i.e. writes are followed by reads in the same cycle) and half-word accesses. Figure 17 shows the RF control and data signal flow. Each write and read port has a 4-bit control input (*wc*[3:0] and *rc*[3:0],

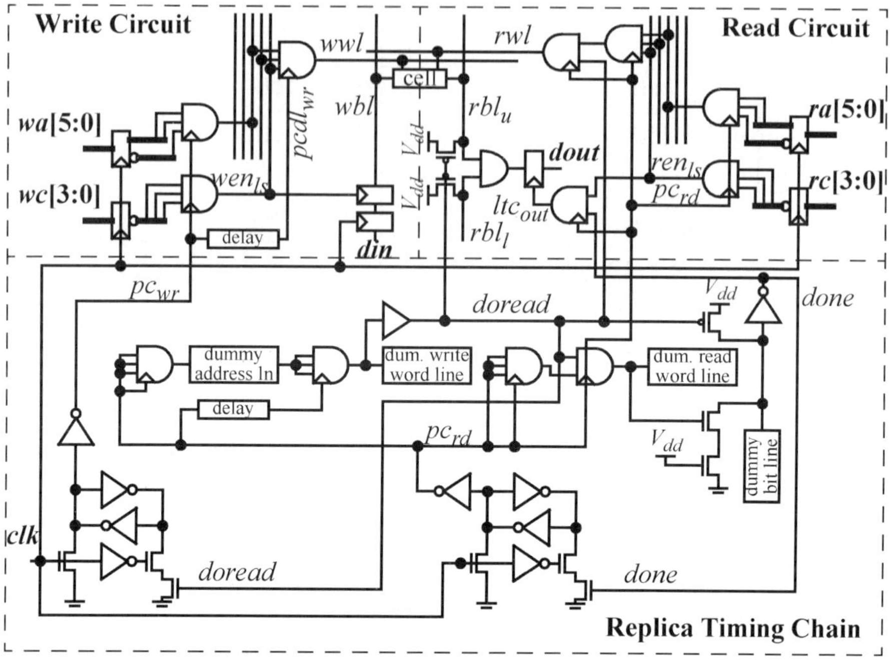

Figure 17. Register file control and data diagram (© 2002 IEEE).

respectively) that enables the port and determines the access width (i.e. LS bits, MS bits, or both) and a 6-bit address (*wa*[5:0] and *ra*[5:0], respectively). Write ports receive a 32- or 64-bit input data and read ports produce a 32- or 64-bit output data. Only one bit is shown in Figure 17; *din* and *dout* respectively. The address and control bits are predecoded (first logic stage) and then drive the final decode stage (second logic stage). Reads have a third stage that is enabled after the write operations have completed. The three logic stages are implemented with precharged dynamic gates. Each word part (i.e. MS and LS) per port is enabled with different control signals (wen_{ls} and wen_{ms} for write ports and ren_{ls} and ren_{ms} for read ports). For write ports, input data drives write bit lines (*wbl* in Figure 17) only when the port is active. For read ports, the output latches are enabled only when the port is active. The RF operation is controlled by a replica timing chain that imitates the sequence of the micro-operations (i.e. write address decoding, data writing, and data reading). The self-timed chain contains dummy predecoded address, write word, read word, and read bit lines. These are placed along the real ones. Dummy word and bit lines are 50% and 25% of their respective actual load.

A timing diagram (Figure 18) from a circuit simulator shows the sequence and dependency of control and data signals. Some signals (i.e. wen_{ls}, ren_{ls}, and ltc_{out}) are omitted for simplicity. Operation is fired on the positive clock

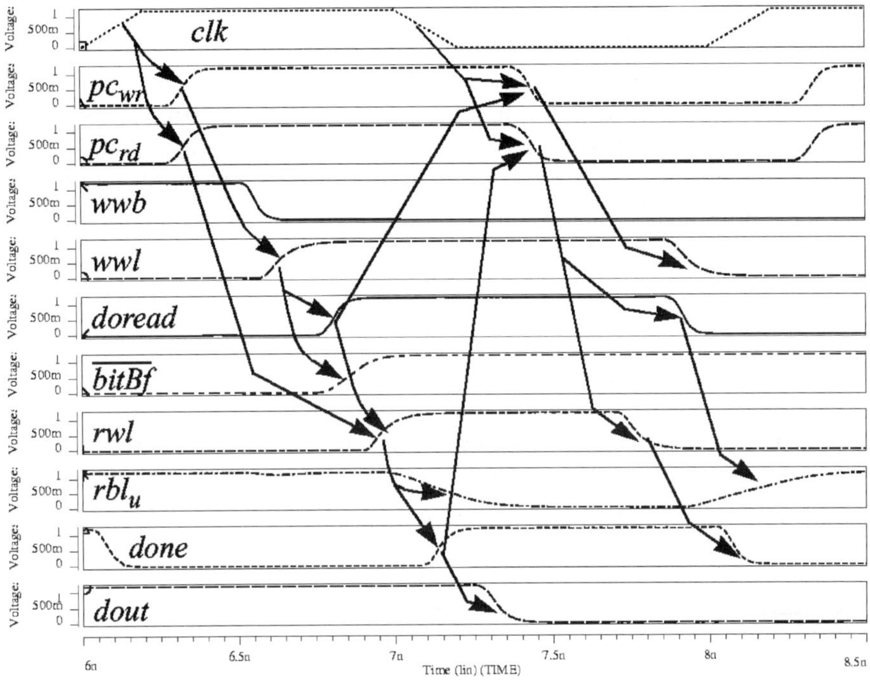

Figure 18. Register file timing diagram (© 2002 IEEE).

edge, which sets the set/reset latches on the bottom of Figure 17. These latches generate two precharge signals: pc_{wr} for the write decoder and pc_{rd} for the read decoder. Setting the precharge signals high initiates the self-timed operation. The dummy decoder generates the *doread* signal which enables read word lines. *Doread* is also used to precharge the read bit lines, which are actively pulled up while the write bit lines switch. The last part of the self-timed logic generates *done*, which indicates the end of the read operation and enables the ltc_{out} signal for read ports. *Doread* in conjunction with the negative edge of the clock reset the latch that generates pc_{wr}. Likewise, *done* resets the latch that generates pc_{rd}.

4. Summary

In this chapter, we presented some basic guidelines for designing efficient single-port SRAM and multi-port RF. In general, design decisions depend greatly on the specifications of the target static memory as well as the CMOS technology process. We covered a variety of design approaches for the key SRAM and RF circuitry. We also addressed design challenges that are related to scaled CMOS technology. As was shown, SRAM static noise margin degrades with decreasing supply and FET threshold voltages. Additionally, lowering the FET threshold voltage decreases the bit-line noise immunity and causes excessive static power dissipation. We presented advanced design techniques to overcome or mitigate these problems. Finally, process, voltage, and temperature variations become key design issues with increasing SRAM sizes and CMOS process scaling. Operating second- or third-level caches asynchronously has been shown to address this problem.

5. Acknowledgements

The author is grateful to Fujitsu Laboratories of America for providing the time to complete this manuscript, and to W. Walker and H. Ando for their constructive comments and suggestions.

References

[1] Archer, D. *et al.* "A 32 b CMOS microprocessor with on-chip instruction and data caching and memory management", *ISSCC Dig. Tech. Papers*, San Francisco, CA, Feb. **1987**, 32–33.

[2] Horowitz, M. *et al.* "A 32 b microprocessor with on-chip 2Kbyte instruction cache", *ISSCC Dig. Tech. Papers*, San Francisco, CA, Feb. **1987**, 30–31.

[3] Weiss, D.; Wuu, J.J.; Chin, V. "The on-chip 3 MB subarray based 3rd-level cache on an Itanium microprocessor", *ISSCC Dig. Tech. Papers*, San Francisco, CA, Feb. **2002**, 112–113.

[4] Chang, J. *et al.* "A 0.13 μm triple-Vt 9 MB third level on-die cache for the Itanium® 2 processor", *ISSCC Dig. Tech. Papers*, San Francisco, CA, Feb. **2004**, 496–497.

[5] Wuu, J. *et al.* "The asynchronous 24 MB on-chip level-3 cache for a dual-core Itanium®-family processor", *ISSCC Dig. Tech. Papers*, San Francisco, CA, Feb. **2005**, 488–489.

[6] Bradley, D.; Mahoney, P.; Stackhouse, B. "The 16 kB single-cycle read access cache on a next-generation 64 b Itanium microprocessor", *ISSCC Dig. Tech. Papers*, San Francisco, CA, Feb. **2002**, 110–111.

[7] Riedlinger, R.; Grutkowski, T. "The high-bandwidth 256 kB 2nd level cache on an Itanium microprocessor", *ISSCC Dig. Tech. Papers*, San Francisco, CA, Feb. **2002**, 418–419.

[8] Lohstroh, J.; Seevinck, E.; de Groot, J. "Worst-case static noise margin criteria for logic circuits and their mathematical equivalence", *IEEE J. Solid-State Circuits*, **1983**, *SC-18*(6), 803–807.

[9] Bhavnagarwala, A.J.; Tang, X.; Meindl, J.D. "The impact of intrinsic device fluctuations on CMOS SRAM cell stability", *IEEE J. Solid-State Circuits*, **2001**, *36*(4), 658–665.

[10] Seevinck, E.; List, F.J.; Lohstroh, J. "Static-noise margin analysis of MOS SRAM cells", *IEEE J. Solid-State Circuits*, **1987**, *SC-22*(5), 748—754.

[11] Hirose, T. *et al.* "A 20-ns 4-Mb CMOS SRAM with hierarchical word decoding architecture", *IEEE J. Solid-State Circuits*, **1990**, *25*(5), 1068–1074.

[12] Rusu, S. *et al.* "A 1.5-GHz 130-nm Itanium® 2 processor with 6-MB on-die L3 cache", *IEEE J. Solid-State Circuits*, **2003**, *38*(11), 1887–1895.

[13] Amrutur B.S.; Horowitz, M.A. "Fast low-power decoders for RAMs", *IEEE J. Solid-State Circuits*, **2001**, *36*(10), 1506–1515.

[14] Sasaki, K. *et al.* "A 9-ns 1-Mbit CMOS SRAM", *IEEE J. Solid-State Circuits*, Oct. **1989**, *24*(5), 1219–1225.

[15] Sasaki, K. *et al.* "A 23-ns 4-Mb CMOS SRAM with 0.2-μA standby current", *IEEE J. Solid-State Circuits*, **1990**, *25*(5), 1075–1081.

[16] Zhao, C. *et al.* "An 18-Mb, 12.3-GB/s CMOS pipeline-burst cache SRAM with 1.54 Gb/s/pin", *IEEE J. Solid-State Circuits*, **1999**, *34*(11), 1564–1570.

[17] Flannagan, S.T. *et al.* "8-ns CMOS 64 K × 4 and 256 K×1 SRAM's", *IEEE J. Solid-State Circuits*, **1990**, *25*(5), 1049–1056.

[18] Seevinck, E.; van Beers, P.J.; Ontrop, H. "Current-mode techniques for high-speed VLSI circuits with application to current sense amplifier for CMOS SRAM's", *IEEE J. Solid-State Circuits*, **1991**, *26*(4), 525–536.

[19] Blalock T.N.; Jaeger, R.C. "A high-speed clamped bit-line current-mode sense amplifier", *IEEE J. Solid-State Circuits*, **1991**, *26*(4), 542–548.

[20] Ho, R.; Mai, K.; Horowitz, M. "The future of wires," *Proc. IEEE*, **2001**, *89*(4), 490–504.

[21] Tzartzanis N.; Walker, W.W. "A differential current-mode sensing method for high-noise immunity, single-ended register files", *ISSCC Dig. Tech. Papers*, San Francisco, CA, Feb. **2004**, 506–507.

[22] Agawa, K. *et al.* "A bitline leakage compensation scheme for low-voltage SRAMs", *IEEE J. Solid-State Circuits*, **2001**, *36*(5), 726–734.

[23] Hsu, S. *et al.* "A 4.5-GHz 130-nm 32-kB L0 cache with a leakage-tolerant self reverse-bias bitline scheme", *IEEE J. Solid-State Circuits*, **2003**, *38*(5), 755–761.

[24] Chappell, T.I. *et al.* "A 2-ns Cycle, 3.8-ns access 512-kb CMOS ECL SRAM with a fully pipelined architecture", *IEEE J. Solid-State Circuits*, **1991**, *26*(11), 1577–1585.

[25] Nakamura, K. *et al.* "A 220 MHz pipelined 16 Mb BiCMOS SRAM with PLL proportional self-timing generator", *ISSCC Dig. Tech. Papers*, San Francisco, CA, Feb. **1994**, 258–259.

[26] McIntyre, H. *et al.* "A 4-MB on-chip L2 cache for a 90-nm 1.6-GHz 64-bit microprocessor", *IEEE J. Solid-State Circuits*, **2005**, *40*(1), 52–59.

[27] Schuster, S.E. *et al.* "A 15-ns CMOS 64 K RAM", *IEEE J. Solid-State Circuits*, **1986**, *SC-21*(5), 704–712.

[28] Amrutur, B.S.; Horowitz, M.A. "A replica technique for wordline and sense control in low-power SRAM's", *IEEE J. Solid-State Circuits*, **1998**, *33*(8), 1208–1219.

[29] Sekiyama, A. *et al.* "A 1 V operating 256-Kbit FULL CMOS SRAM", *Symp. of VLSI Circuits*, Honolulu, HI, June **1990**, 53–54.

[30] Ishibashi, K. *et al.* "A 1-V TFT-load SRAM using a two-step word-voltage method", *IEEE J. Solid-State Circuits*, **1992**, *27*(11), 1519–1524.

[31] Mai, K.W. *et al.* "Low-power SRAM design using half-swing pulse-mode techniques", *IEEE J. Solid-State Circuits*, **1998**, *33*(11), 1659–1671.

[32] Alowersson, J.; Andersson, P. "SRAM cells for low-power write in buffer memories", *Symp. on Low Power Electronics Dig. Tech. Papers*, San Jose, CA, Oct. **1995**, 60–61.

[33] Hattori, S.; Sakurai, T. "90% write power saving SRAM using sense-amplifying memory cell", *Symp. of VLSI Circuits*, Honolulu, HI, June **2002**, 46–47.

[34] Tzartzanis, N.; Athas, W.; Svensson, L. "A low-power SRAM with resonantly powered data, address, word, and bit lines", *Proc. Eur. Solid-State Circuits Conf.*, Stockholm, Sweden, Sept. **2000**, 336–339.

[35] Mizuno, H.; Nagano, T. "Driving source-line cell architecture for sub-1-V high-speed low-power applications", *IEEE J. Solid-State Circuits*, **1996**, *31*(4), 552–557.

[36] Yamauchi, H. *et al.* "A 0.8 V/100 MHz/sub-5 mW-operated mega-bit SRAM cell architecture with charge-recycle offset-source driving (OSD) scheme", *Symp. of VLSI Circuits*, Honolulu, HI, June **1996**, 126–127.

[37] Amrutur, B.S.; Horowitz, M.A. "Techniques to reduce power in fast wide memories", *Symp. on low power electronics Dig. Tech. Papers*, San Diego, CA, Oct. **1994**, 92–93.

[38] Matsumiya, M. *et al.* "A 15-ns 16-Mb CMOS SRAM with interdigitated bit-line architecture", *IEEE J. Solid-State Circuits*, **1992**, *27*(11), 1497–1503.

[39] Agarwal, A.; Li, H.; Roy, K. "A single-V_t low-leakage gated-ground cache for deep submicron", *IEEE J. Solid-State Circuits*, **2003**, *38*(2), 319–328.

[40] Osada, K. *et al.* "16.7-fA/cell tunnel-leakage-suppressed 16-Mb SRAM for handling cosmic-ray-induced multierrors", *IEEE J. Solid-State Circuits*, **2003**, *38*(11), 1952–1957.

[41] Nii, K. *et al.* "A 90-nm low-power 32-kB embedded SRAM with gate leakage suppression circuit for mobile applications", *IEEE J. Solid-State Circuits*, **2004**, *39*(4), 684–693.

[42] Agarwal, A. *et al.* "Effectiveness of low power dual-V_t designs in nano-scale technologies under process parameter variations", *Proc. Int. Symp. on Low Power Electronics and Design*, San Diego, CA, Aug. **2005**, 14–19.

[43] Kawaguchi, H.; Itaka, Y.; Sakurai, T. "Dynamic leakage cut-off scheme for low-voltage SRAM's", *Symp. of VLSI Circuits*, Honolulu, HI, June **1998**, 140–141.

[44] Itoh, K. *et al.* "A deep sub-V, single power-supply SRAM cell with multi-V_T, boosted storage node and dynamic load", *Symp. of VLSI Circuits*, Honolulu, HI, June **1996**, 132–133.

[45] Nii, K. *et al.* "A 90 nm dual-port SRAM with 2.04 μm^2 8 T-thin cell using dynamically-controlled column bias scheme", *ISSCC Dig. Tech. Papers*, San Francisco, CA, Feb. **2004**, 508–509.

[46] Yamaoka, M. *et al.* "A 300-MHz 25-μA/Mb-leakage on-chip SRAM module featuring process-variation immunity and low-leakage-active mode for mobile-phone application processor", *IEEE J. Solid-State Circuits*, **2005**, *40*(1), 186–194.

[47] Bhavnagarwala, A.J. *et al.* "A pico-joule class, 1 GHz, 32 KByte × 64 b DSP SRAM with self reverse bias", *Symp. of VLSI Circuits*, Kyoto, Japan, June **2003**, 251–252.

[48] Parke, S.A. *et al.* "Design for suppression of gate-induced drain leakage in LDD MOSFET's using a quasi-two-dimensional analytical model", *IEEE Trans. Electron Dev.*, **1992**, *39*(7), 1694–1703.

[49] Schroder, D.K.; Babcock, J.A. "Negative bias temperature instability: road to cross in deep submicron silicon semiconductor manufacturing", *J. Appl. Phys.*, **2003**, *94*(1).

[50] Yamaoka, M.; Osada, K.; Ishibashi, K. "0.4-V logic library friendly SRAM array using rectangular-diffusion cell and delta-boosted-array-voltage scheme", *Symp. of VLSI Circuits*, Honolulu, HI, June **2002**, 170–173.

[51] Bhavnagarwala, A.J. *et al.* "A transregional CMOS SRAM with single, logic V_{DD} and dynamic power rails", *Symp. of VLSI Circuits*, Honolulu, HI, June **2004**, 292–293.

[52] Zhang, K. *et al.* "A 3-GHz 70 Mb SRAM in 65 nm CMOS technology with integrated column-based dynamic power supply", *ISSCC Dig. Tech. Papers*, San Francisco, CA, Feb. **2005**, 474–475.

[53] Hidaka, H. *et al.* "Twisted bit-line architectures for multi-megabit DRAM's", *IEEE J. Solid-State Circuits*, Feb. **1989**, *24*(1), 21–27.

[54] Rabaey, J.M. *Digital Integrated Circuits: A Design Perspective*, Prentice Hall, Upper Saddle River, NJ, **1996**.

[55] Takeda, K. *et al.* "A 16-Mb 400-MHz loadless CMOS four-transistor SRAM macro", *IEEE J. Solid-State Circuits*, **2000**, *35*(11), 1631–1640.

[56] May, T.; Woods, M. "Alpha-particle-induced soft errors in dynamic memories", *IEEE Trans. Electron Dev.*, Jan. **1979**, *ED-26*(1), 2–9.

[57] Ziegler, J.F. *et al.*, Special Issue on "Terestrial cosmic rays and soft errors", *IBM J. Res. Dev.*, **1996**, *40*(1), 2–129.

[58] Tosaka, Y. *et al.* "Simple method for estimating neutron-induced soft error rates based on modified BGR model", *IEEE Electron Dev. Letters*, **1999**, *20*(2), 89–91.

[59] Sato, H. *et al.* "A 500-MHz pipelined burst SRAM with improved SER immunity", *IEEE J. Solid-State Circuits*, **1999**, *34*(11), 1571–1579.

[60] Osada, K. *et al.* "Cosmic-ray multi-error immunity for SRAM, based on analysis of the parasitic bipolar effect", *Symp. of VLSI Circuits*, Kyoto, Japan, June **2003**, 255–258.

[61] Suzuki, T. *et al.* "0.3 to 1.5 V embedded SRAM with device-fluctuation-tolerant access-control and cosmic-ray-immune hidden-ECC scheme", *ISSCC Dig. Tech. Papers*, San Francisco, CA, Feb. **2005**, 484–485.

[62] Kalter, H.L. *et al.* "A 50-ns 16-Mb DRAM with a 10-ns data rate and on-chip ECC", *IEEE J. Solid-State Circuits*, **1990**, *25*(5), 1118–1128.

[63] Henkels, W.H.; Hwang, W.; Joshi, R.V. "A 500 MHz 32-word × 64-bit 8-port self-resetting CMOS register file and associated dynamic-to-static latch", *Symp. of VLSI Circuits*, Kyoto, Japan, June **1997**, 41–42.

[64] Murabayashi, F. *et al.* "3.3-V BiCMOS circuit techniques for a 120-MHz RISC microprocessor", *IEEE J. Solid-State Circuits*, **1994**, *29*(3), 298–302.

[65] Fetzer, E.S.; Orton, J.T. "A fully-bypassed 6-issue integer datapath and register file on an itanium microprocessor", *ISSCC Dig. Tech. Papers*, San Francisco, CA, Feb. **2002**, 420–421.

[66] Golden, M.; Partovi, H. "A 500 MHz, write-bypassed, 88-entry, 90-bit register file", *Symp. of VLSI Circuits*, Kyoto, Japan, June **1999**, 105–108.

[67] Tzartzanis, N. *et al.* "A 34 word × 64 b 10R/6W write-through self-timed dual-supply-voltage register file", *ISSCC Dig. Tech. Papers*, San Francisco, CA, Feb. **2002**, 416–417.

[68] Tzartzanis, N.; Walker, W.W. "A transparent voltage conversion method and its application to a dual-supply-voltage register file", *Proc. Int. Conf. on Computer Design*, San Jose, CA, Oct. **2003**, 107–110.

[69] Lev, L.A. *et al.* "A 64-b microprocessor with multimedia support", *IEEE J. Solid-State Circuits*, **1995**, *30*(11), 1227–1238.

[70] Asato, C. "A 14-port 3.8-ns 116-word 64-b read-renaming register file", *IEEE J. Solid-State Circuits*, **1995**, *30*(11), 1254–1258.

[71] Tang, S. *et al.* "A leakage-tolerant dynamic register file using leakage bypass with stack forcing (LBFS) and source follower NMOS (SFN) techniques", *Symp. of VLSI Circuits*, Honolulu, HI, June **2002**, 320–321.

[72] Krishnamurthy, R.K. *et al.* "A 130-nm 6-GHz 256 × 32 bit leakage-tolerant register file", *IEEE J. Solid-State Circuits*, **2002**, *37*(5), 624–632.

Chapter 5

LARGE-SCALE CIRCUIT PLACEMENT

Ameya R. Agnihotri[1], Satoshi Ono[1], Mehmet Can Yildiz[2], and Patrick H. Madden[1]
[1]*SUNY Binghamton, Computer Science Department*
E-mail: pmadden@binghamton.edu
[2]*IBM, Austin Research Laboratories*

Abstract: Modern computing systems contain staggering numbers of transistors and interconnecting wires, and blocks of widely varying size and shape. The placement problem is intractable for even small problems and simple metrics; heuristic methods, and in particular, multi-level formulations, are commonplace across the industry. In this chapter we survey modern techniques for circuit placement, with an emphasis on how placement interacts with logic synthesis and routing. Stability of placement methods is now a key concern: to allow timing closure it is essential that gate sizing, buffer insertion, and routing can be completed without large disruptions to the overall physical structure of a circuit. We also discuss fundamental aspects of computing circuits that have made the placement problem progressively more difficult – resulting in a recent shift toward "multi-core" microprocessors. We argue that this change is an extremely significant event, as it fundamentally changes how microprocessor design must be pursued.

Key word: Placement.

1. Introduction

Circuit placement – the physical location of all logic elements – is a fundamental part of integrated circuit design, and has been studied for many years. The first circuits could be designed by hand; with a small number of logic elements a human design can easily arrange them in a compact and efficient way. As design sizes have increased, the problems have become far too complex

Vojin G. Oklobdzija and Ram K. Krishnamurthy (eds.),
High-Performance Energy-Efficient Microprocessor Design, 121–146.

for easy comprehension. Even at the highest levels of abstraction there can be hundreds of objects and constraints to consider.

Microprocessors are at the leading edge of Moore's Law; they are large and complex, and provide the most challenging problems to placement tools. In this chapter we consider two different placement formulations: floorplanning, which is used at the highest levels of abstraction, and mixed size placement, which combines both low-level logic elements and large functional blocks. Algorithms for circuit placement have become more sophisticated over the years. For floorplanning, simulated annealing has been dominant; in mixed size placement one commonly finds annealing, analytic methods, and recursive bisection.

In this chapter we will use a typical abstraction: we will approach placement as the embedding of a hypergraph into a plane. The vertices of the hypergraph correspond to logic elements – these are normally rectangular, with either a fixed outline, or a range of feasible outlines. The hyperedges correspond to the interconnecting wires, or "signal nets." Because an interconnect wire may attach to several logic elements, the hypergraph formulation is appropriate – hyperedges model the "multiple connection" concept directly.

We first consider clustering techniques, and in particular, multi-level optimization techniques. Modern designs are simply too large to be handled in a "flat" way. With hundreds of millions, or even billions, of logic elements, representing a circuit down to its lowest element is simply not practical. By grouping related logic functions in a hierarchical manner the placement problem becomes more tractable.

Next, we discuss placement at the highest level of abstraction – the floorplanning phase. While modern circuits may contain huge numbers of transistors, it is almost always possible to reduce them to a few hundred distinct logic blocks. By restricting these blocks to be rectangular, an arrangement that has little wasted space and low wire length can be found. Our discussion will revolve around an early floorplan representation, the sequence pair. A good floorplan is an essential component in the production of a good microprocessor design; a poor floorplan may doom a design.

Below the floorplanning level is mixed size placement, in which macro blocks and large numbers of standard cells are intermixed. In mixed size placement research, one finds annealing, analytic methods, and recursive bisection. Mixed size placement is relevant to modern designs, as it allows the reuse of predesigned blocks of logic, as well as the customization freedom of standard cell design.

Modern designs normally require a great deal of gate sizing and buffer insertion to meet performance constraints. The "stability" of a placement algorithm is of great concern, and we discuss recent methods and metrics to quantify this. If the overall structure of a placement changes significantly after the insertion of a buffer, the timing behavior may also change. During tuning, this erratic behavior can prevent design convergence.

We conclude our discussion by presenting theoretical limits to circuit placement. Soaring power demands have resulted in the introduction of "dual core" microprocessors; this trend is likely to continue. By examining geometric constraints on the embedding of circuits, this change is not in the least surprising – and there are interesting implications for the future of microprocessor design.

2. Multi-level Circuit Representation

As mentioned in the introduction, a circuit is frequently modeled as a hypergraph $G(V, E)$. The graph is composed of a set of vertices V and a set of hyperedges E.

Many of the circuit details – the functionality of each logic element or IP macro block, the size of each element, signal directions on each net, and so on, are abstracted away to simplify the placement problem. This simplification is necessary for large designs: there is simply too much information to be processed effectively. Issues such as timing criticality are usually considered only indirectly in placement: computing the "true" critical path is too costly for use within the inner loop of a placement engine.

The most common and effective simplification is *clustering* (also known as "coarsening"); logic elements are grouped together, so that fewer objects must be manipulated. In recent years *multi-level clustering* has become an essential part of placement research.

2.1. Basic Clustering Techniques

Clustering reduces the size of the input graph $G(V, E)$ by creating a new graph G'. This new graph contains fewer vertices and edges, normally with pairs (or groups) of vertices being condensed into one.

If a pair of vertices v_i and v_j are merged, both can be replaced by a single new vertex $v_{i,j}$. If all vertices of a signal net e_i are merged, the net can be eliminated from consideration. If two nets e_i and e_j share an identical set of vertices, the nets can be replaced by a single new net $e_{i,j}$, which may have increased weight. There are many methods to cluster vertices and edges – we mention a few here.

- While multi-level partitioning had been performed previously [1], the *hMetis* partitioner [2] developed by Karypis achieved a leap in performance. *hMetis* supports a variety of clustering methods; the *first-choice*

approach simply groups connected vertices together, and merges them in a greedy fashion. A variety of hybrid methods modify the coarsening scheme, to allow a preference for heavily weighted nets, or for merge operations that eliminate hyperedges. Perhaps most significant is the use of multiple V-cycles, which are described below.
- The *edge separability* approach developed in Cong and Lim [3] pursues clustering with a more sophisticated objective function. The authors developed a "CAPFOREST" algorithm, which distinguishes between vertices with a high degree of local connectivity (good candidates for clustering) and vertices that have less local connectivity.
- A third method, *best-choice* [4], mixes more complex (and time consuming) objective functions with a lazy update scheme.

There are quite literally dozens of different methods to perform clustering. They are all heuristic in nature, and there is no clear definition of "optimality" for clustering. While many complex (and computationally expensive) methods have been explored, for many applications this is not necessary – the use of "multiple V-cycles" can frequently overcome the shortcomings of a poor clustering method.

2.2. The V-cycle

Clustering alone simply reduces the complexity of a hypergraph. For physical design there is obviously a need for a solution to the original "flat" problem. This is achieved by a "V-cycle," as shown in Figure 1.

In a typical multilevel approach, a hypergraph is repeatedly clustered until a manageable problem size is obtained; this is the "downward" portion of a V-cycle. A solution is found for the coarse graph – in the case of partitioning, one typically finds the hill-climbing approach of Fidducia and Mattheyses [5].

The solution for the coarse graph is then projected onto a slightly less coarse version of the graph. Given a high-quality initial configuration the graph can be refined, and the solution quality improved. Repeated uncoarsening and refinement forms the "upward" portion of a V-cycle.

A significant contribution of Karypis *et al.* [2] is the introduction of multiple V-cycles. Rather than repeatedly uncoarsening a graph, the graph may be reclustered, producing a zig-zag pattern. With multiple V-cycles, excellent solutions can be found for partitioning; most feel that modern partitioners are close to optimal for even very large problems.

Multiple V-cycles address the shortcomings of clustering algorithms. It is entirely possible for a clustering to make an optimal solution unreachable. Consider the partitioning problem shown in Figure 2; if the signal net indicated is clustered, a balanced partition with minimum cut cannot be obtained. By using

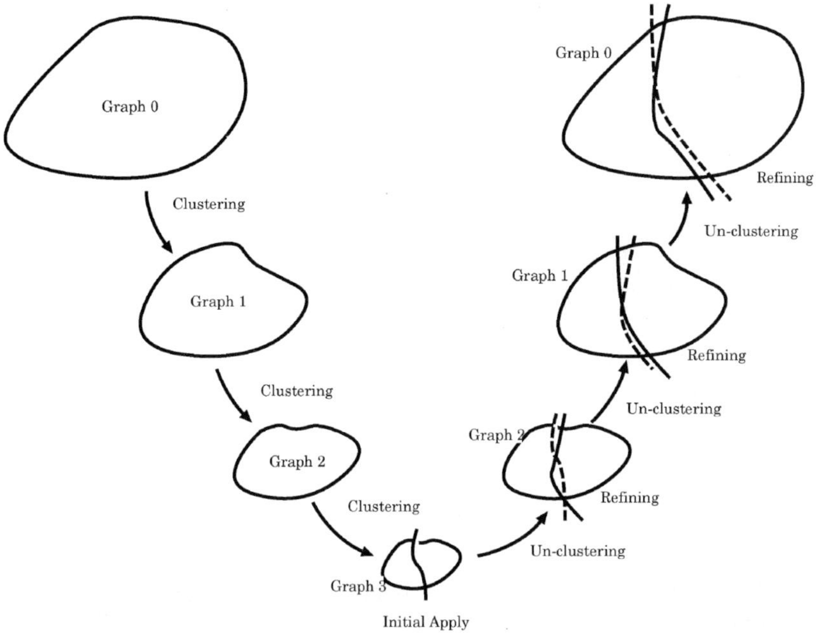

Figure 1. In both placement and partitioning, clustering is normally performed with a "V-cycle." An initial circuit structure is repeatedly clustered, to simplify optimization. The circuit is then progressively unclustered, with optimization at each step using the prior solution as an initial starting point.

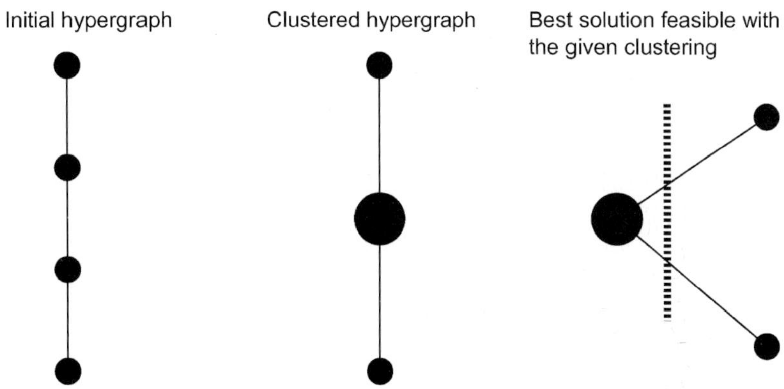

Figure 2. A poor clustering may bound a solution away from optimality; in this case, contracting the center edge results in balanced partitionings having a cut of two (compared to the optimal solution of one). Multiple V-cycles are an effective means to overcome poor clustering choices.

multiple V-cycles an algorithm may recover from a poor clustering choice. As theoretical results described in Section 6 indicate, it may be impossible to avoid such choices.

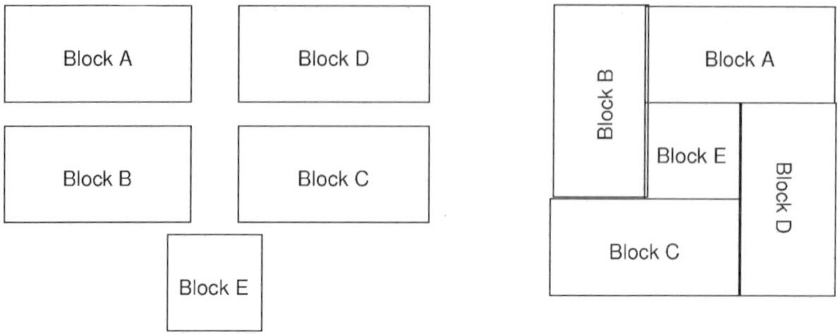

Figure 3. A simple floorplanning problem; blocks of varying sizes must be packed into a rectangular space.

3. Floorplanning

At the highest level of abstraction for placement is *floorplanning*. The objects to be placed are frequently referred to as "blocks." The number of objects that must be placed is typically relatively small – from a dozen or so, up to perhaps a few thousand. Each block may represent a very large, and very complex, functional unit.

The placement problem at this stage may simply be making things *fit* into the allowed space; there is a classic (and NP-complete) block packing problem that is well known in the computer science community. Figure 3 shows a simple example, in which four 2×1 and one 1×1 block must be placed. The total area of the logic elements is nine; fitting them into a three-by-three area, however, is not a simple matter.

Optimization of a floorplan (to minimize either interconnect length or chip area) requires both a way to represent the arrangement of blocks, and a method to modify that arrangement. While there are a variety of floorplan representations, we focus on *sequence pairs* [6] here.

3.1. Slicing vs. Non-slicing Floorplans

An important issue to note is that many good floorplans are not "slicing:" it is not possible to divide the floorplan into two sections with either a vertical or horizontal line. The example in Figure 3 illustrates this. The non-slicing nature of the floorplanning problem makes the use of divide-and-conquer algorithms more difficult.

There is a puzzle-like avor to floorplanning. In addition to being concerned with the relative positions of the logic blocks, we must also consider how the blocks interlock.

3.2. The Sequence Pair

A primary constraint of a valid floorplan is that no two blocks overlap. The sequence pair representation, which has been incorporated into many modern design automation tools, is an elegant method to achieve this.

Consider the "non-overlap" constraint. In a non-overlapping floorplan, if a pair of blocks A and B are vertically aligned, then either A must be above B, or vice-versa. Similarly, if the blocks are horizontally aligned, then A must be to the left of B, or vice-versa. If at least one of these conditions holds for every pair of blocks, then the floorplan has no overlaps; the sequence pair representation ensures that this is true.

A sequence pair consists of two strings; each string is simply one permutation of a list of the blocks to be placed. The overlap constraints are derived from the relative positions of the blocks within each string. If A appears before B in both strings, we interpret this as a constraint that A be to the *left* of B. If A appears before B in the first string, but their relative order is reversed in the second, then A is below B. The other possible constraints (*right* or *above*) are variations of the first two – the relationships are reflexive.

With n blocks there are $n! \times n!$ different possible sequence pair combinations; the possible solution space is extremely large, and covers a wide range of possibilities.

Translation from a sequence pair to an actual floorplan (with specific locations for each block) is relatively simple; the horizontal and vertical positions can be determined separately. If one considers the "to the left of" constraints, a directed graph can be constructed, as shown in Figure 4. Traversing the graph with a longest path algorithm provides horizontal positions for the blocks such that they cannot overlap horizontally. A similar vertical traversal provides positioning along the other axis.

Note that, in many cases, it is possible for a block to slide horizontally or vertically; this is frequently performed as a post-processing step. In floorplanning, blocks may also be mirrored, rotated, or flipped into one of eight different orientations; these operations are normally kept separate from the sequence pair organization, and are optimized separately.

3.3. Annealing Based Optimization

If the number of blocks to be planned is relatively small, it may be possible to explore all sequence pairs to find the optimal configuration. In practice, however, problems are large enough to prohibit brute force search, and heuristic methods are applied.

Given the floorplan representation, almost all modern floorplanners perform optimization with an annealing [7] based approach. Starting from an initial

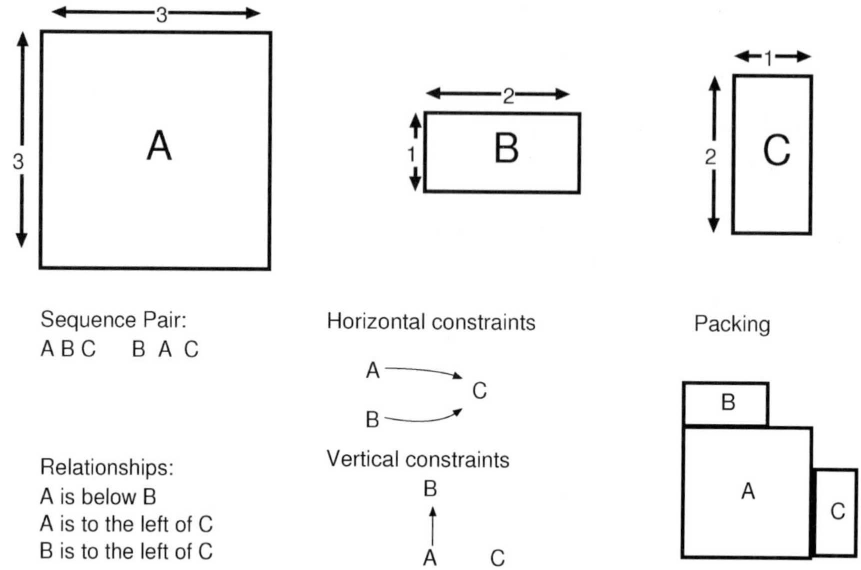

Figure 4. A small floorplanning problem, with a sequence pair representation. The ordering of blocks in each sequence has either a horizontal or vertical layout constraint. The x and y coordinates for each block are determined by traversing the constraint graphs with a longest path algorithm.

random configuration, the placement can be improved – by swapping the sequence positions of two blocks, rotation of a block, and so on. The annealing approach allows for exploration of the solution space while avoiding becoming stuck within a local minima.

In early floorplanning work the primary objective was area minimization. With each perturbation of the sequence pair the total (rectangular) area needed to enclose the floorplan is computed; acceptance or rejection of a move is based on the total amount of excess space required, the annealing temperature, and the random variable.

As interconnect delay has increased, wire length has also become a common optimization objective during floorplanning. By combining both area and length (with the total solution cost normally being a weighted sum), the annealing process can achieve both low area and low wire length.

3.4. The Human Touch

While annealing based floorplanners are effective at packing blocks together, and can do an acceptable job with wire length minimization, it is common for human designers to need some degree of control. These experts, who have developed the circuit architecture with specific critical paths, busses, and configurations in mind, can commonly produce results that are far superior to automatically generated floorplans.

Much of the recent work in floorplanning has focused on addressing user-supplied constraints. Alignment of blocks (to support a signal bus), or matched distances between blocks, are two common themes. Obstacle avoidance has also received attention: many large designs have I/O pads within the boundary of the placement area, or blocks in locations fixed by a designer.

4. Mixed Size and Standard Cell Placement

Below the floorplanning level is *mixed size placement*, which combines large numbers of standard cells with macro blocks. The problem is sometimes referred to as "boulders and dust" – to highlight the difference in size of the blocks and cells. A small mixed size placement problem is shown in Figure 5; all objects, both large and small, may be moved.

Almost all tools divide the placement problem into two steps; *global* and *detailed* placement, to simplify the problem. The global placement step finds

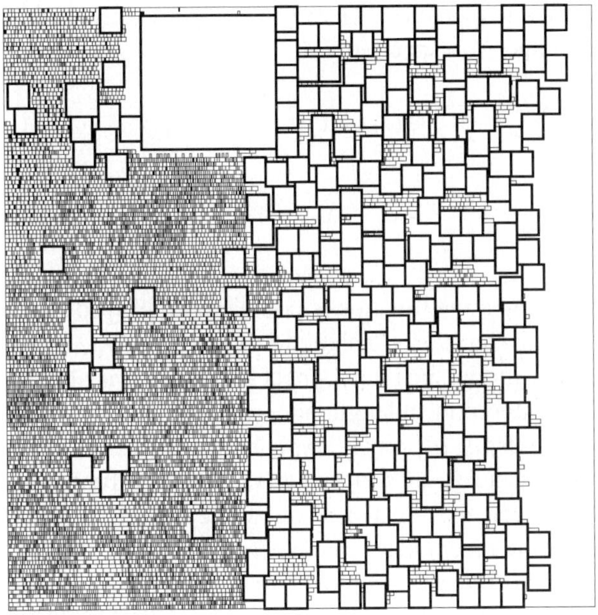

Figure 5. A mixed size placement problem. Modern microprocessor blocks may contain hundreds of thousands of standard cells, and thousands of macro blocks – some of which may be moved, and some which may be fixed in place. By increasing the number of objects that can be handled with mixed size placement, the number of blocks addressed by floorplanning can be reduced.

a rough distribution of cells in the placement region with the primary objective of wire length minimization. The detailed placement step removes overlaps from the global placement and performs local optimization to further improve wire length. A uniform distribution of cells by the global placer is very important to make the job easier for the detailed placer.

4.1. Global Placement Techniques

Several competitive approaches have been developed over the years for global placement and we survey them here.

4.1.1. Analytic methods

Analytic methods formulate the circuit placement problem as a mathematical optimization function, typically *quadratic* or *linear*. An excellent overview of the approach is given by Vygen [8]. If logic elements v_i and v_j are connected to each other, then expressions (1) and (2) represent quadratic and linear objective functions respectively. Here (x_i, y_i) and (x_j, y_j) are the geometric coordinates of v_i and v_j respectively.

$$\text{minimize} \sum_{\forall Nets} \sum_{\forall i,j \in Net_v} \{(x_i - x_j)^2 + (y_i - y_j)^2\} \tag{1}$$

$$\text{minimize} \sum_{\forall Nets} \sum_{\forall i,j \in Net_v} \{|x_i - x_j| + |y_i - y_j|\} \tag{2}$$

The quadratic objective function is popularly employed as it can be solved efficiently using sparse matrix techniques. One problem, however, is that it does not accurately model the interconnect length. Linear objective function is a true measure of interconnect length but cannot be efficiently solved for large problem size. Doll *et al.* [9] nicely explains this trade off. Recently, Naylor *et al.* [10] proposed a *log-sum-exponential* objective function that balances the trade-offs between the linear and quadratic objective. This is employed successfully by some recent academic placers [11, 12].

Algorithm 1 illustrates the typical framework of an analytic placement engine. The well-known analytic placement tool *GORDIAN* [13] could be considered a template for many modern methods. It recursively utilized a quadratic formulation to find a relative ordering, and then used quadrisection to separate logic elements into different regions. Without some method to force logic elements apart, analytic tools may produce degenerate solutions in which all logic elements are located on top of each another.

Algorithm 1. The framework of a typical analytic placement approach. For large designs the problem is typically simplified through a multilevel clustering approach. Clustering effectively reduces the numbers of variables and constraints, allowing faster convergence with relatively little loss in solution quality.

```
Create the initial placement region R_init containing all components;
List of regions, L_r = {R_init};
for each region R_i in L_r do
    L_r = L_r − {R_i};
    if |R_i| > k then
        Use an analytic solver to solve the optimization function;
        Partition the placement based on the locations derived from previous
        step into two or four regions;
        Add new regions to the list L_r;
    end if
end for
```

4.1.2. Recursive bisection

Capitalizing on the dramatic improvements of partitioning algorithms, a number of tools based on recursive bisection have produced excellent results with remarkably low run times. The placement approach is remarkably simple and intuitive; modern placers such as *Feng Shui* [14] and *Capo* [15] have much in common with the early work by Breuer [16] or Dunlop and Kernighan [17].

A typical recursive bisection placement flow is presented in Algorithm 2. The process begins by creating a single *region* which represents the initial placement area. A multi-level partitioning algorithm splits the circuit into two components; the region is split into two halves (with either a vertical or horizontal cut line), and portions of the circuit are assigned to each half. The lengths of nets that span the two partitions are minimized through the use of terminal propagation [17].

The direction of cut lines is typically determined by the aspect ratio of a region [18]. If a region is tall, cut lines are horizontal, and vice-versa. With each partitioning the number of regions doubles, until each logic element is assigned to a single region.

If the number of components in a region R_i (denoted by $|R_i|$ in the algorithm) is one, then the component is placed at the center of the region and the region is removed from the list of regions (L_r) to be partitioned. The algorithm terminates when L_r is empty.

Unlike analytic methods, there is little difficulty with overlap removal. The partitioning objective function (minimum cut) does not model the wire length

Algorithm 2. Recursive bisection placement. With each partitioning, a multi-level algorithm such as *hMetis* performs clustering and multiple v-cycles. The "placement" view of a large circuit may be flat, but a key component utilizes a multi-level approach.

Create the initial placement region R_{init} containing all components;
List of regions, $L_r = \{R_{init}\}$;
for each region R_i in L_r **do**
  $L_r = L_r - \{R_i\}$;
  if $|R_i| > 1$ **then**
    Partition R_i into R_x and R_y;
    $L_r = L_r \cup \{R_x; R_y\}$;
  else
    Place the only component in R_i at the center of R_i;
  end if
end for

objective as accurately, however, and solution quality from modern bisection based tools typically trails the leading analytic tools.

4.1.3. Simulated annealing.

A typical annealing based placement flow is presented in Algorithm 3. An initial placement is first generated either randomly or using some heuristic; for example, the recursive bipartitioning-based technique described above. An initial annealing temperature is then set.

The optimization consists of two main loops: outer and inner loop. The outer loop runs till a termination condition is not true. This condition is usually satisfied if a certain number of iterations of the inner loop pass without any improvements to the placement quality. The inner loop is run for a predefined number of iterations. During each of these iterations, several perturbations are tried on the placement. Each change is accepted if it reduces the placement cost (condition ($\Delta cost < 0$) in Algorithm 3) or if the annealing cost function value ($e^{-\Delta cost/T}$) falls above a random number between 0 and 1; otherwise the move is rejected. At the end of each outer iteration the annealing temperature is reduced.

Placement cost function is typically a combination of wire length and overlaps generated by placement perturbations.

The perturbations could be: swapping locations of a pair of cells, moving a cell to a new location, changing the orientation of a cell. During each of these changes, two things need to be taken care of: first, the overlaps created

Algorithm 3. Simulated annealing-based placement flow. The termination condition of the outer *while* loop is usually a passage of a few iterations of the inner *for* loop with little or no improvements to the placement quality.

```
Generate an initial random placement;
Initialize current temperature for cooling schedule, T = T_init;
while Termination condition is not satisfied do
  for A predefined number of iterations do
    Save current placement P_i;
    Perturb current placement P_i to get placement P_j;
    Change in cost due to the perturbation, Δcost = cost_j − cost_i;
    if (Δcost < 0)||(e^(−Δcost/T) > random[0, 1]) then
      Accept placement P_j;
    else
      Roll back to placement P_i;
    end if
  end for
  Decrease temperature T;
end while
```

by such changes; secondly, a limit on the cell displacement. The first goal is typically achieved by penalizing changes that create overlaps. The second goal is achieved by defining windows for cell displacement which are reduced toward the end of the optimization.

In the presence of macros, certain additional rules could be enforced on the kind of moves generated. For example, the placement tool *TimberWolf* [19] prevents swapping of objects if they do not belong to the same *exchange class*, where cells, macros and pads belong to three different exchange classes. Another example is fixing the orientation of big macros, especially the ones with a large aspect ratio, as this would sweep a huge number of standard cells from their initial locations.

The annealing cost function is such that initially, when the temperature is high, its value is large. So the probability of accepting a move that worsens the cost is high. This makes sure that we do not get stuck in *local minima*. As we reduce the temperature, the value of the annealing cost function also reduces and the chances of accepting a bad move reduce. This makes sense, because as our solution starts converging, we would not want to accept too many moves that make the solution worse.

Given sufficient time by making sure that the temperature is reduced *slowly*, annealing-based heuristics are known to produce high-quality results. The negative exponential function in Algorithm 3 indeed complies with this idea.

While annealing is clearly the dominant method for floorplanning, it has fallen out of favor with mixed-size placement. A primary concern is a lack of scalability; run times increase rapidly, making it impractical for largeproblems.

For small problems, the results of early annealing based placers such as *TimberWolf* [20] are still impressive. To handle large problems, recent methods such as *Dragon* [21] apply a hybrid of bisection and annealing.

4.2. Detailed Placement

Detailed placement involves removal of overlaps between logic elements followed by local improvements to the placement to further minimize wire length. This is explained in the following subsections.

4.2.1. Legalization

Placements produced by the techniques presented above can have overlaps between logic elements. These overlaps need to be removed in order to make the placement legal, a process called legalization. This involves the following steps.

- Overlap removal.
- Alignment of y coordinates of movable elements with placement row boundaries.
- Alignment of x coordinates of movable elements with legal site locations within a placement row.

For legalization, *dynamic programming* [22, 23] and *network flow* [24]-based heurisics have been applied effectively.

For mixed size legalization, many current placers rely on a simple but extremely fast and effective *greedy* technique proposed in Agnihotri *et al.* [14] which is based on the patent in Hill [25]. This method is presented in Algorithm 4. The cost function minimizes the displacement between the cell's initial

---

Algorithm 4. A placement legalization algorithm. Input is a list of cells C where each cell has a location assigned by the global placement algorithm. Output is a legal placement.

Sort C by *left-edge* locations of cells;
for each cell c_i in C, in sorted order **do**
  Find a location for c_i in a row that minimizes the cost function $(\Delta x + \Delta y)$;
  Fix c_i and mark the corresponding placement sites as occupied;
end for

---

and final (legal) location. Once fixed, each legalized cell acts as an obstacle for subsequent non-legalized cells.

4.2.2. *Local improvements*

Post-legalization, many tools use sliding window techniques to further improve interconnect length. In this approach a small group of cells is selected and optimal permutation of cells is found, most commonly using a *branch and bound* technique. Sliding the window over the entire placement improves the wire length considerably. Due to run time limitations, these techniques can usually be applied only to a small number of cells, typically six to eight.

An effective dynamic programming technique called *optimal interleaving* was presented in Hur and Lillis [26]; the approach is polynomial time, allowing optimization of larger groups than can be handled with branch and bound. A small number of cells are selected in a window and divided into two sets A and B. Then, using a simple dynamic programming formulation, an optimal permutation of these cells is found that minimizes the wire length. The relative ordering of cells from the subsets A and B is preserved. For example, if A and B contain $\{a_1, a_2\}$ and $\{b_1, b_2\}$ respectively in that order from left to right, then $\{a_1, b_1, a_2, b_2\}$ is a valid placement whereas $\{a_2, b_1, a_1, b_2\}$ is not, as the ordering of cells a_1 and a_2 is reversed.

5. Synthesis and Stability

In modern designs, interconnect delay dominates system delay. As a result the critical path is not known until placement is nearly complete – and at this point it may be impossible to address performance problems. For large circuits the need to run placement tools multiple times, and to adjust the size of logic elements within the circuit, is unavoidable.

Commonly, in a physical synthesis flow, placement is applied to a given netlist several times iteratively and all logic changes such as repowering, buffer insertion, cloning, swapping, etc., are performed after each placement step. As logic changes provide some improvement on the given objective (commonly timing and/or power) they are mostly local and not capable of solving the problems which involve a large number of gates. If placement changes are not limited to small areas, the disruption of the overall circuit structure can significantly change the state of the design. While placement is a powerful tool in the physical synthesis flow, the changes introduced may increase the time needed for convergence. A stable placement algorithm, which generates similar solutions even with changes in the input netlist and placement parameters, is important for a fast convergence in a physical synthesis flow.

Increasing design size and complexity and decreasing feature size makes timing closure more difficult. Today, a tool can spend a week or more on a several-million-gate design. If a designer makes a small change in the input netlist and runs the tool again he or she should expect to see a result similar to the previous run. If the underlying placement algorithm is not stable, it can produce a totally different solution with a new set of problems for the designer to fix.

High run times, and the need for convergence without a great deal of designer effort, has made stability a key concern. Many groups would gladly sacrifice wire length minimization for increased stability.

5.1. Stability Metrics

In order to improve stability of a placement algorithm we first should be able to measure the stability; otherwise it is not possible to come up with a solution and say that this increase the stability. Alpert *et al.* [4] defined two stability metrics: object movement and net clusters. Both the metrics can measure the stability of placement solutions if the netlist is the same in both the runs.

Quantifing stability requires defining the similarity metric between two placements. We seek to answer; "what does it mean for a placement A to be similar to a placement B?" Let $S(A, B)$ represent a stability metric between two placements A and B. A desirable characteristic would be measure stability on a 0 to 1 scale, i.e. if $S(A, B) = 0$, then A and B are identical, and if $S(A, B) = 1$ they are completely different. One can think of this as a percentage, e.g. if $S(A, B) = 0.35$, then they are 35% different. This criterion allows one to measure algorithm stability for a range of designs and still intuitively understand the behavior. In addition, the following properties of metrics should hold.

- Reflexive property: $S(A, A) = 0$.
- Associative property: $S(A, B) = S(B, A)$.
- Triangle inequality: $S(A, B) + S(B, C) \leq S(A, C)$.

5.1.1. *Measuring object movement*

Object movement metric measures how much the cell locations changed from one placement to the next. The total object movement of cells in terms of Manathan distance is given by:

$$OM(A, B) = \sum_{i=1}^{n} a_i \left(|x_i^A - x_i^B| + |y_i^A - y_i^B| \right) \quad (3)$$

The following equation, the object movement stability metric, gives us the total amount of location changes weighted by area and scaled by the expected value of the location differences.

$$S_{OM}(A,B) = \frac{3\sum_{i=1}^{n} a_i \left(\left|x_i^A - x_i^B\right| + \left|y_i^A - y_i^B\right|\right)}{A_t(W+H)} \tag{4}$$

a_i is the area of cell v_i, and $A_t = a_i + \ldots + a_n$ is the total area of all cells. (x_i^A, y_i^A) is the location of cell v_i in placement A. W and H are the width and the height of the placement area respectively.

This metric measures the linear movement between cells. Another object movement metric is the squared object movement metric. This metric penalizes the objects that move a long distance while reducing the effect of short movements. As is the practice in analytical placement the squared object movement can be measured keeping the horizontal and vertical components separate. Equation (5) gives the squared object movement.

$$S_{OMS}(A,B) = \frac{6\sum_{i=1}^{n} a_i \left((x_i^A - x_i^B)^2 + (y_i^A - y_i^B)^2\right)}{A_t(W^2+H^2)} \tag{5}$$

[Alpert] has both the details of the metrics and all the proofs.

5.1.2. *Measuring net clusters*

Both object movement metrics measure what happens to the cells, but they tell us nothing about what happens to the nets or structure of the placement. Consider the example in Figure 6. The placement A has a natural structure in which there are two dominant clusters of logic as indicated by the coloring of cells. Placement B is the mirror image of A while placement C is random.

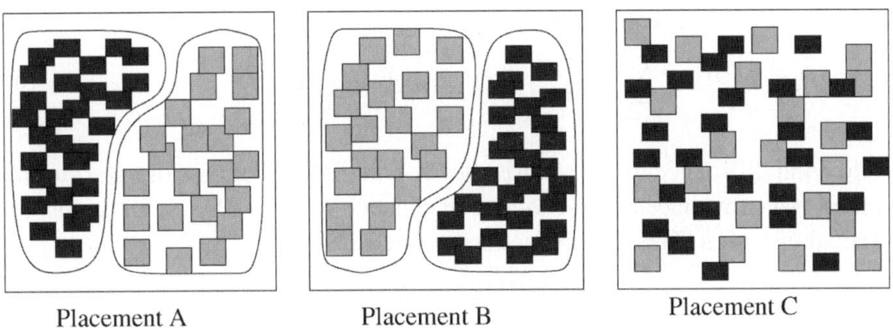

Figure 6. Placement A has two natural clusters, placement B is a mirror image of A and placement C is random. Arguably, A and B are similar, especially when compared to C, yet this is not captured by object movement stability metrics.

According to object movement metrics, it is likely that $S(A, B)$ would actually be larger than one, since the average cell location moves quite a bit from A to B. However, clearly A and B are quite similar, especially when compared to C. Both A and B preserve the natural connectivity based clustering of cells. Net center displacement metrics seek to quantify how well cells connected to each net stay grouped together from one placement to the next.

Let n_j be the number of pins connected to e_j , (x_{cj}, y_{cj}) be the coordinates of the center of the net and let

$$HD_j(A, B) = \sum_{v_i \in e_j} \left| |x_i^A - x_{cj}^A| - |x_i^B - x_{cj}^B| \right| \tag{6}$$

be the total difference of the horizontal distance of pins of net e_j to the center of e_j for placement A to that of placement B. In other words, consider the sum of the cumulative distances from each point to its net center. This is equivalent to total net length in the star topology (for one-dimension). Similarly let $VD_j(A, B)$ be the same as $HD_j(A, B)$ but only for vertical direction. The total net center displacement over all nets is defined as:

$$NC(A, B) = \sum_{j=1}^{m} (HD_j(A, B) + VD_j(A, B)) \tag{7}$$

To create a stability metric $NC(A, B)$ needs to be normalized with the expected value $E[NC(A, B)]$. So the net center metric is

$$S_{NC}(A, B) = \frac{\sum_j^m K(n_j)(HD_j(A, B) + VD_j(A, B))}{P(W + H)} \tag{8}$$

where $K(n_j)$ is the constant coeffecient for the nets with size of n_j , P is the the total number of pins in the netlist and W and H are the width and height of the image area respectively.

The squared net center metric which reflects the distance in the placements more accurately is given below.

$$HDS_j(A, B) = \sum_{v_i \in e_j} \left| (x_i^A - x_{cj}^A)^2 - (x_i^B - x_{cj}^B)^2 \right| \tag{9}$$

is the total difference of the squared horizontal distance of pins of net e_j to the center of e_j for placement A to that of placement B. $HDS_j(A, B)$ is for the vertical direction. The squared net center metric is

$$S_{NCS}(A, B) = \frac{\sum_j^m K_S(n_j)(HDS_j(A, B) + VDS_j(A, B))}{P(W^2 + H^2)} \tag{10}$$

where $K_S(n_j)$ is the constant coeffecient for the squared metric.

Empirically derived values for K and K_S for each net of size n_j is given in Table 1.

Table 1. Net degree constants for net center metrics

Size	$K(n)$	$K_S(n)$	Size	$K(n)$	$K_S(n)$
2	7.50	20.00	14	6.05	12.23
3	6.66	15.24	16	6.04	12.19
4	6.39	13.89	18	6.03	12.15
5	6.27	13.27	20	6.03	12.13
6	6.20	12.94	30	6.02	12.07
7	6.16	12.72	40	6.02	12.06
8	6.12	12.58	60	6.01	12.03
9	6.10	12.48	80	6.00	12.01
10	6.08	12.40	100	6.00	12.00
12	6.07	12.30	∞	6.00	12.00

5.2. Achieving Timing and Power Closure on Real Designs

Companies always look for better and cheaper products in order to compete well against others. This situation usually forces a chip designer to design aggressively (i.e. using more transistors for better functionality as aiming for shorter delay and less power). It is almost always true that a physical synthesis tool cannot close on timing or give better power during the first run on a new design. A tool needs to be tuned for each design to some extent based on to its needs. Therefore, a physical synthesis tool should be as flexible as possible to deliver a better result for variety of design.

6. Fundamental Limits for Placement

Microprocessor architectures have undergone a profound change in the past few years. Starting with a revision of the PowerPC architecture in 2001, the use of multiple processor cores has become common. The manufacturers of large microprocessors have all moved toward multiple-core designs, while single-core designs have all but been abandoned. Increasing clock rates (a traditional method to improve performance) can no longer be done, as the frequency f directly impacts power consumption. Similarly, lowering supply and threshold voltages is difficult because of insufficient noise margins.

Parallelism (through multiple cores) allows for increased numbers of instructions to be executed, without increasing the clock rate. While this might appear to be a "free" way to improve performance, it is by no means clear if these designs will be practical for consumers. Parallel computing (in various forms) has long been a staple of the scientific community, but attempts to broaden the market have failed repeatedly [27].

While power constraints were by no means unexpected, one might wonder why the shift to multiple cores has come about so abruptly. There have been many prior "barriers" to increased complexity in circuit design, and with great regularity the industry has overcome these. By considering theoretical aspects of the placement problem, we show that there are other factors at work – making the placement problem more difficult with each generation, and accelerating increases in power consumption. In this section we bring Moore's Law, Rent's Rule, and classic computational complexity together, to provide better insight into factors which have forced the move to multi-core designs.

6.1. A Theoretic Approach to Placement

As a first step we will assume we are working at some level within the hierarchy of a computing system. We have a number of components (possibly transistors, standard cells, macro blocks, etc.), and these are interconnected in some way. To simplify the discussion we will further assume that these components are square and of uniform size.

We abstract the concept of "length" by defining the length that a signal can travel without buffering, or within a single clock cycle, to be some constant factor of the average width or height of the components.

We will refer to this distance as L. If a connection must be made that is longer than L, then buffers, repeaters, or latches become necessary. The delay along this connection is assumed to be linear with distance, or possibly worse.

If our component represents a standard cell, a larger cell has higher drive strength (and can drive a longer wire between repeaters). Inserting additional repeaters allows the delay of the wire to be linearized.

If our component represents a block within a hierarchy, we assume that higher portions of the hierarchy operate at lower clock rates – and thus, the distance that can be traveled between latches increases with block size. In a similar manner, L can be linearized by insertion of additional latches.

We next define the concept of "adjacency" in a formal way. We again assume equal-sized square components. For any given component in a planar layout there are four "immediately adjacent" locations – above, below, and to the left and right.

If we consider a square group of n components, the number of immediately adjacent locations to this group is $4 \times n^{0.5}$. Extending further, within a distance L from the group, there are $4 \times L \times n^{0.5} + 2 \times L^2$ locations. This is illustrated in Figure 7.

This property holds no matter what the components are: it is irrelevant if they represent individual transistors, FPGA lookup tables, macro blocks, or microprocessor cores. Further, it is irrelevant what the value of L is; it can

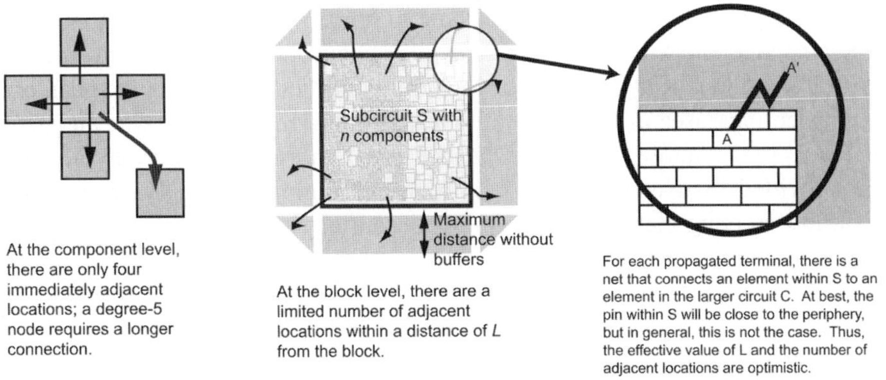

Figure 7. The number of physical locations adjacent to a logic block grows with the size of the block. In the first example, a single block has four neighboring locations; if the block has five connections, at least one of these connections must have a length of at least two units. As the number of components in a block increases, the number of adjacent locations (or locations within a distance of L) grows at the square root of the number of elements within the block.

correspond to either the distance that can be traveled before buffering, or a distance at which a latch must be inserted.

6.2. Rent's Rule

Rent's Rule [28, 29] was observed in the late 1960s – essentially, the observation was that as the number of components in a portion of computing hardware increased, so did the number of electrical connections to that hardware. Specifically, Rent's Rule is commonly formulated at follows.

$$T = k \times n^p$$

The number of terminals T on a portion of circuitry grows with the number of components n within that circuit; the value k is a constant which addresses the typical fanout of logic elements. p is the "Rent parameter" which models the complexity of a circuit. For a pipeline (with low complexity), $p = 0$; if a subcircuit is extracted randomly, p is at worst 1.

For a regular mesh, $p = 0.5$; if one considers a circuit structure such as memory or a programmable logic array, the number of terminals on the perimeter of the circuit is the square root of the number of logic elements. For "general-purpose" circuitry, repeated studies over the past 40 years have found p in the range of 0.6–0.8.

While Rent's Rule has been typically used for the estimation of interconnect lengths or routing congestion, it also sheds light on why multicore processors have emerged. In short, it is not possible to embed a large circuit into a two-dimensional plane without having many long connections. Higher performance was obtained in part by increasing clock frequency; compounding the problem have been increases in the length of average and critical path wires, and also increases in the number of buffers and the size of drivers. While device scaling has provided increasing numbers of transistors, more of these are diverted to simply moving signals down wires – consuming additional power without providing computational advantage.

6.3. Rent Limit on System Size

If one has a computing architecture similar to those designed in the past 40 years, the number of terminals required for a block with n components is $k \times n^p$ for $p \geq 0.6$; this is $O(n^p)$. The number of locations in which an "external" computing element can be placed grows at $L \times n^{0.5} + L^2$, which is $O(n^{0.5})$.

Suppose we decompose a very large computing system into subcircuits of size n. Without question, for some value of n, there will be external connections from a subcircuit that *must* travel a distance of greater than L. As n increases, the percentage of these "long connections" increases. This is true for any value of L, and for any degree of component fan-in or fan-out. This is illustrated in Figure 8.

The increase in n is tied to Moore's Law – and thus, we have a direct conict with Rent's Rule. We will refer to the necessity of long wires as the **Rent Limit** on system size. At some point the benefit gained by increasing the number of computing elements within a system is surpassed by the increased delay of the

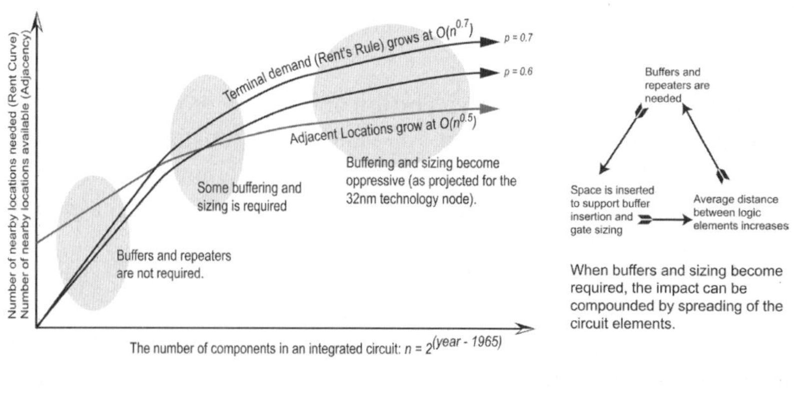

Figure 8. Computational complexity should make the following obvious: if the number of adjacent locations increases at $O(n^{0.5})$, and the terminal demand for the block increases at $O(n^p)$. for $p > 0.5$, the two curves *will* cross.

interconnecting wires. *As circuit sizes increase, the percentage of interconnect wires that can be considered "worst case" increases towards 100%.*

We wish to avoid digressing into specific process technology and device details; as with the work of Hartmanis and Stearns [30], we feel that this clouds the discussion and distracts from the fundamental nature of the problem. Improved device technologies can shift where this "crossover point" occurs – but we see no way in which it can be avoided using a traditional computing architecture.

With the theoretic formulation of the Rent Limit, a number of observations can be made.

- As the number of logic elements increases, the average length of an interconnect wire must also increase. This supports the arguments made in Sylvester and Keutzef [31, 32] that the size of a logic block should be limited to avoid the need for large numbers of repeaters.
- Buffer insertion, which has become essential, in fact compounds the problem. The Rent parameter has traditionally been tied to components that "do something." By inserting buffers into a design, we do not increase the complexity of computations performed, but we do increase the total area.

  In the start of this section we assumed that components were uniformly sized squares. If we amortize the area demand for buffers and repeaters, the "average size" of each component increases, thereby decreasing the effective value of L. Additionally, the increased average size of components means that wires internal to the block also increase in length – and may themselves need additional buffers and repeaters.

  Buffer and repeater insertion, as well as gate sizing, can be viewed as having diminishing returns. As we increase the "average" size of a computing element, the buffer demand increases as well. This matches with the projections of Saxena *et al.* [33].
- At the architectural level it should be obvious that increasing interconnect lengths imply that signals must go "further" during a clock period. This, coupled with increasing clock rates, mean that many paths must have high performance (and thus consume large amounts of power).

6.4. The Future of Microprocessor Placement

In many respects the physical design of microprocessors has been an uphill battle. With each technology generation, required interconnect lengths have increased – this is a fundamental requirement for the planar embedding of a circuit. With increased clock rates there was the simultaneous need to transmit signals over longer distances, and in less time. The only practical method to achieve this is with buffering and gate sizing, resulting in exploding power demand.

As power had become a limiting factor for microprocessors, clock rates have stopped increasing. To avoid increased critical path lengths, the number of logic elements in a processor core – the n used in Rent's Rule – have also become stable.

This is an interesting and somewhat surprising situation; with modern fabrication, designers have access to more devices than can be used effectively. The only apparent solution is to implement a mesh of independent processing elements. A mesh structure as one might find with a multi-core design has a Rent parameter $p = 0.5$ at the top level, matching the physical space perfectly.

Over the next few years it is reasonable to expect that the number of cores on a microprocessor will increase. This may occur much more quickly than traditional scaling – the repeater explosion projected by Saxena *et al.* [33] may force smaller blocks in designs – and this may cause a ripple effect throughout the architecture, requiring lower transistor counts in each core.

It is by no means clear how the consumer market will react to these designs. While scientific computing has utilized parallel machines for many years, consumers have given the approach a very cool reception. Microprocessor manufacturers have thrived on a vast market for high performance components; multi-core designs represent a fundamental change in the product, and the next few years look to be extremely challenging for the industry.

References

[1] Cong, J.; Smith, M. "A parallel bottom-up clustering algorithm with applications to circuit partitioning in VLSI design", *Proc. Design Automation Conf.*, **1993**, 755–780.

[2] Karypis, G.; Aggarwal, R.; Kumar, V.; Shekhar, S. "Multilevel hypergraph partitioning: application in VLSI domain", *Proc. Design Automation Conf.*, **1997**, 526–529.

[3] Cong, J.; Lim, S.K. "Edge separability-based circuit clustering with application to multilevel circuit partitioning", *IEEE Trans. on Computer-Aided Design of Integrated Circuits and Systems*, **2004**, *23*(3), 246–357.

[4] Alpert, C.J.; Nam, G.-J.; Villarrubia, P.G.; Yildiz, M.C. "Placement stability metrics", *Proc. Asia South Pacific Design Automation Conf.*, **2005**, 1144–1147.

[5] Fiduccia C. M.; Mattheyses. R.M. "A linear-time heuristic for improving network partitions", *Proc. 19th IEEE Design Automation Conf.*, **1982**, 175–181.

[6] Murata, H.; Fujiyoshi, K.; Nakatake, S.; Kajitani, Y. "VLSI module placement based on rectangle-packing by the sequence pair", *IEEE Trans. on Computer-Aided Design of Integrated Circuits and Systems*, **1996**, *15*(12), 1518–1524.

[7] Kirkpatrick, S. "Optimization by simulated annealing: quantitative studies", *J. Stat. Phys.*, **1984**, *34*, 975–986.

[8] Vygen, J. "Algorithms for large-scale at placement", *Proc. Design Automation Conf.*, **1997**, 746–751.

[9] Doll, K.; Sigl, G.; Johannes, F.M. "Analytical placement: a linear or a quadratic objective function?", *Proc. Design Automation Conf.*, **1991**, 427–431.

[10] Naylor, W. *et al.* US patent 6,301,693: Non-linear optimization system and method for wire length and delay optimization for an automatic electric circuit placer, **2001**.

[11] Sze, K.; Chan, T.; Cong, J. "Multilevel generalized force-directed method for circuit placement", *Proc. Int. Symp. on Physical Design*, **2005**, 185–192.

[12] Kahng, A.B.; Wang, Q. "Implementation and extensibility of an analytic placer", *Proc. Int. Symp. on Physical Design*, **2004**, 18–25.

[13] Kleinhans, J.; Sigl, G.; Johannes, F.; Antreich, K. "GORDIAN: VLSI placement by quadratic programming and slicing optimization", *IEEE Trans. on Computer-Aided Design of Integrated Circuits and Systems*, **1991**, *10*(3), 356–365.

[14] Agnihotri, A.R.; Ono, S.; Li, C. *et al.* "Mixed block placement via fractional cut recursive bisection", *IEEE Trans. on Computer-Aided Design of Integrated Circuits and Systems*, **2005**, *24*(5), 748–761.

[15] Adya, S.N.; Chaturvedi, S.; Roy, J.A.; Papa, D.A.; Markov, I.L. "Unification of partitioning, placement, and floorplanning", *Proc. Int. Conf. on Computer Aided Design*, **2004**, 550–557.

[16] Breuer, M.A. "A class of min-cut placement algorithms", *Proc. Design Automation Conf.*, **1977**, 284–290.

[17] Dunlop, A.E.; Kernighan, B.W. "A procedure for placement of standard-cell VLSI circuits", *IEEE Trans. on Computer-Aided Design of Integrated Circuits and Systems*, January **1985**, *CAD-4*(1), 92–98.

[18] Yildiz, M.C.; Madden, P.H. "Improved cut sequences for partitioning based placement", *Proc. Design Automation Conf.*, **2001**, 776–779.

[19] Sechen, C.; Sangiovanni-Vincentelli, A. "The timberwolf placement and routing package", *IEEE J. Solid-State Circuits*, **1985**, *20*(2), 510–522.

[20] Swartz, W.; Sechen, C. "Timing driven placement for large standard cell circuits", *Proc. Design Automation Conf.*, **1995**, 211–215.

[21] Yang, X.; Choi, B.-K.; Sarrafzadeh, M. Routability driven white space allocation for fixed-die standard cell placement. *Proc. Int. Symp. on Physical Design*, **2002**, 42–50.

[22] Agnihotri, A.; Yildiz, M.C.; Khatkhate, A.; Mathur, A.; Ono, S.; Madden. P.H. "Fractional cut: improved recursive bisection placement", *Proc. Int. Conf. on Computer Aided Design*, **2003**, 307–310.

[23] Brenner, U.; Pauli, A.; Vygen, J. "Almost optimum placement legalization by minimum cost flow and dynamic programming", *Proc. Int. Symp. on Physical Design*, **2004**, 2–9.

[24] Doll, K.; Johannes, F.M; Antreich, K.J. "Iterative placement improvement by network flow methods", *IEEE Trans. on Computer-Aided Design of Integrated Circuits and Systems*, **1994**, *13*(10), 1189–1200.

[25] Hill, D., US patent 6,370,673: Method and system for high speed detailed placement of cells within an integrated circuit design, **2002**.

[26] Hur, S.-W.; Lillis, J. "Mongrel: hybrid techniques for standard cell placement", *Proc. Int. Conf. on Computer Aided Design*, **2000**, 165–170.

[27] Furht, B. "Parallel computing: glory and collapse", *IEEE Computer*, **1994**, *27*(11), 74–75.

[28] Radke, C.E. "A justification of, and an improvement on, a useful rule for predicting circuit-to-pin ratios", *Proc. Design Automation Conf.*, **1969**, 257–267.

[29] Landman, B.; Russo, R. "On a pin versus block relationship for partitioning of logic graphs", *IEEE Trans. on Computers*, **1971**, *C-20*, 1469–1479.

[30] Hartmanis, J.; Stearns, R.E. "On the computational complexity of algorithms", *Trans. AMS*, **1965**, *117*, 285–306.

[31] Sylvester, D.; Keutzer, K. "Getting to the bottom of deep submicron", *Proc. Int. Conf. on Computer Aided Design*, **1998**, 203–211.

[32] Sylvester, D.; Keutzer, K. "Getting to the bottom of deep submicron II: a global wiring paradigm", *Proc. Int. Symp. on Physical Design*, **1999**, 193–200.

[33] Saxena, P.; Menezes, N.; Cocchini, P.; Kirkpatrick, D.A. "Repeater scaling and its impact on CAD", *IEEE Trans. on Computer-Aided Design of Integrated Circuits and Systems*, **2004**, *23*(4), 451–463.

[34] Alpert, C.; Kahng, A.; Nam, G.-J.; Reda, S.; Villarrubia, P. "A semipersistent clustering technique for VLSI circuit placement", *Proc. Int. Symp. on Physical Design*, **2005**, 200–207.

Chapter 6

ENERGY-DELAY CHARACTERISTICS OF CMOS ADDERS

Vojin G. Oklobdzija and Bart R. Zeydel
ACSEL Laboratory, University of California Davis

Abstract: Choosing the right algorithm and a corresponding adder topology depends on many factors closely related to the technology of implementation. With the transition to CMOS, where circuit delay has a complex relation to implementation parameters, and the recent transition to deep-submicron technologies which introduce further complexity, it becomes even more difficult to make the right choice. This relationship is even more complicated when power consumption is included. In this chapter we present this complex relationship and highlight the important factors that influence the right selection of algorithm, circuit topology, operating conditions and power consumption.

Key words: adders; digital arithmetic; digital circuits; energy-delay optimization; VLSI arithmetic; fast digital circuits.

1. Introduction

For almost half a century realizations of addition algorithms have been continually refined to improve performance due to changing technology and operating constraints [1]. With each technology generation, the gap between the underlying algorithms for addition and efficient realization of those algorithms has grown. Many of the adders in use today were developed for older technologies and under a different set of constraints than those imposed by current technology, such as energy efficiency. To solve this problem a method for analyzing designs in the energy-delay space was developed [2, 3], which allowed for the energy-delay tradeoffs to be taken into account. In addition, this method provides guidance for algorithm selection and realization. Using

Vojin G. Oklobdzija and Ram K. Krishnamurthy (eds.),
High-Performance Energy-Efficient Microprocessor Design, 147–169.

this method we explore the leading addition recurrence algorithms and their realizations, to identify favorable characteristics of each for the development of efficient adder realizations in modern CMOS technology. A comparison of various schemes in the energy-delay space is presented to demonstrate the relative performance and energy efficiency of the proposed structures.

The most important step in the process of VLSI adder design is selecting the initial adder topology which is expected to yield desired performance in the allotted power budget. However, the performance and power will be known only after a time-consuming design and simulation process is completed. Therefore, the validity of the initial selection will not be known until the later stages of the design process or even after several schemes under consideration have been designed and completed. Going back and forth between several designs is often prohibited by the design schedule, making it impossible to correct initial mistakes. Thus, an uncertainty always remains as to whether a higher performance or lower power was possible with a more appropriate choice of topology or simply more effort. This problem is aggravated by a lack of proper delay and power estimation techniques that are guiding the development of computer arithmetic algorithms. The majority of algorithms in use today are based on outdated methods of counting the number of logic gates on the critical path, producing inaccurate and misleading results. The importance of transistor sizing, load effects and power are not taken into account by most.

Different adder topologies may influence fan-out and wiring density, thus influencing design decisions and yielding better area/power trade-offs than known cases [4]. This emphasizes the disconnect existing between algorithms and implementation. The importance of fan-in and fan-out effects on the critical path was demonstrated at the time CMOS technology started replacing nMOS [5]. Similar conclusions were expressed later in the logical effort (LE) method of Sutherland and Sproull [6] regarding critical-path delay estimation, which was later introduced into common practice by Harris and co-workers [7]. Comparison of delay estimates of various VLSI adders obtained via LE, to simulation results obtained using H-SPICE [8] demonstrates good matching and confirms the validity of LE. This matching is under 10% in most cases (Table 1).

Table 1. Delay comparison of 64-bit adders using logical effort

Circuit family	Adder topology	HSPICE (F04)	LE estimate (F04)
Static	Kogge-Stone [8]	11.8	10.9
	Mux based adder [9]	11.4	12.8
	Han-Carlson [10]	12.8	13.3
Dynamic	Kogge-Stone [8]	8.7	9.2
	Ling [11]	9.0	9.5
	Han-Carlson [10]	9.8	9.9

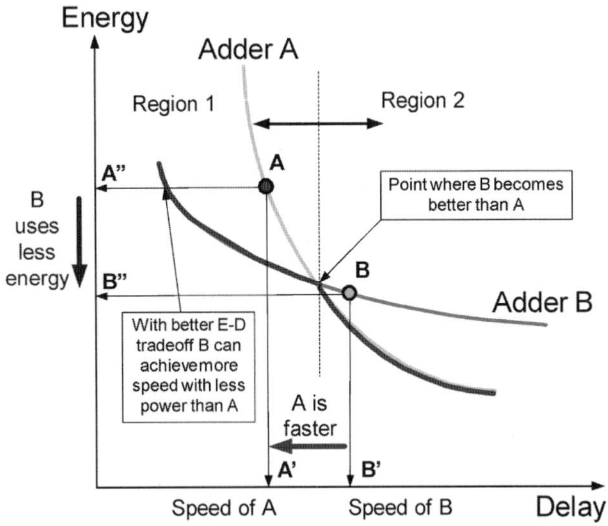

Figure 1. Energy-delay dependency.

However, this is still an incomplete picture, because delay and energy can be traded against each other; thus the energy aspect of this analysis is missing.

A method for analyzing the energy-delay tradeoffs of an adder was developed following the LE guidelines [3]. Using this method it is possible to compare different adders in the energy-delay space (see Figure 1). This method satisfies two requirements: it is simple and quick, yet sufficiently accurate to guide correct selection of the appropriate algorithm and realization (topology).

In this chapter we elaborate on the problem of energy-delay design trade-offs and their estimation, showing how different technology parameters affect the performance of different algorithms. Further, we show how the best algorithm and topology should be selected, and point to the most important factors in their selection. Finally, we show the best topology for an energy-efficient high-performance adder that was obtained in such a way.

2. Comparison of VLSI Adders

The most common approach in comparing VLSI adders is to use a single delay point [4, 5]. An example of such a comparison (based solely on delay) of high-performance 64-bit adders is shown in Table 1. The comparison shows LE delay estimates and H-SPICE pre-layout simulation results in 130 nm technology.

The comparison shows a significant speed difference between static CMOS and dynamic CMOS implementations. This fact has been well known to practitioners: high-speed processors use dynamic CMOS logic [12–14]. However, while the delay difference between different circuit families is more apparent, the delay difference between topologies using the same circuit family is relatively small, making it difficult to know which design can be improved further in terms of speed. Energy is also important because if too much power is used in order to achieve a target delay, hot spots can be created [15].

To illustrate the problem, suppose that two adders A and B were compared against each other based on delay only. Such a hypothetical comparison is illustrated in Figure 1, where the delays of adder A and adder B are shown as points A and B respectively. From the single point comparison, adder A appears faster than adder B, leading to a conclusion that the topology of the adder A is better. However, such a comparison provides an incomplete and potentially misleading picture. If we consider that energy can be traded for delay, it is clear that further analysis is needed. Hypothetical energy-delay dependencies of two designs, A and B, optimized under the same constraints are illustrated in Figure 1.

As the curves show, adder B has more room for delay improvement, it uses less energy in the high-performance region (Region 1) as compared to the adder A.

On the other hand, if lower computational energy is the design objective, adder A is the better choice as it uses less energy in the low performance region (Region 2) compared to adder B.

The challenge is to make such a comparison early in the design process without significant time overhead. A method for estimating energy and delay with relatively low effort and in a short amount of time has been developed in ref. 2. This method provides sufficient accuracy to make appropriate choices for algorithm and circuit topology.

3. Delay and Energy Estimation

The speed of a VLSI adder depends on several factors: technology, circuit family, adder topology, transistor sizes, wires, leakage currents and second-order effects. As a result there are no simple rules to be applied when estimating delay. Skilled engineers are capable of fine-tuning the design to obtain the best performance and lowest energy through transistor sizing. However, this is often an *ad-hoc* process not leading to the best solution. Thus, it is difficult, if not impossible, to predict the best topology.

3.1. Delay Estimation

The introduction of LE was a significant step forward because it provided a better way to estimate delay. Further, LE provides an optimal sizing for delay.

There is a tradeoff in delay estimation where improved accuracy is paid for by complexity or resorting to CAD tools. LE simplifies the delay model to a single parameter referred to as *stage effort*, f, which is used during optimization and modeling. The LE model for gate delay is $t_d = (f + p)\tau$, where $f = gh$ [5]. Each gate has a *logical effort*, g, which represents its drive capability relative to an inverter. The term h represents the electrical effort or effective fan-out of the gate, $h = C_{out}/C_{in}$. The parasitic delay, p, corresponds to the delay associated with parasitic capacitance. The term τ is the per fan-out delay increment of an inverter, and is used to introduce technology-independent estimation of delay.

The accuracy of LE can be improved by obtaining the coefficients g and p through H-SPICE characterization of logic gates in the chosen technology. This step incorporates characteristics of a technology, slopes, and layout estimates into the LE parameters. Gate characterization is performed under the constraint of fixed input-to-output slope relationship to obtain the best matching between estimation and simulation data. This step improves the accuracy of LE estimation considerably and typically brings it within 10% of H-SPICE simulation, as shown in Table 1.

The effect of gate-to-gate wiring is not accounted for using basic LE modeling, and is often ignored in comparisons. However, we have observed that in 130 nm technology, for example, wire resistance and capacitance can contribute up to 1F04 delay degradation in 64-bit adders. The wire capacitance introduces a constant load at the output of each gate, which can be estimated from the wire length. The impact of wire resistance can be estimated using the approximation $T_{wire} = 0.38 R_{wire} C_{load}$, which provides reasonable matching versus H-SPICE. A comparison of the impact of for the worst case impact of wire resistance on a 64-bit adder is shown in Table 2.

Application of LE to simple path delay optimization is straightforward; however, it is often difficult to apply the analysis to complex paths due to branching (b). LE defines branching as $b = (C_{on} + C_{off})/C_{on}$, where the terms C_{on} and C_{off} must be determined relatively. This analysis becomes prohibitively complex when branches have differing gate types and number of stages. In addition, constant loads, such as wires, require iterative computation. As a result, the optimization of a complex path using the LE gate delay model must be performed by changing individual stage efforts (f's) of each gate to achieve

Table 2. Worst-case delay impact of wire resistance in 130 nm 64-bit adders

Wire length (bits crossed)	HSPICE (no resistance)	HSPICE (with resistance)	Estimate (with resistance)
80 μm (8 bits)	54.7 ps	58.5 ps	58.9 ps
160 μm (16 bits)	57.7 ps	66.0 ps	66.8 ps
320 μm (32 bits)	64.0 ps	84.7 ps	84.2 ps

minimal delay. Instead of using a simple paper-and-pencil method (as suggested by LE) the use of numerical optimization such as the built-in gradient based optimization of Microsoft®-Excel or Matlab is required for delay optimization. The use of these does not require considerable overhead compared to paper-and-pencil analysis.

3.2. Energy Estimation

The use of LE for delay optimization provides a delay estimate and corresponding gate sizing. However, it does not account for energy. A model for estimating the energy of a gate based on its LE sizing was presented in refs 1 and 2. By estimating the energy of each gate from its LE sizing, an estimate for the total energy of a design can be obtained.

The energy of a gate is primarily a function of the output load, C_L, and parasitic capacitance which is proportional to gate size. The ratio of the energy associated with C_L and the energy due to parasitic capacitance varies depending on the electrical effort (h). For small values of h, the parasitic energy is comparable to the energy associated with the output load, while for larger values of h the energy associated with the output load increases relative to the parasitic energy (Figure 2).

Gate energy parameters can be extracted from H-SPICE simulation by varying C_L and gate size. A linear dependence of energy on C_L and size is observed in ref. 1, which results in the following energy model for a gate:

$$E = E_p \cdot gate\ size + E_g \cdot C_L + E_{internal\text{-}wire}$$

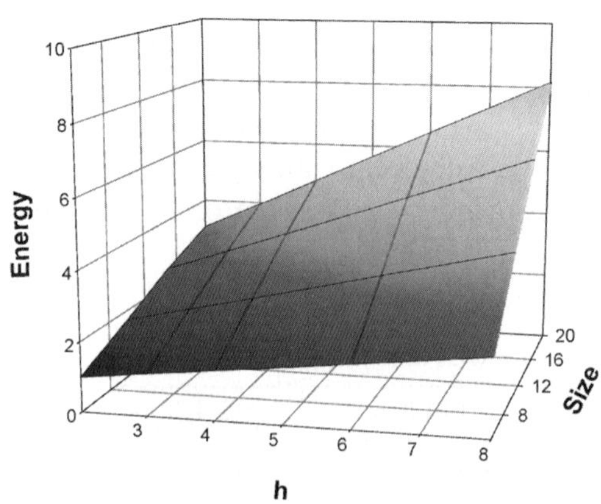

Figure 2. Dependence of gate energy on its size and electrical effort (h).

where E_p is the energy per unit size, E_g is energy per unit load, and $E_{internal\text{-}wire}$ is an offset due to internal wiring introduced by layout estimation. This energy model directly accounts for parasitics, local wiring, and output load dependence, while performing a best fit for crowbar current and leakage. The parameters, E_g, E_p, and $E_{internal\text{-}wire}$, are obtained using the same gate characterization setup as in LE with the slight overhead of performing the characterization for multiple gate sizes.

4. Energy-delay Estimation Method

The objective of the energy-delay estimation method (EDE) is to provide a relatively simple and quick way to compare designs in the energy-delay space [3] so that it can be used before design decisions are made and committed.

LE provides reasonable delay results and sizing, but it does not account for wiring. To improve the estimate, wires and the correct handling of complex branch dependencies are included in our analysis. As we are interested in comparing designs over a range of performance targets, each design is compared over the same range of path gain (H), where $H = C_{out}/C_{in}$. After characterizing a technology to find parameters g, p, E_g, E_p, and $E_{internal\text{-}wire}$ for each gate, the following steps are performed to obtain delay and energy estimates of a design for each H:

1. Determine the critical path of the design.
2. Optimize the delay of the critical path to determine f_{opt}.
3. Use f_{opt} to size the gates on the critical path.
4. Estimate the energy of the entire design.

Using the sizing from Step 3 we can estimate the energy of the critical path. However, Step 4 requires an estimate for the energy of the entire design and not just the critical path. The energy of gates within a design can be estimated according to two cases: gates on paths with the same number of stages as the critical path, and gates on paths with fewer stages than the critical path.

For paths with the same number of stages as the critical path, the size of each gate is proportional to the size of the gates on the critical path, allowing for the energy of each gate to be computed directly. To facilitate this analysis the energy of each gate is assumed to be the same as the energy of the equivalent gate on the critical path.

For paths with a different number of stages than the critical path, the size of each gate is not directly proportional to the gates on the critical path. Instead, to obtain an energy estimate, the path must first be sized to have the same delay as the critical path. Once the sizing is obtained, the energy of each gate can be estimated. Similar to first case, the energy of any paths with the same number of stages can be computed proportionally to this path.

The energy of each gate depends on its switching activity. In application of EDE to VLSI adders it is common to use a 15% switching activity factor for each gate in a static path and a 50% switching activity for each gate in the dynamic path. These switching factors were obtained as an average of designs that were analyzed based on experience and are consistent with the "rule-of-thumb" used in the industry.

5. Energy-delay Estimation of Adders

Energy-delay estimation is a useful tool in comparing various tradeoffs in adder design, such as the algorithm or circuit topology. The choice of the adder topology and design style is also dependent on the required performance and pressures to meet the critical path. In some instances the adder may not be the critical path and the speed requirements may be relaxed, or it may be possible to improve the speed of the clocked storage elements (flip-flops and latches) to meet the required timing.

The analysis of these tradeoffs was performed by Zyuban and Strenski, who termed these design characteristics "hardware intensity" [16, 17]. Hardware intensity defines the design point in terms of the trade-off between energy and delay. A tangent on the energy-delay curve represents the percentage of delay reduction being paid for by the percentage increase in energy. Thus, several algorithms can be examined in various design regions and the best one can be chosen. Also, various circuit design styles can be examined. For example, the analysis of three different circuits design styles: static CMOS, dynamic CMOS and compound (dynamic-static) CMOS design, revealed that compound CMOS achieves speed of dynamic CMOS design while maintaining the low energy of static CMOS. An EDE comparison of 64-bit adders implemented in these three design styles is shown in Figure 3.

5.1. Domino and Compound Domino CMOS Analysis

In order to improve adder performance, domino CMOS logic is often used for implementation of an adders carry-merge (CM) blocks resulting in the circuit shown in Figure 4(a). The static CMOS inverter is necessary after each dynamic block in order to make the logic behave in a "domino" fashion. The signal inversion, which is necessary in the domino CMOS logic block, can be achieved using a more complex static gate instead of inverter. This is often referred to as compound-domino, or dynamic-static CMOS. Thus, two domino carry-merge stages can be merged into one by replacing the inverter with an AOI, as shown in Figure 4(b).

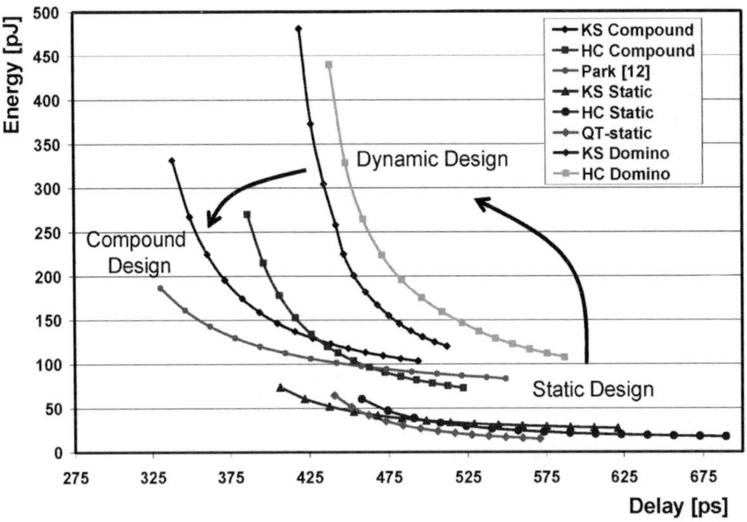

Figure 3. EDE comparison of circuit design styles on 64-bit adders: compound CMOS shows benefits of static and dynamic CMOS circuits.

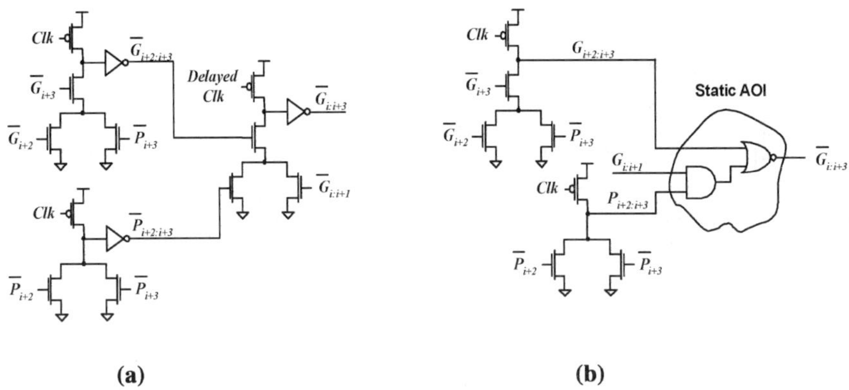

Figure 4. (**a**) Carry-merge: domino implementation; (**b**) carry merge: compound domino implementation.

Comparison of the 64-bit Kogge–Stone (KS) [9] and Han–Carlson (HC) [11] adders implemented in dynamic-domino CMOS and compound-domino CMOS is shown in Figure 5.

Comparing dynamic-domino to compound-domino, EDE helps us to observe the benefits obtained by utilizing compound-domino logic. For the same energy budget (e.g. 200 pJ) compound-domino KS yields 20% delay improvement over domino KS. EDE provides a clear picture of the impact compound-domino can have on adder design.

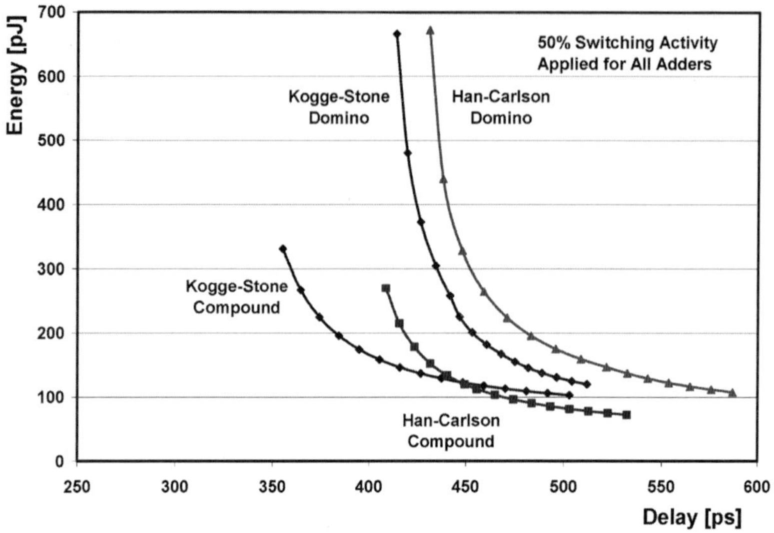

Figure 5. EDE Comparison of 64-bit HC and KS domino and compound-domino adders in 130 nm technology.

6. Adder Comparison

The ability, provided by EDE, to observe differences between implementations of the same adder using different circuit families, is beneficial in selecting the appropriate circuit design style. However, it is also important to see tradeoffs between adder topologies implemented using the same circuit family. The accuracy of EDE for demonstrating tradeoffs in the energy-delay space is shown by comparing optimized H-SPICE results for 32-bit compound-domino KS [9] and QT [13] adders versus EDE results in 100 nm technology. A comparison of EDE estimation with simulation results adopted for a 100 nm technology is shown in Figure 6 [3]. EDE estimates demonstrate the same tradeoffs as observed in simulation. These results confirmed the validity of the QT adder as a viable option for reducing energy without sacrificing performance.

Compound-domino and static 64-bit adders were analyzed using EDE to see what tradeoffs exist (Figure 7). Different points on the energy-delay curve were obtained by varying the size of the input gates for each adder. The output of each adder was loaded with a 1 mm wire. A range of path gains (H's) was chosen from $H = 2$ to the maximum H (i.e. where minimum input size occurs).

The results show the benefits associated with sparse designs: HC and QT. However, the benefits of compound-domino QT are lesser at 64 bits than for the 32-bit design. This is a result of the one extra stage that the QT implementation used versus the KS implementation. In the 32-bit design the QT implementation

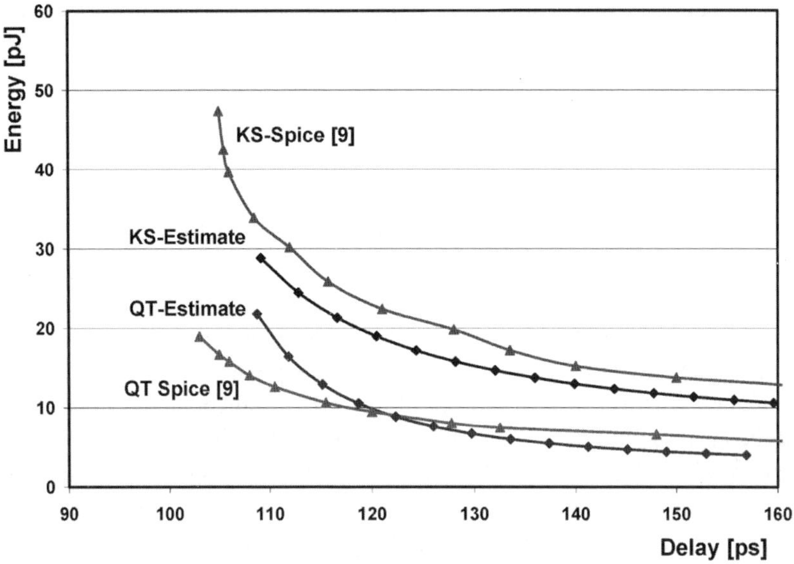

Figure 6. Comparison of 32-bit QT and KS adders: EDE vs. simulation in 100 nm technology.

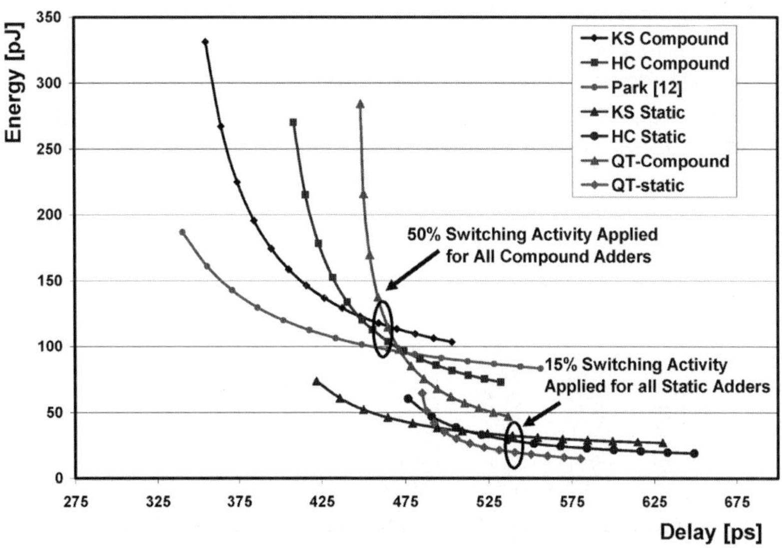

Figure 7. EDE analysis of 64-bit compound-domino and static adders in 130 nm technology.

used the same number of stages as KS. We also observe that the compound-domino KS prefix-4 Park adder [18], which utilizes fewer stages at the cost of increased gate complexity and branching, shows further benefits. The increased gate complexity in the Park adder is offset by a reduction in parasitic delay and number of stages, which allows for the Park adder to achieve lower delay with less energy than the other designs. At lower performance targets the overhead of the more complex gates is to much and designs such as QT and HC achieve lower energy.

6.1. Representative High-performance Adders

In this section we present a comparison of high-performance adders used in leading microprocessors in industry. All of the adders used in the comparison were implemented using compound-domino design style which combines the "best of both worlds": low power and high performance, as has been realized by the design community. The comparison shown in Figure 8 includes: IBM implementation of KS adder [18], Kogge–Stone 4-2 consisting of a prefix 4 dynamic and prefix two static compound-domino stage [9], Quaternary adder (QT) developed by Intel [13], Ling adder used by IBM and the Intel Itanium processor, and Han-Carlson adder (HC) [11]. We first compared the adders using the energy-delay^2 figure of merit (Figure 8). Energy-delay^2 only represents one point on the E-D curve; however, it is often used as a metric for comparing high-performance designs [19]. From Zyuban's analysis, this means that we are willing to trade 2% of energy increase for a 1% improvement in speed [16, 17]. The tangent to the E-D curve at this point has a slope of 2. We see that, even though it is not the fastest, IBM adder developed by Park *et al.* [18] shows the best ED^2 figure of merit.

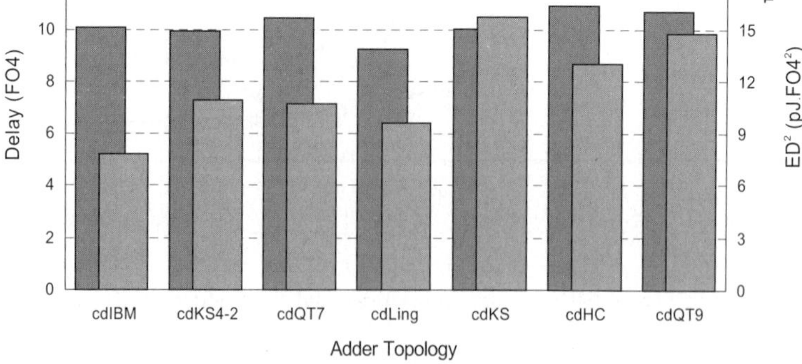

Figure 8. Comparison of representative high-performance adders used in the industry; "cd" designates compound-domino circuit design style in 130 nm technology.

6.2. Contribution of Wire

In order to properly evaluate all of the adder topologies, the impact of wires on adder performance and energy consumption needs to be properly accounted for. Wires contribute in two ways: they add delay to the adder (effect of long wires), and increase energy. In the past these effects were often ignored, but as the technology continues to scale wire effects could make a substantial difference between realizations. Therefore, the impact of wires on energy and delay must be included in the optimization and analysis.

Figure 9 shows the wire effects expressed as the total wire capacitance in the adder. This capacitance contributes to performance and energy deterioration.

Wire energy is shown as a fraction of the total energy of the adder. The tradeoff between wire and delay is best seen on the IBM adder. Given the small delay difference between the fastest and the slowest adders, this difference can potentially change the ranking. A comparison of representative adders, with and without wire, is shown in Figure 10. The results are obtained by applying the EDE method to the entire adder, where the energy of each gate is individually calculated. Depending on the length of wires on the critical path, wire delay can impact the adder delay by up to 1FO4, which is more than 10% of the total delay.

The impact of wire diminishes the differences between the best and the worst designs. This shows that a hidden tradeoff exists in some of the designs between the wiring and logic complexity. We can trade a stage of the adder at the expense of more intense wiring. If wire contribution is not properly accounted for, an unfair advantage may be apparent. Looking at, for example, six-stage KS and seven-stage QT7, in Figure 11(a) and (b), one can notice that they switch order when wire effects are properly accounted for.

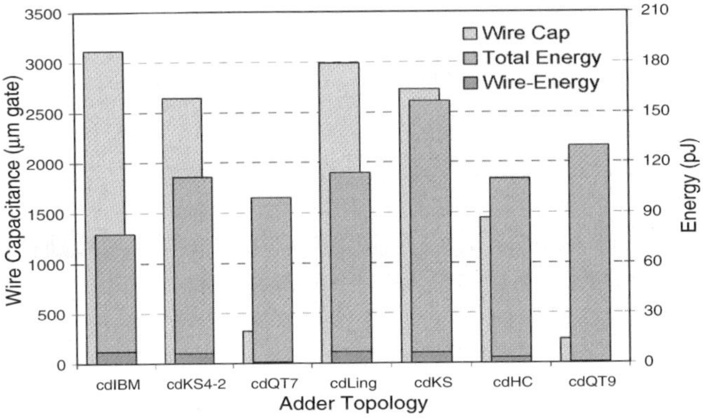

Figure 9. Impact of wires on high-performance adders in 130 nm technology.

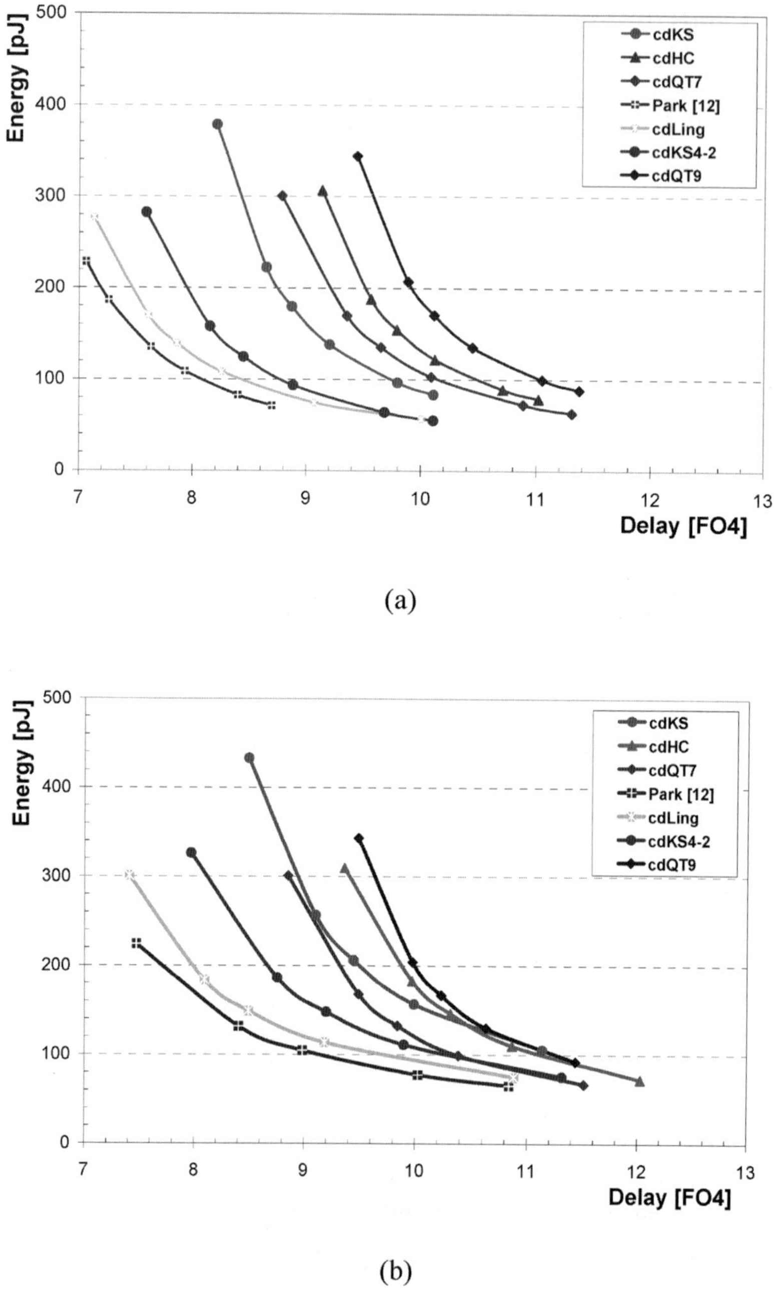

Figure 10. Energy-delay behavior of representative high-performance adders in a 130 nm technology: (**a**) energy-delay with wire excluded from the model, (**b**) energy-delay with wire included. It should be noted that, in general, wire effect diminishes performance differences between designs, while energy increases only slightly.

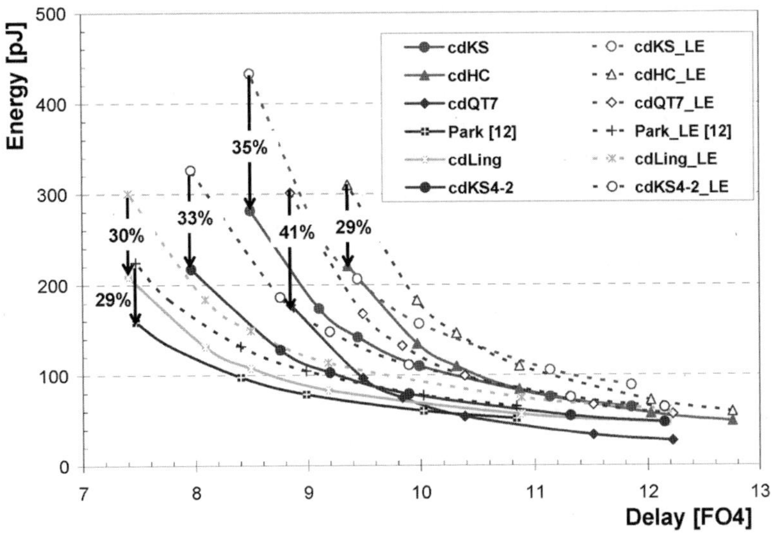

Figure 11. Energy-delay behavior of representative high-performance adders after energy optimization. The best behavior is that of IBM adder designed by Park *et al.* [18].

By applying energy minimization techniques to transistor sizing [20] the energy can be reduced even further. The ultimate energy-delay Figure of the representative adders is shown in Figure 11. This is about the best energy and performance one can achieve [21].

7. The Ultimate Adder Topology

Efficient adder design requires proper selection of a recurrence algorithm and its realization. Using the insight obtained through the application of EDE, we analyzed several algorithms for their flexibility and suitability for realization in CMOS. We found that the use of Ling's algorithm provides up to 12% improvement in performance of 32-bit static adders with the same recurrence trees. Using Ling's algorithm we developed general techniques for efficient realizations based on technology constraints. From these techniques several high-performance realizations of Ling's algorithm are developed that achieve better performance and energy efficiency than existing Ling and Weinberger designs [22].

Technology characteristics limit potential realizations of Weinberger's and Ling's recurrences for addition. The primary constraint in modern CMOS is the fan-in of a gate, which is commonly limited to between 2 and 5. Several realization techniques have been developed to map recurrence algorithms to CMOS under these constraints [22].

7.1. Combined Bit Operator and First Carry Stage

In adder realizations, one stage can be removed by combining the 1-bit operation for g and t into the first prefix computation stage [12, 18]. This technique is more favorable to Ling's recurrence than to Weinberger's. Under the same transistor stack height constraint a Ling realization can use a prefix of 1 more than a realization using Weinberger's. This saving is observed in the prefix-2 equations for Ling's recurrence:

$$\begin{aligned} H_i &= a_i b_i + a_{i-1} b_{i-1} \\ T_{i-1} &= (a_{i-1} + b_{i-1})(a_{i-2} + b_{i-2}) \end{aligned}$$

and for Weinberger's recurrence:

$$\begin{aligned} G_{i+1} &= a_i \cdot b_i + (a_i + b_i)(a_{i-1} b_{i-1}) \\ T_i &= (a_i + b_i)(a_{i-1} + b_{i-1}) \end{aligned}$$

The implementation of H_i requires a stack of two nMOS transistors, while the implementation of G_{i+1} requires a stack of three nMOS transistors. The implementations of transmit for both require a stack of two nMOS transistors.

The fan-in of the first logic stage for Ling's transformation is reduced by 1 in CMOS compared to Weinberger's. Subsequent stages for both have the same fan-in since the recursion is performed using the prefix "•" operator. Ling's transformation is especially useful for static realizations, as the first stage and bit operator stage can be combined, while in Weinberger's such a combination would result in a stack of three nMOS transistors.

7.2. Conditional Computation of Sum

Several adder implementations make use of conditional logic for the computation of sum. Conditional logic allows for the number of gates in the recurrence to be reduced at the cost of increased fan-out in the recurrence tree. Additionally, in dynamic adders the conditional logic is implemented using static gates allowing for the switching activity of the gates to be reduced compared to the dynamic gates on the carry path. Both Weinberger's and Ling's recurrence fit well into conditional computation. The issue with conditional computation is determining how many bits to compute conditionally and using what structure. The number of gates on the critical path consists of the carry structure, the bit operator gate (if not combined into the first carry stage), and the sum. The conditional sum must be computed prior to the sum selection by the carry path. For example, in a realization with a four-gate carry tree the conditional sum must be completed in the same time as the four-gate carry tree used to select sum. As the number of stages in the carry tree increases, conditional computation is a viable solution for reducing energy. If, however, the number of stages in the carry tree decreases, the possibility that the conditional sum becomes the critical portion of the design increases. The optimal number of bits to compute conditionally, as well as the implementation of the conditional computation, either by rippling the recurrence or through the use of separate recurrence trees, is dependent on delay target and technology.

7.3. Ling Realizations and their Alternatives

7.3.1. Static adders

For static adders designers are often limited to a two stack of nMOS transistors and a two stack of pMOS transistors. Knowles described [4] how to create minimum depth carry trees for static adders using Weinberger's recurrence. The same trees can be constructed with Ling's transformation by combining the first stage of the carry tree as described in Section 7.1. This construction allows for one stage to be removed from the critical path of the adder.

7.3.2. Dynamic adders

Ling's transformation shows advantages of reduced logic complexity of the critical path and should therefore yield good structures for addition in CMOS technology. Several types of 64-bit dynamic realizations of Ling's transformation are proposed in the following sections.

7.3.3. Fast parallel prefix ling adders

Ling's transformation only displays advantages over Weinberger's when the factored t_i term can be removed from the critical path. As shown in refs 22 and 23, this can be accomplished through the use of conditional logic for the sum. The fastest Ling implementations are dependent on CMOS technology limitations. In modern technology the nMOS transistor stack height is commonly limited between two and five, while pMOS transistors are typically limited to a stack height of two.

7.3.4. A three-stage ling adder (TSL)

A three-stage adder can be constructed using a fully parallel prefix tree with Ling's transformation. A technology limitation of five stack nMOS for dynamic stage allows for prefix-4 gates to be used in dynamic stages, while a limitation of two stack pMOS limits static gates to prefix-2. Under these constraints a full

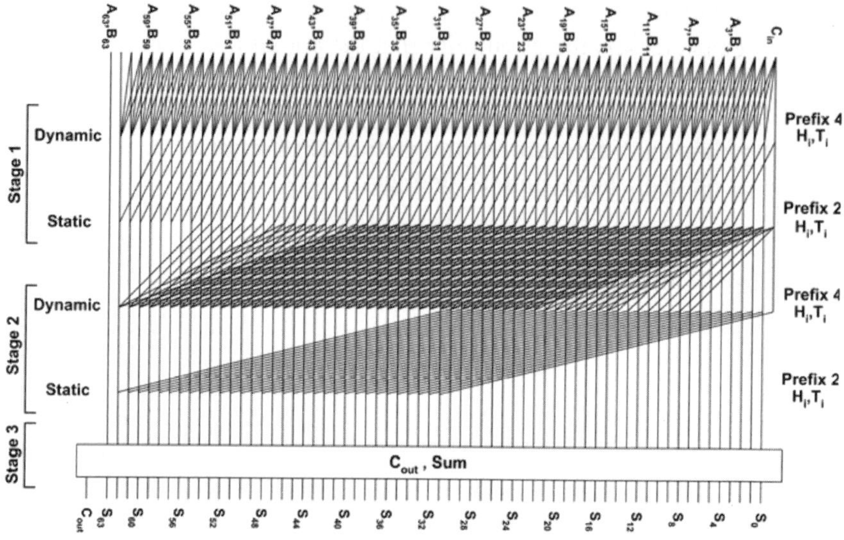

Figure 12. A 64-bit three-stage Ling adder (TSL).

prefix tree with prefix 4, 2, 4, and 2 for the first, second, third and fourth gates respectively can be constructed (Figure 12).

The equations for the first level H_i and T_i are:

$$\begin{aligned} H_i &= A_iB_i + A_{i-1}B_{i-1} + (A_{i-1} + B_{i-1})A_{i-2}B_{i-2} \\ &\quad + (A_{i-1} + B_{i-1})(A_{i-2} + B_{i-2})A_{i-3}B_{i-3} \\ T_{i-1} &= (A_{i-1} + B_{i-1})(A_{i-2} + B_{i-2})(A_{i-3} + B_{i-3})(A_{i-4} + B_{i-4}) \end{aligned}$$

Both equations result in a worst-case stack height of four nMOS transistors. These gates must be footed, because they are in the first stage of the adder, bringing the worst case stack height to the technology limit of 5. The second, third and fourth level H_i and T_i computations follow traditional prefix "•" product operations for prefix-2 and prefix-4 and can be implemented without violating stack height limitations. Weinberger's recurrence can be used with the same full parallel prefix tree for the carry recurrence at an increase of one in the stack height of the first stage for the carry recurrence relative to Ling's. However, this would violate the limitation of stack of five nMOS transistors in the first stage of the recurrence.

7.3.5. *Three-stage conditional sum ling adder (CSL)*

The amount of wiring in an adder realization can be reduced without increasing the number of stages by generating every other H_i and performing a 2-bit

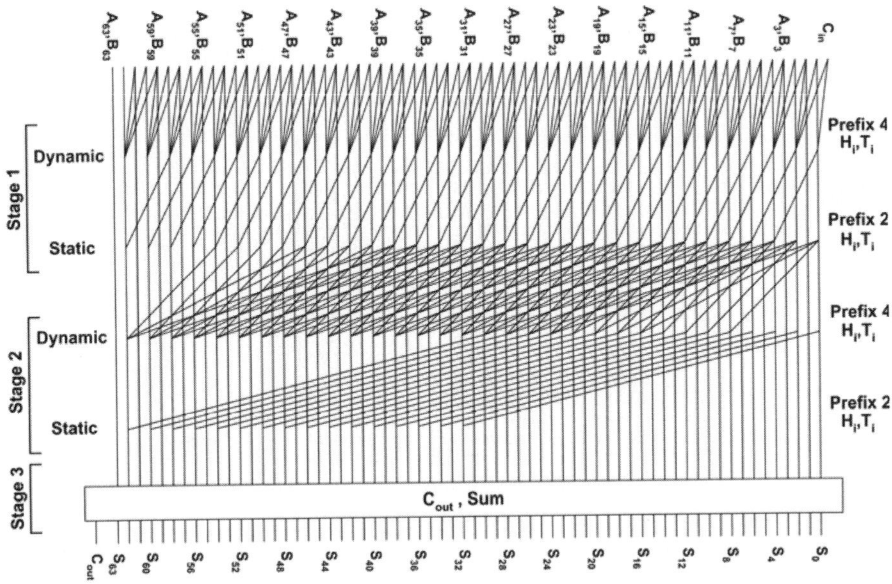

Figure 13. A 64-bit three-stage conditional sum Ling adder (CSL).

conditional sum to be selected by a prefix 4-2-4-2 carry tree. The conditional sum length was chosen based on the limitations on the number of conditional sum bits described in Section 7.2 (Figure 13) and does not add complexity to the carry path.

7.4. Results

All results are obtained using estimates for 130 nm technology by applying the energy-delay estimation method (EDE) we developed in ref. 2 to the entire adder. A comparison of 32-bit static adder implementations between Weinberger's recurrence and Ling's transformation is shown in Figure 14.

Ling's transformation demonstrates a delay improvement of up to 12%, confirming the benefit that Ling can achieve in static implementations limited to a stack height of two nMOS and two pMOS transistors. For dynamic implementations, technology constraints and adder size determine whether the advantage of using Ling's transformation is a logic stage or a reduction in transistor stack height.

A comparison of 64-bit Ling dynamic adders with and without conditional sum is shown in Figure 15.

The results show an energy savings for the 2-bit conditional sum variants. This is primarily due to the reduced switching activity of the static gates on

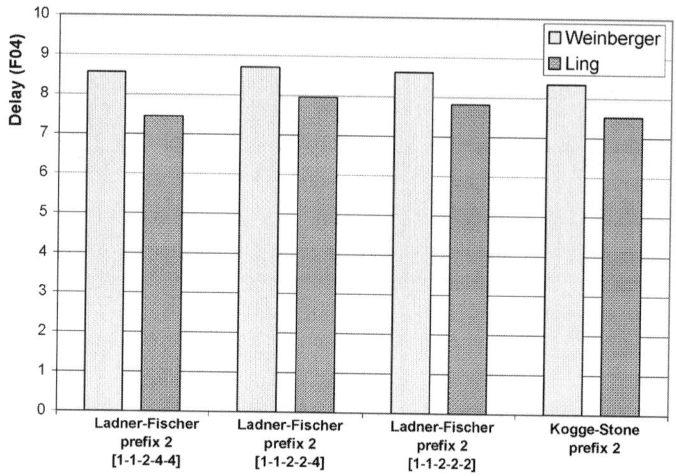

Figure 14. Comparison of 32-bit static Weinberger and Ling adders.

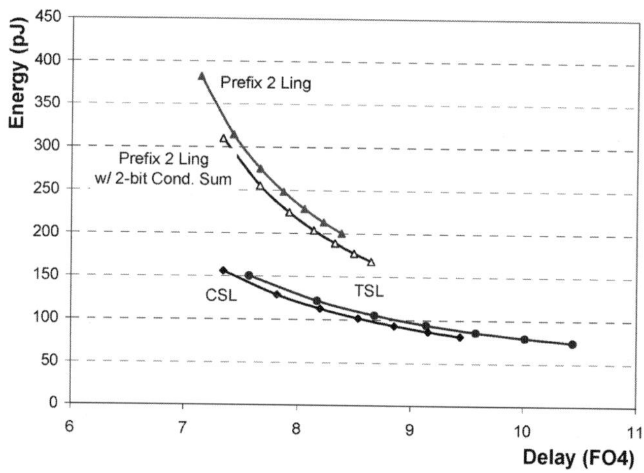

Figure 15. Comparison of conditional sum in high-performance 64-bit dynamic adders.

the conditional path. In the fully parallel prefix-2 Ling carry tree, applying a 2-bit conditional sum improves energy at only a slight increase in delay. The delay penalty is due to increased loading of the adder input caused by the static gates on the conditional sum path. The CSL adder results in a slight energy savings and improved performance compared to the TSL design. In contrast to the prefix-2 design, the static gates of the conditional sum path reduce the loading of the inputs to the adder due to their reduced complexity compared to the prefix-4 gates of the carry path. A comparison of the best 64-bit

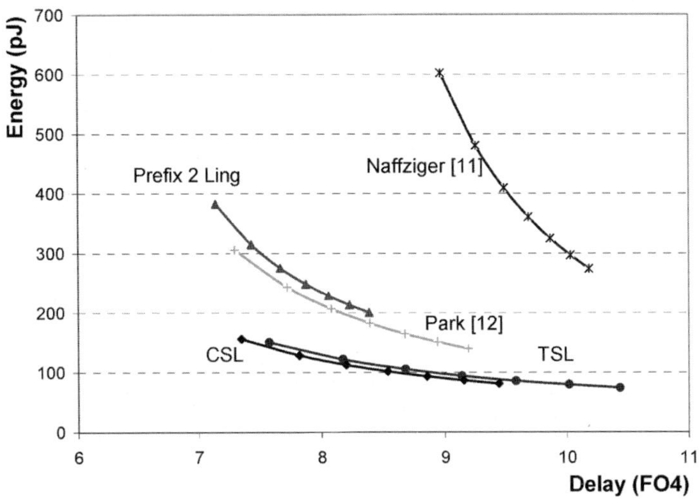

Figure 16. Energy-delay comparison of high-performance 64-bit dynamic adders.

adder implementations for Ling's [25] and Weinberger's recurrence [26] and the proposed realizations are shown in Figure 16.

The results show the significant advantage obtained by the proposed realizations. These realizations demonstrate better performance than the Weinberger adder and the previous best implementation of a Ling adder. While the best delay is obtained by the fully parallel prefix-2 Ling design, the most energy-efficient design is the proposed CSL adder.

8. Conclusion

We presented an energy-delay estimation (EDE) method that extends logical effort (LE) and its application to the analysis and selection of high-performance VLSI adders. The EDE method has proven to be a much-needed and effective tool in design space exploration, in particular when comparing high-performance adders in the early stages of design. Further, the sufficient accuracy of the method for adder selection in the energy-delay space was demonstrated when comparing designs implemented in 130 nm and 100 nm CMOS technologies using static, domino and compound-domino circuit styles. The method described here brings a new perspective to comparing arithmetic circuits and advances the analysis and design of VLSI-oriented computer arithmetic algorithms.

Ling's and Weinberger's addition recurrence algorithms demonstrate favorable characteristics for efficient CMOS mapping. Ling's algorithm demonstrates a fundamental advantage for high-performance addition in CMOS by reducing the complexity of the first stage of the carry tree. Guidelines are presented which aid in the selection of efficient realizations of Ling's transformation for both prefix selection for the recurrence and selection of conditional computation size. The proposed Ling structures, TSL and CSL, demonstrate up to 50% savings in energy at the same delay when compared to the fastest previous designs [18].

References

[1] Davari, B.; Dennard, R.H.; Shahidi, G.G. "CMOS scaling for high performance and low power – the next ten years", *Proc. IEEE*, *83*, April **1995**.

[2] Oklobdzija, V.G.; Zeydel, B.R.; Dao, H.Q.; Mathew, S.; Krishnamurthy, R. "Energy-delay estimation technique for high-performance microprocessor VLSI adders", *Proc. 16th Int. Symp. on Computer Arithmetic*, Santiago de Compostela, Spain, June **2003**.

[3] Oklobdzija, V.G.; Zeydel, B.R.; Dao, H.Q.; Mathew, S.; Krishnamurthy, R. "Comparison of high-performance VLSI adders in energy-delay space", *Transaction on VLSI Systems*, **2005**, *13*(6), 754–758.

[4] Knowles, S. "A family of adders", *Proc. 14th Symp. on Computer Arithmetic*, Adelaide, Australia, April **1999**.

[5] Oklobdzija, V.G.; Barnes, E.R. "On implementing addition in VLSI technology", *IEEE J. Parallel and Distributed Computing*, **1988**, *5*, 716–728.

[6] Sutherland, E.; Sproull, R.F. "Logical effort: designing for speed on the back of an envelope", *IEEE Advanced Research in VLSI*, C. Sequin (editor), MIT Press, **1991**.

[7] Sutherland, I.E.; Sproull, R.F.; Harris, D. *Logical Effort Designing Fast CMOS Circuits*, Morgan Kaufmann, **1999**.

[8] Dao, H.Q.; Oklobdzija, V.G. "Application of logical effort techniques for speed optimization and analysis of representative adders", 35th Annual Asilomar Conference on Signals, Systems and Computers, **2001**.

[9] Kogge, P.M.; Stone, H.S. "A parallel algorithm for the efficient solution of a general class of recurrence equations", *IEEE Trans. Computers*, **1973**, *C-22*(8), 786–793.

[10] Farooqui, A.A.; Oklobdzija, V.G.; Chehrazi, F. "Multiplexer based adder for media signal processing", *Int. Symp. on VLSI Technology, Systems, and Applications*, Taipei, Taiwan, June **1999**.

[11] Han, T.; Carlson, D.A.; Levitan, S.P. "VLSI design of high-speed low-area addition circuitry", *Proc. IEEE Int. Conf. on Computer Design: VLSI in Computers and Processors*, **1987**, 418–422.

[12] Naffziger, S. "A sub-nanosecond 0.5 μm 64-b adder design", *1996 IEEE International Solid-State Circuits Conf., Dig. Tech. Papers*, Feb. **1996**, 362–363.

[13] Mathew, S.K. *et al.*, "Sub-500-ps 64-b ALUs in 0.18 μm SOI/bulk CMOS: design and scaling trends", *IEEE J. of Solid-State Circuits*, Nov. **2001**, *36*, 1636–1646.

[14] Mathew, S.K. *et al.*, "A 4 GHz 130 nm address generation unit with 32-bit sparse-tree adder core", *IEEE J. of Solid-State Circuits*, **2003**, *38*, 689–695.

[15] Harris, D.; Naffziger, S. "Statistical clock skew modeling with data delay variations", *IEEE Trans. VLSI Systems*, Dec. **2001**, *9*, 888–898.

[16] Zyuban, V.; Strenski, P. "Balancing hardware intensity in microprocessor pipelines", *IBM J. Res. Dev.*, **2003**, *47*(5/6).

[17] Zyuban, V.; Strenski, P. "Unified methodology for resolving power-performance tradeoffs at the microarchitectural and circuit levels", *Proc. Int. Symp. on Low Power Electronics and Design*, Aug. **2002**, 166–167.

[18] Park, J. *et al.* "470 ps 64-bit parallel binary adder", *Symposium on VLSI Circuits, Dig. Tech. Papers*, **2000**.

[19] Nowka, K. IBM Austin Research Lab, Austin, TX, private communication, August **2001**.

[20] Dao, H.Q.; Zeydel, B.R.; Oklobdžija, V.G. "Energy minimization method for optimal energy-delay extraction", *ESSCIRC 2003, Estoril*, Portugal, 16–18 September **2003**.

[21] Oklobdzija, V.G. "Energy-delay tradeoffs in CMOS digital circuits design", *Presentation at Dallas IEEE CAS Workshop*, Richardson, Texas, 10 October **2005**.

[22] Zeydel, B.R.; Kluter, T.T.J.H.; Oklobdžija, V.G. "Efficient mapping of addition recurrence algorithms in CMOS", *Int. Symp. on Computer Arithmetic*, ARITH-17, Cape Cod, Massachusetts, USA, 27–29 June **2005**.

[23] Sklanski, J. "Conditional-sum addition logic", *IRE Trans. on Electronic Computers*, **1960**, *EC-9*(2), 226–231.

[24] Bedrij, O.J. "Carry-select adder", *IRE Trans. on Electronic Computers*, **1962**, *EC-11*, 340–346.

[25] Ling, H. "High-speed binary adder", *IBM J. Res. Dev.*, **1981**, *25*(3), 156–166.

[26] Weinberger, A.; Smith, J.L. "A logic for high-speed addition", *Nat. Bur. Stand. Circ.*, **1958**, *591*, 3–12.

Chapter 7

HIGH-PERFORMANCE ENERGY-EFFICIENT DUAL-SUPPLY ALU DESIGN

Sanu K. Mathew, Mark A. Anders, and Ram K. Krishnamurthy
Circuits Research Laboratories, Intel Corporation, Hillsboro, OR, USA

Abstract: This chapter describes the design of a single-cycle 64-bit integer execution ALU fabricated in 90 nm dual-Vt CMOS technology, operating at 4 GHz in the 64-bit mode with a 32-bit mode latency of 7 GHz (measured at 1.3 V, 25° C). The lower- and upper-order 32-bit domains operate on separate off-chip supply voltages, enabling conditional turn-on/off of the 64-bit ALU mode operation and efficient power-performance optimization. High-speed single-rail dynamic circuit techniques and a sparse-tree semi-dynamic adder core enable a dense layout occupying $280 \times 260\,\mu m^2$ while simultaneously achieving (i) low carry-merge fan-outs and inter-stage wiring complexity, (ii) low active leakage and dynamic power consumption, (iii) high DC noise robustness with maximum low-Vt usage, (iv) single-rail dynamic-compatible ALU write-back bus, (v) simple 2Φ 50% duty-cycle timing plan with seamless time-borrowing across phases, (vi) scalable 64-bit ALU performance up to 7 GHz measured at 2.1 V, 25° C, and (vii) scalable 32-bit ALU performance up to 9 GHz measured at 1.68 V, 25° C.

Key words: arithmetic and logic unit (ALU); sparse-tree architecture; semi-dynamic design; dual-supply voltage design.

1. Introduction

Fast 32-bit and 64-bit arithmetic and logic units (ALU) with single-cycle latency and throughput are essential ingredients of high-performance superscalar integer and floating-point execution cores. Furthermore, in a typical

Vojin G. Oklobdzija and Ram K. Krishnamurthy (eds.),
High-Performance Energy-Efficient Microprocessor Design, 171–187.

ALU operation, the lower-order 32 bits of the ALU output are required early for address generation and rapid back-to-back operations [1]. These constraints require a high-performance ALU with a compact layout footprint that minimizes interconnect delays in the core. ALU also contribute to one of the highest power-density locations on the processor, resulting in thermal hotspots and sharp temperature gradients within the execution core. The presence of multiple execution engines in current-day processors [1] further aggravates the problem, severely impacting circuit reliability and increasing cooling costs. Therefore, this strongly motivates energy-efficient ALU designs that satisfy the high-performance requirements, while reducing peak and average power dissipation. Traditional dense-tree adder architectures such as Kogge–Stone [2] use full binary carry-merge trees that result in large transistor sizes because of their high fan-outs and require wide routing channels for inter-stage wiring. Adder architectures such as Ladner–Fischer [3] address the wiring problem by reducing the number of inter-stage interconnects at the expense of exponentially increasing carry-merge fan-outs.

In this chapter a single-cycle 64-bit integer execution ALU [4] fabricated in 90 nm dual-Vt CMOS technology [5] is described. A sparse-tree adder architecture is employed to address the high fan-out issue mentioned above, as well as to reduce the inter-stage wiring complexity by up to 80% [6]. High-speed single-rail dynamic circuit techniques and migration of non-critical paths to fully static CMOS enable low carry-merge fan-outs, low active leakage and dynamic power consumption, high DC noise robustness, and a dense layout. The complete 64-bit ALU operates at 4 GHz in the 64-bit mode measured at 1.3 V, 25° C and consumes 300 mW total power. The corresponding 32-bit mode latency is 7 GHz, also measured at 1.3 V, 25° C. The lower- and upper-order 32-bit domains operate on separate off-chip supply voltages, enabling conditional turn-on/off of the 64-bit ALU mode operation and efficient power-performance optimization, resulting in up to 22% power savings. The 64-bit ALU performance is scalable up to 7 GHz measured at 2.1 V, 25° C, and the 32-bit ALU performance is scalable up to 9 GHz measured at 1.68 V, 25° C. Burn-in tolerant conditional keepers are inserted on all dynamic gates to enable full functionality under worst-case noise conditions and elevated supply/temperature stress tests [7].

The remainder of this chapter is organized as follows: Section 2 describes the organization of the ALU; Sections 3 and 4 present the key circuits that enable an energy-efficient single-rail ALU design; the ALU clocking scheme and timing plan are described in Section 5; Section 6 discusses the benefits of this design over a conventional dual-rail dynamic implementation; Section 7 presents the 90 nm dual-Vt CMOS implementation and silicon measurement results; the dual-supply voltage operation of the ALU is described in Section 8. Finally the chapter is summarized in Section 9.

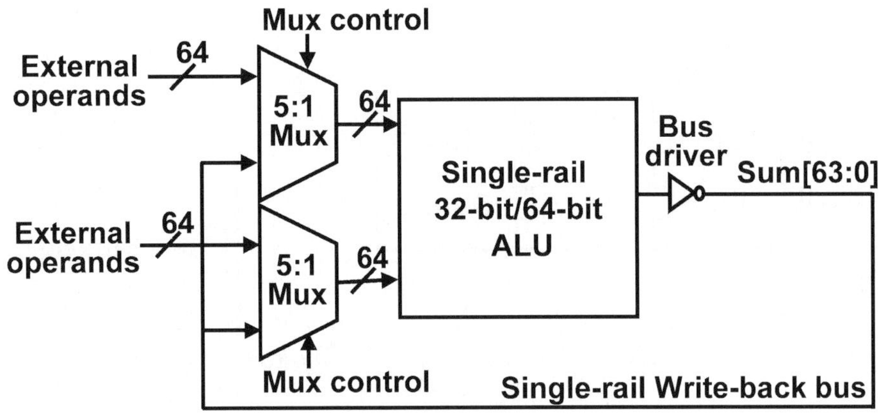

Figure 1. 32-bit/64-bit ALU organization.

2. ALU Organization

The integer execution ALU operates on inputs that originate from the integer register file, the L0 cache, bypass cache, and ALU write-back bus results. A 5:1 source multiplexer selects a pair of these operands and delivers them as inputs to the single-rail adder/logical unit core. The output of the adder shares a bus driver with the logical unit, and drives a 110 μm write-back bus that sends the ALU outputs back to its inputs. The ALU is designed for operation in both 32-bit and 64-bit modes. This organization (Figure 1) enables single-cycle, back-to-back execution of 32/64-bit Add, Subtract, Accumulate and Logic instructions.

3. Single-rail Source Multiplexer

During Add/Logic operations, the source multiplexer selects a pair of ALU operands from four sources, which include the integer register file, L0 cache, bypass cache and the single-rail write-back bus. However, subtract operations will also require a dynamic compatible complementary version of the write-back bus result. This is obtained using the single-to-dual rail converting source multiplexer, shown in Figure 2. This circuit has two dynamic pull-down paths, both of which are precharged with the same Φ1 clock [8]. The first dynamic node (Sum#.En4) directly drives a dynamic inverter, whose output computes (Sum#.En4)#. Depending on the logic state of the inputs (Sum and En4#), either of these two nodes will discharge to ground, and will hold the complementary node at Vcc through a cross-coupled PMOS network.

Due to the non time-borrowable nature of this gate, its inputs must be set up before the clock switches to avoid false evaluation. This constraint is

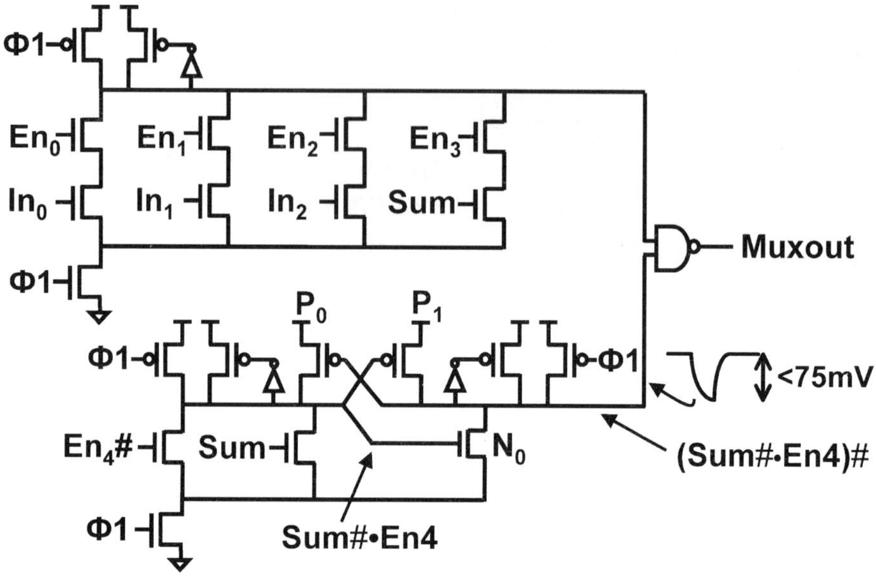

Figure 2. Single-rail to dual-rail converting dynamic source multiplexer.

required to prevent the turned on NMOS device (N0) from falsely discharging the output node. Careful sizing of the cross-coupled PMOS keepers (P0 and P1) limits the switching noise on the output node. Figure 3 shows the impact of clock skew on peak noise-glitch at the non-switching dynamic output node. Noise constraints of this design require the maximum noise-glitch to be less than 75 mV. Simulations in 90 nm CMOS technology show that an 18 ps setup time at the Φ1 boundary enables this circuit to tolerate up to ±15 ps clock variation about the nominal point, while adequately meeting all propagated noise constraints. The single-to-dual-rail converting source multiplexer circuit eliminates the need for a differential write-back bus, thereby contributing to the energy-efficiency and compact layout of this design.

4. Sparse-tree Adder Core

4.1. Critical Sparse-tree: 32-bit Mode of Operation

The performance-setting and power-limiting block in the ALU is the adder core, which is designed to operate in both 32-bit and 64-bit modes. In the 32-bit mode the upper-order 32-bit section is powered down, leaving the lower-order section to execute a 32-bit instruction (Figure 4). The first stage of the adder is the Propagate-Generate (PG) block that outputs *propagate* ($P_i = A_i$#

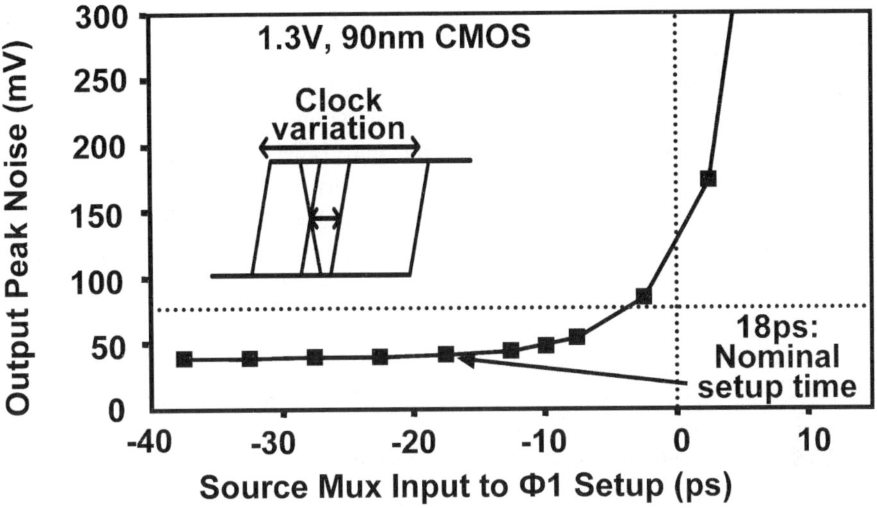

Figure 3. Source-multiplexer peak output noise vs. input data arrival time trade-off.

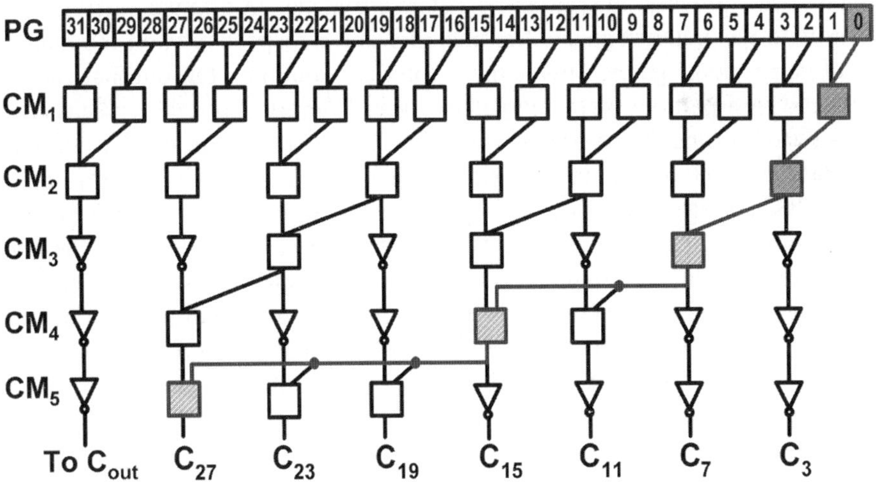

Figure 4. Critical sparse carry-merge tree: lower 32 bits.

NAND B_i#) and *generate* (G_i = A_i# *NOR* B_i#) signals from the adder inputs A_i# and B_i#. Complementary version of all ALU inputs, except for the write-back, bus are available at the front-end multiplexer. As described in Section 3, the complementary version of the writeback bus result can be easily generated using the single-to-dual rail converter circuit. The use of complementary adder inputs results in single-transistor pull-up (GP_i# = P_i *NAND* P_{i-1}) and pull-down (GP = P_i# *NOR* P_{i-1}#) evaluation paths in the static and dynamic group-propagate gates, leading to a lower average transistor size on these paths. Other implementations of this design using true inputs are also possible. The adder core, implemented in single-rail dynamic logic, has a worst-case evaluation stack of 2-NMOS and is followed by five stages of carry-merge (footless dynamic gate [CM2] and footed dynamic gate [CM4] interspersed between three static gates [CM1, CM3, CM5]) that perform a radix-2 carry-merge operation in both the static and dynamic stages. The static carry-merge block CM1 outputs the two-way group-generate ($GG_i = G_i + P_iG_{i-1}$) and group-propagate ($GP_i = P_iP_{i-1}$) signals. The output of CM1 will pre-discharge low (since its inputs are precharged high) and has a worst-case 2-PMOS pull-up evaluation path. Thus, the carry-merge tree has a worst-case evaluation path of 2N-2P-2N-2P-3N-2P in order to generate the carry. The 3-NMOS evaluation stack in the source multiplexer and CM4 is due to the presence of the footed clock evaluation transistors required at phase boundaries.

This reduced-fanout carry-merge tree is the key advantage of the sparse-tree adder architecture [6]. The sparse-tree generates every fourth carry (C_3, $C_7, \ldots, C_{23}$ and C_{27}), unlike the Kogge–Stone architecture that generates carries for every bit using a dense tree, with generate and propagate fan-outs of 2 and 3, respectively. In contrast, the sparse-tree has 73% fewer carry-merge gates with generate/propagate fan-outs of 1 and 2 respectively, on the majority of carry-merge gates. This is the theoretical minimum fan-out that can be achieved in a parallel carry-merge structure. Consequently, the critical path reduces to a pruned carry-merge tree with 33/50% reduction in P/G fan-outs per stage and 25% reduction in maximum inter-stage interconnect length (spans 12 bits vs. 16 bits in the Kogge–Stone design). Furthermore, the 80% reduction in wiring complexity permits the use of wider/shielded wires on the few performance-critical inter-stage "group generate/propagate" signals. The fan-out reduction and relaxed wiring results in 20% improvement in performance and 56% reduction in power consumption of the sparse-tree design [6]. Thus the critical path of the adder is improved by retaining the same number of gate stages as the Kogge–Stone architecture with reduced fan-outs/stage and reduced interconnect delay between stages.

4.2. Non-critical Conditional Sum-generators

In parallel to the critical sparse-tree, four-bit sum generators speculatively generate conditional sums corresponding to a carry-in of 0 and 1. The non-criticality of the sum-generator permits the usage of a ripple carry-merge scheme to generate the conditional carries. Thus, as shown in Figure 5, the carry-in at the first level of each conditional carry rail is tied to 0 and 1 respectively, generating two rails of conditional carries. An "XOR" of the partial sum with the conditional carries generates the conditional sums. The carries from

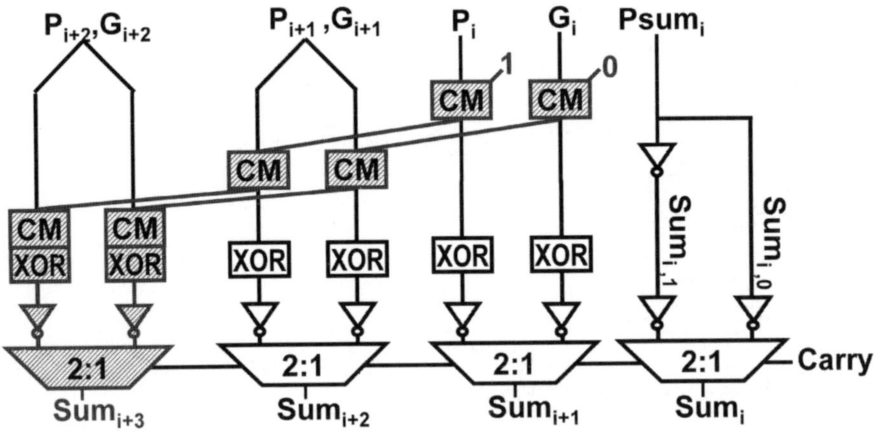

Figure 5. Non-critical four-bit conditional sum-generators.

the sparse-tree ($C_3, C_7, \ldots, C_{23}$ and C_{27}) then select the appropriate four-bit conditional sums using a 2:1 multiplexer, delivering the final 32-bit sum. In this way logic traditionally implemented in the main carry-tree using expensive parallel prefix logic is implemented in the sparse-tree design using an energy-efficient architecture. Such an approach results in smaller area, reduced energy consumption and lower leakage.

The conditional sum-generator block can be further optimized by exploiting correlation of the inputs to the first-level conditional carry gates. As shown in Figure 6, both first-level conditional carry-merge gates reduce to inverters, which can be merged with the next gate in the ripple carry chain, reducing the number of stages in the non-critical path from five to four. This additional slack between the critical and non-critical paths may be further exploited to save power.

4.3. Semi-dynamic Implementation

The performance criticality of the ALU demands a dynamic adder implementation. Partitioning the carry-merge tree into critical and non-critical sections enables an energy-efficient implementation by leveraging dynamic and static techniques. Figure 7 shows the critical and non-critical sections of the adder core. The critical path, implemented in single-rail dynamic logic, begins with the two-stage source multiplexer (A dynamic 5:1 mux with a static (S) NAND gate merging the true and complementary signals, as shown in Figure 2), followed by the PG block that outputs the P_i and G_i signals from the inputs A_i# and B_i#. At this point the critical and non-critical paths fork, with the five-stage

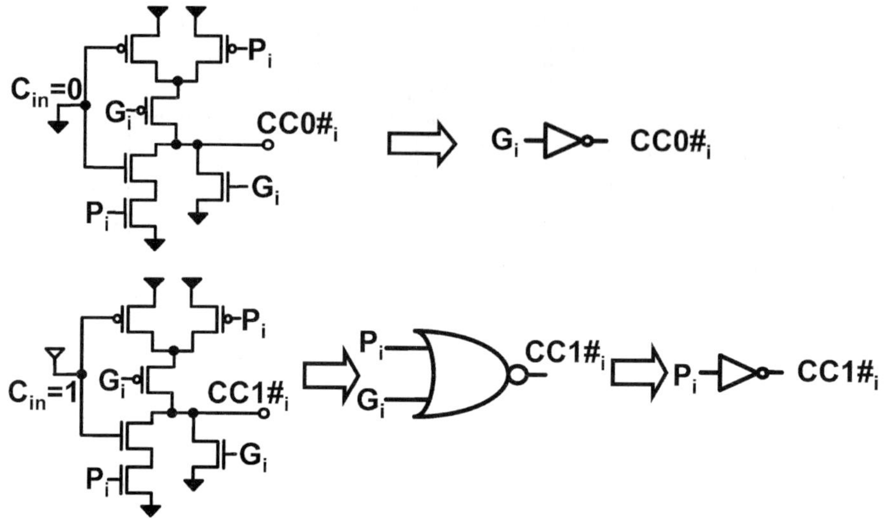

Figure 6. Optimizing conditional-carry rails.

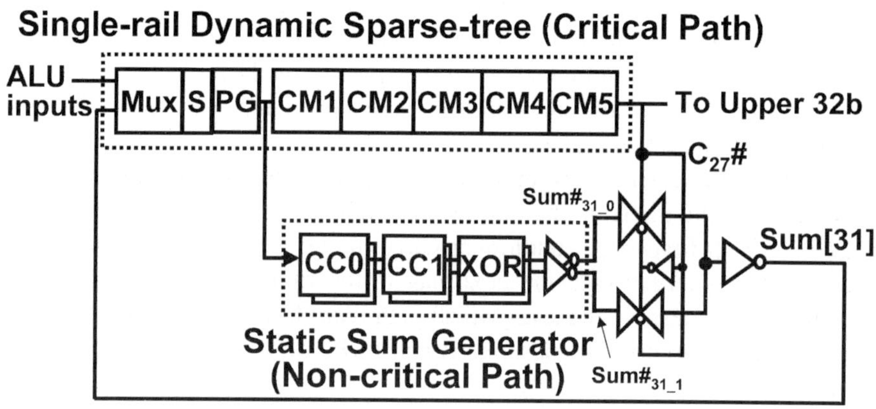

Figure 7. Critical and non-critical paths in sparse-tree adder.

sparse-tree (CM1-CM5) in the critical path and the four-stage conditional sum generator (two stages of conditional ripple-carry gates, a sum XOR and an inverter) in the non-critical path. The final 1 in 4 carry (C_{27}#) from the critical path selects between the two conditional sums (Sum#$_{31\_0}$ and Sum#$_{31\_1}$) using a 2:1 transmission gate multiplexer. The one-stage slack enables reduced transistor sizes in the non-critical path, resulting in additional power savings.

To meet the performance requirement of the ALU, the critical path is implemented in single-rail dynamic logic. Compared to other dual-rail dynamic implementations [9], this 50% reduction in transistor count and interconnect

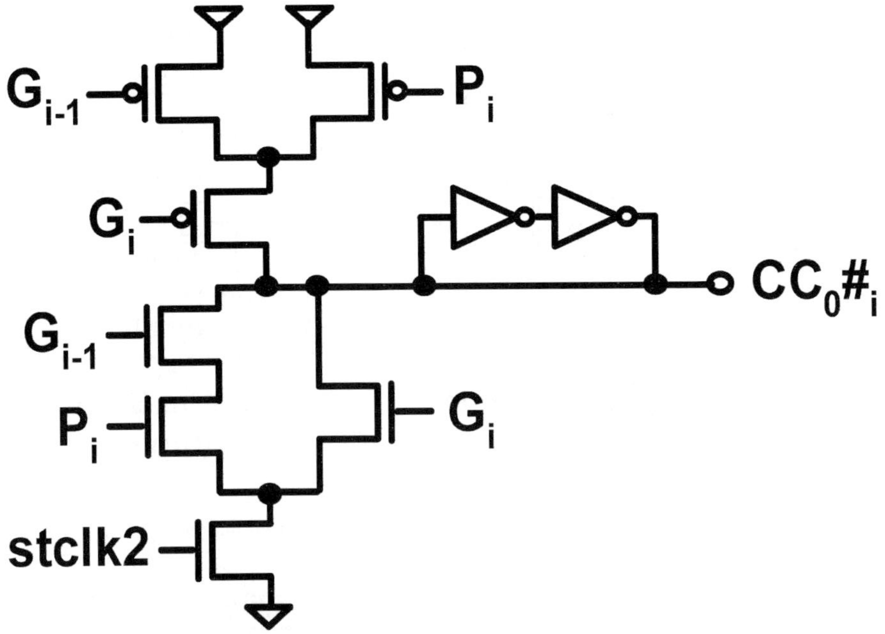

Figure 8. Stage 1 of sum-generator: set-dominant latch.

complexity results in reduced leakage power and higher performance. The critical sparse-tree path includes the PG block and carry-merge gates implemented in dynamic logic. The non-criticality of the sum generator allows a completely static CMOS logic implementation. The low switching activity of static gates reduces the average power consumption in the ALU.

The inputs (G_i, P_i, and G_{i-1}) to the conditional carry gate (Figure 8) of the static sum generator are dynamic signals that go high during precharge. To prevent this precharge activity from propagating through the sum generators when the clock goes low, the first gate in the sum generator is converted to a set-dominant latch (Figure 8) by the addition of the clocked footer NMOS device to the static carry-merge gate, clocked with stclk2, a staggered Φ 1 clock, as described in Section 5. This transistor cuts off the discharge path for the output during the precharge phase, while a full keeper added to the output node holds state. Thus switching activity in the downstream static blocks is reduced, resulting in a semi-dynamic design that reduces average power consumption without impacting performance.

The interface between the static and dynamic signals occurs at the 2:1 transmission-gate multiplexer, with the gates CM4 and CM5 representing the fourth and fifth level carry-merge gates of the sparse tree (Figure 9). The Carry# signal is a dynamic signal (pre-discharged low) that drives the select node of the multiplexer. Inputs to the transmission-gate multiplexer are static conditional

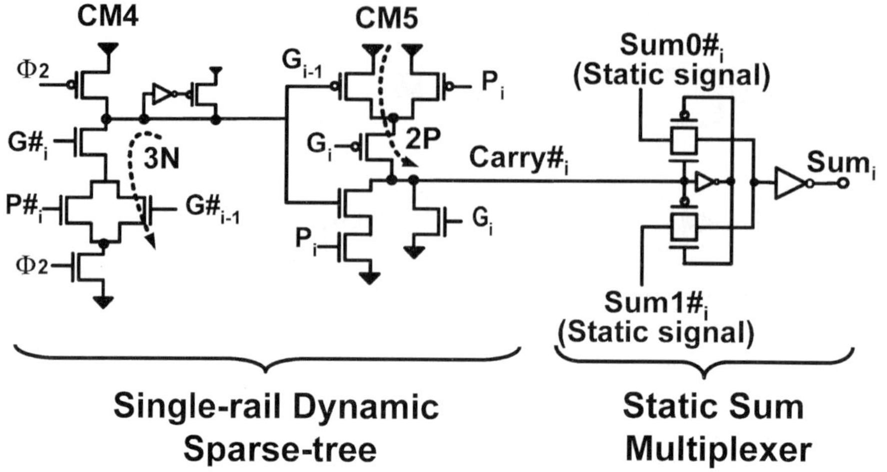

Figure 9. Semi-dynamic design: interfacing static and dynamic signals.

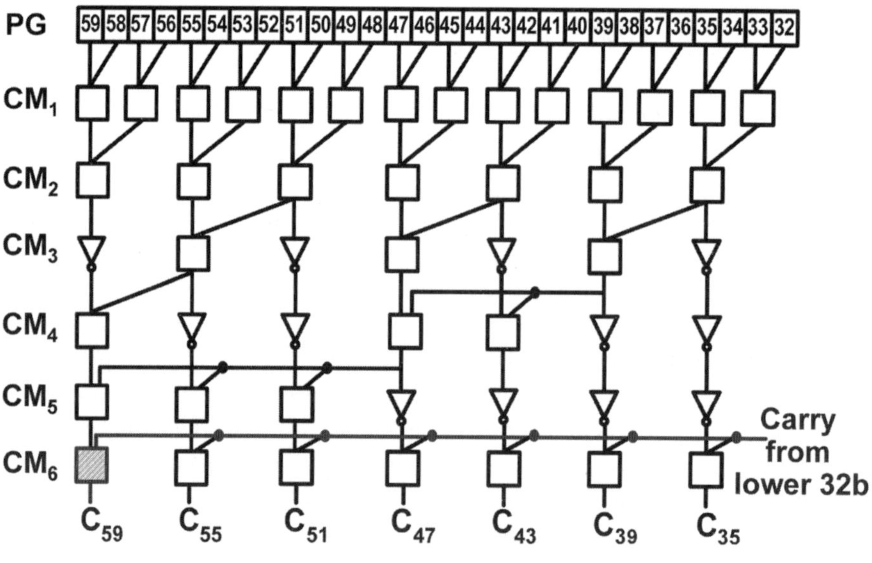

Figure 10. Sparse carry-merge tree: upper 32 bits.

sum signals from the side-paths. During precharge, the lower transmission gate is turned ON, and the output sum is equal to Sum1. During evaluation, Carry# may remain low or go high in response to activity in the dynamic carry-merge gates. The multiplexer will select either of the conditional sums (Sum#0 or Sum#1) to deliver the final sum, enabling seamless time-borrowing at the dynamic–static interface with no race conditions between the two paths.

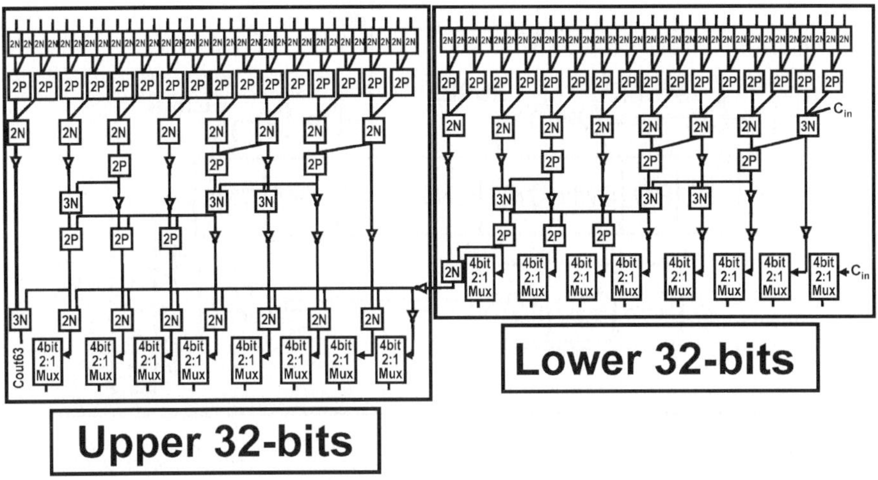

Figure 11. Sparse carry-merge tree: 64 bits.

4.4. 64-bit mode of Operation

In the 64-bit mode the supply to the upper 32 bits of the adder is activated (Figure 10). To enable layout reuse of the 32-bit adder cores, the upper 32-bit section uses a similar sparse-tree to generate 32-way group-generates and propagates. The carryout from the lower-order 32 bits is merged with these signals in the additional carry-merge stage (CM6), generating every fourth carry of the upper 32 bits (Figure 11). This architecture reduces design/layout effort and produces the final 64-bit ALU outputs with only three additional gate stages while minimally affecting the performance of the lower 32 bits of the adder.

5. ALU Clocking Scheme

The ALU loop operates on a simple 50% duty cycle 2Φ dynamic timing scheme with seamless time-borrowing at locally generated clock boundaries [10]. Φ1 begins at the dynamic source multiplexer and terminates at the third carry-merge gate (CM3). Φ2 begins at the fourth carry-merge gate (CM4) and terminates at the end of the bus (Figure 12). By locally generating the Φ2 clock from an inversion of the incoming Φ1 clock, the falling and rising edges of Φ1 and Φ2 are locked, enabling race-free debug at slow frequencies. The conversion of the first stage of the non-critical path (CC0) to a latch, which is clocked with a staggered Φ1 clock (stclk2) , reduces switching activity in the non-critical paths.

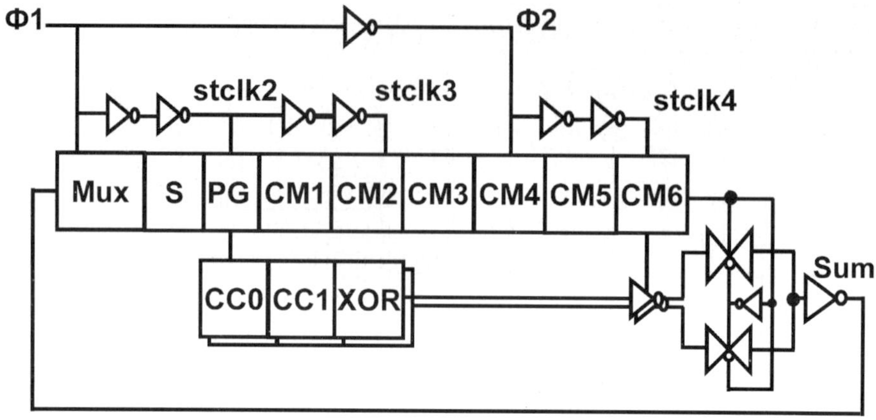

Figure 12. ALU clocking scheme.

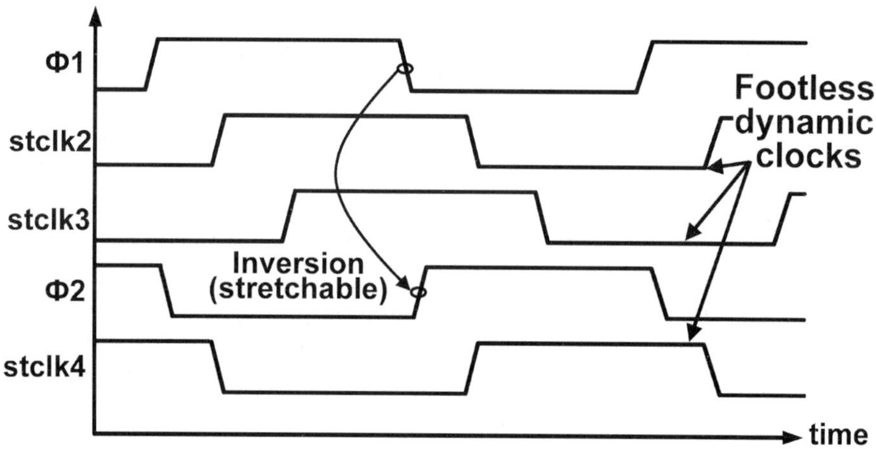

Figure 13. ALU timing plan.

The timing plan (Figure 13) illustrates the relative positioning of the ALU clocks. The phase clocks Φ1 and Φ2 drive footed dynamic gates in the source multiplexer and the fourth level carry-merge gate respectively. The remaining carry-merge gates are implemented using footless dynamic circuits driven by locally generated staggered clocks. Generation of these staggered clocks from the phase clocks limits precharge races. Seamless time-borrowing occurs at all clock boundaries except at the Φ1 boundary, which requires a setup time of 18 ps to minimize noise generated at the single-to-dual-rail converting multiplexer's output.

6. Benefits over Dual-rail Dynamic Implementation

A dual-rail dynamic implementation and its inherent costs are avoided by the use of a single-to-dual rail converting multiplexer. This circuit allows generation of dynamic-compatible true/complementary versions of the ALU inputs from a single-rail write-back bus, thereby enabling a single-rail ALU design with single-rail conditional carry circuits and simplified sum XOR gates. This enables 50% reduction in area due to elimination of the dual-rail paths, 10% reduction in delay due to shorter critical path interconnects, and 40% reduction in active leakage power due to reduction in total transistor count.

7. Measurement Results

Figure 14 shows a microphotograph of the die implemented in 90 nm dual-Vt CMOS technology, with the lower and upper 32-bit sections of the ALU in the middle, a total die dimension of 1622 μm × 294 μm, and the ALU occupying 280 μm × 260 μm. Frequency and power measurements of this ALU (Figure 15) are obtained by sweeping the supply voltage from 0.8 V to 2.1 V in a temperature-stabilized environment of 25°C. At 1.3 V (nominal supply voltage for this technology), the 64-bit ALU operates at a maximum frequency of 4 GHz with worst-case switching power of 300 mW and a leakage component of 9.6 mW. Average power consumption for a switching activity of 10% is 89 mW. ALU performance in the 64-bit mode is scalable to 7 GHz at 2.1 V. In the 32-bit mode the ALU operates at a maximum frequency of 7 GHz with a power consumption of 238 mW at the nominal design point, with performance scalable to 9 GHz at 1.68 V.

8. Dual Supply Voltage Operation

Depending on the mode of operation, the upper 32 bits of the ALU can be selectively turned on or turned off (Figure 16). In the 32-bit mode the supply to

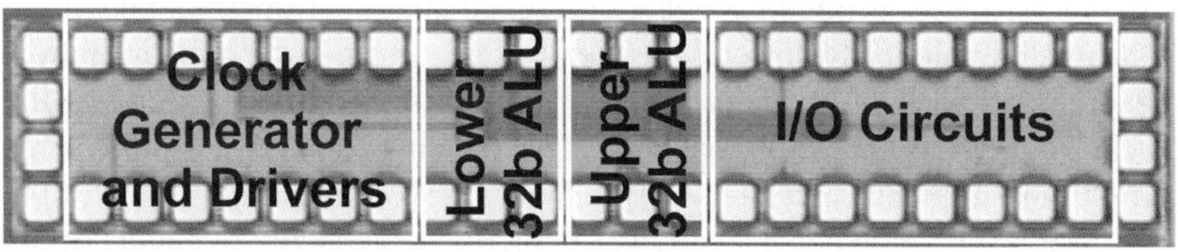

Figure 14. Die microphotograph.

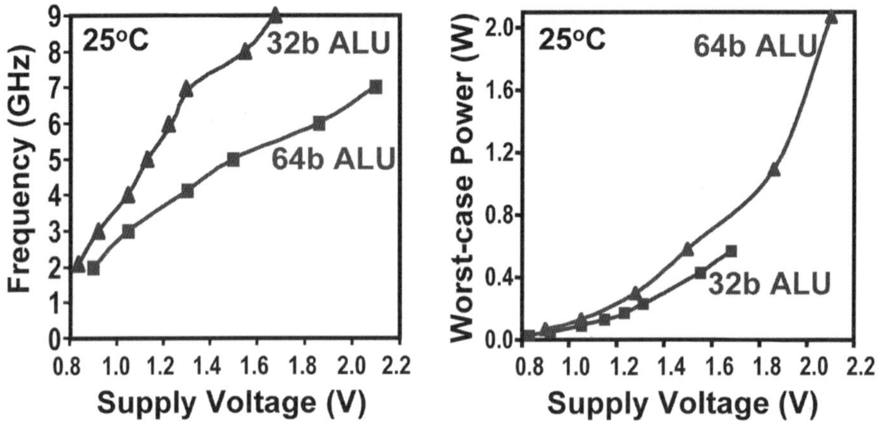

Figure 15. 32-bit and 64-bit ALU maximum frequency and power measurements.

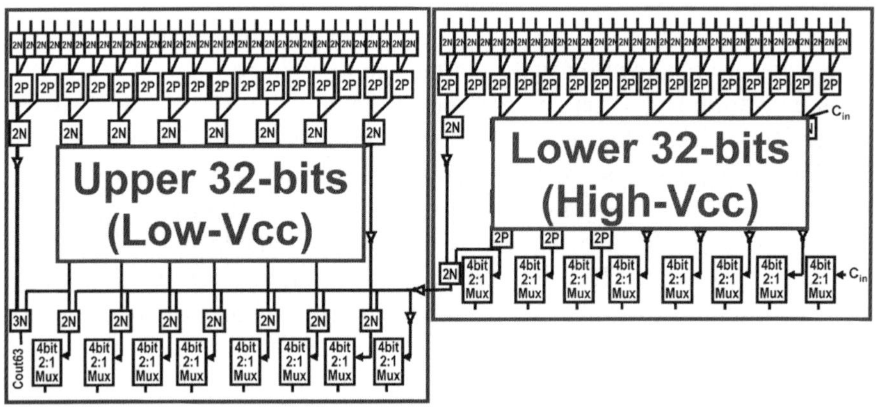

Figure 16. Dual-supply operation of 64-bit ALU.

the upper section is turned off, thereby reducing active leakage power by 50%. In the 64-bit mode the three-stage slack present in the upper-order 32-bit section of the tree is exploited by lowering its supply, resulting in lower dynamic power consumption, without impacting performance. Low-voltage operation of the ALU also provides the additional benefit of reducing leakage power. Leakage current measurements of a 64-b ALU in 90 nm CMOS technology (Figure 17) show that a 66% active leakage power reduction is achieved by reducing the supply voltage from 1.3 V down to 1 V. By operating the upper 32-bit section of the ALU at this lower supply in the 64-bit mode we can obtain a 33% leakage

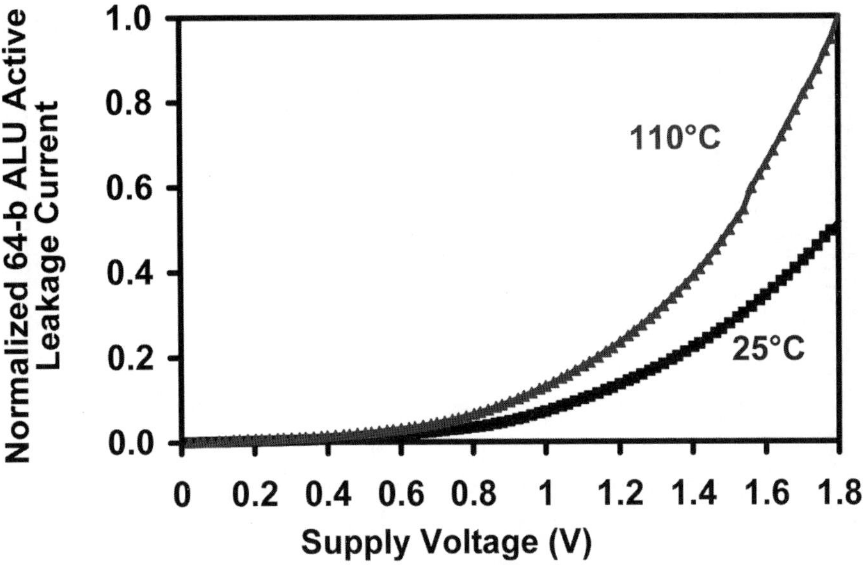

Figure 17. 90 nm measured results: scaling of 64-bit ALU active leakage with supply voltage.

power reduction without impacting performance. No level converter is required at the interface of the two supplies, since the carryout signal at a higher supply voltage drives a carry-merge gate at a lower supply.

Figure 18 illustrates the power-performance tradeoffs in the 64-bit mode of operation. The total power savings and the associated delay penalty are shown as the supply voltage to the upper 32 bits is lowered from 1.3 V down to 0.9 V. The lower 32-bit section of the ALU operates at the nominal supply voltage of 1.3 V and therefore its performance remains unaffected. As the supply voltage is reduced from 1.3 V to 1.05 V the performance-setting path of the ALU includes the lower 32-bit carry-merge tree (operating at the nominal supply voltage) and the final three stages of the upper 32-bit ALU (operating at the lower supply). A super-linear savings in power is obtained, while in this voltage range the total delay penalty is minimal since only three stages of the performance-setting path operate at the lower supply. Below 1.05 V the slack between the upper and lower 32-bit sections is exhausted and the performance-setting path of the ALU shifts entirely to the upper 32-bit section of the ALU, which operates at the lower supply. Consequently, the delay penalty overtakes the percentage power savings. Thus we have demonstrated that, with appropriate choice of a lower supply voltage, efficient power–performance optimization can be achieved, enabling up to 22% total power savings.

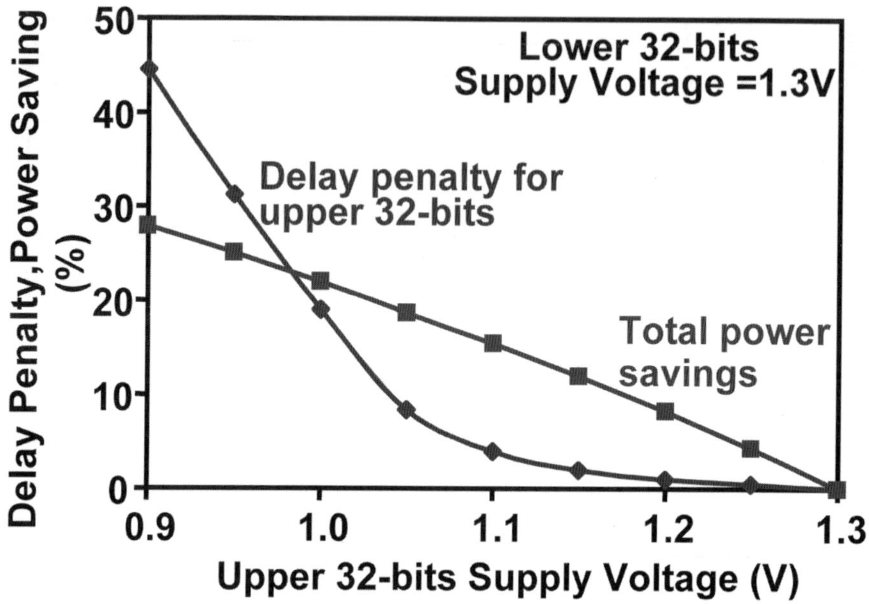

Figure 18. Total power savings and performance penalty vs. supply trade-off for upper-order 32 bits.

9. Summary and Conclusions

The design of an energy-efficient 64-bit integer execution ALU operating at 4 GHz in a 1.3 V, 90 nm CMOS technology is described, with performance scalable to 7 GHz at 2.1 V (measured at 25° C). The single-rail sparse-tree adder core design enables an energy-efficient ALU design with a worst-case power of 300 mW, measured at 4 GHz with a low leakage component of 9.6 mW. In the dual-supply mode of operation up to 22% power savings are demonstrated. Finally, the semi-dynamic design results in reduced switching activity in the adder core with a low average power consumption of 89 mW. This mitigates the power density issues and thermal hotspot challenges within the execution core, increasing reliability and reducing cooling costs.

Acknowledgments

The authors thank R. Saied, S. Wijeratne, D. Chow, M. Kumashikar, C. Webb, G. Taylor, I. Young, H. Samarchi, G. Gerosa, for discussions; Desktop Platforms Group, Folsom, CA for layout support; and S. Borkar, M. Haycock, J. Rattner and S. Pawlowski for encouragement and support.

References

[1] Sager, D. *et al.* "A 0.18 μm CMOS IA32 microprocessor with a 4 GHz integer execution unit", *Digest of Tech. Papers, IEEE Intl. Solid-State Circuits Conf.*, February **2001**, 324–325.

[2] Kogge, P.; Stone, H.S. "A parallel algorithm for the efficient solution of a general class of recurrence equations", *IEEE Trans. on Computers*, **1973**, *c22*, 786–793.

[3] Knowles, S. "A family of adders", *Proc. 14th IEEE Intl. Symp. on Computer Arithmetic*, April **1999**, 277–281.

[4] Mathew, S.; Anders, M.; Bloechel, B.; Nguyen, T.; Krishnamurthy, R.; Borkar, S. "A 4 GHz 300 mW 64-bit integer execution ALU with dual supply voltages in 90 nm CMOS", *Digest of Tech. Papers, IEEE Int. Solid-State Circuits Conf.*, February **2004**, 162–163.

[5] Thompson, S. *et al.* "A 90 nm logic technology featuring 50 nm strained silicon channel transistor, 7 layer of Cu interconnects, low-k ILD, 1 μm^2 SRAM cell", *IEDM Tech. Dig.*, December **2002**, 61–64.

[6] Mathew, S.; Anders, M.; Krishnamurthy, R.; Borkar, S. "A 4 GHz 130 nm address generation unit with 32-bit sparse-tree adder core", *IEEE J. Solid State Circuits*, **2003**, *38*, 689–695.

[7] Alvandpour, A.; Krishnamurthy, R.; Borkar, S. "A sub-130 nm conditional keeper technique", *IEEE J. Solid State Circuits*, **2002**, *37*, 633–638.

[8] Anders, M.; Mathew, S.; Bloechel, B. *et al.* "A 6.5 GHz 130 nm single-ended dynamic ALU and instruction scheduler loop", *Dig. Tech. Papers, IEEE Int. Solid-State Circuits Conf.*, February **2002**, 410–411.

[9] Naffziger, S. "A sub-nanosecond 0.5 μm 64-bit adder design", *Dig. Tech. Papers, IEEE Int Solid-State Circuits Conf.*, February **1996**, 362–363.

[10] Alvandpour, A.; Krishnamurthy, R.; Eckerbert, D.; Apperson, S.; Bloechel, B.; Borkar, S. "A 3.5 GHz 32 mW 150 nm multiphase clock generator for high-performance microprocessors", *Dig. Tech. Papers, IEEE Int Solid-State Circuits Conf.*, February **2003**, 112–113.

Chapter 8

BINARY FLOATING-POINT UNIT DESIGN:

The fused multiply-add dataflow

Eric M. Schwarz
IBM Corp., MS:P310, 2455 South Road, Poughkeepsie, NY 12601

Abstract: Since 1990 many floating-point units have been designed using a fused multiply-add dataflow. This type of design has a huge performance advantage over a separate multiplier and adder. With one compound operation, effectively two dependent operations per cycle can be achieved. Even though a fused multiply-add dataflow is now common in today's microprocessors, there are many details which have never been discussed in papers. This chapter shows the implementation of the different parts of the fused multiply-add dataflow including the counter tree, suppression of sign extension encoding, leading zero anticipation, and end around carry adder design. This chapter illustrates algorithms and implementation details used in today's floating-point units that have been passed down from designer to designer, becoming the folklore of floating-point unit design.

Key words: binary floating-point; fused multiply-add; computer arithmetic; end around carry adder; leading zero anticipation.

1. Introduction

Floating-point units have been developed for many radices including binary, hexadecimal, and even decimal. The most popular radix is binary, though the future may change with the introduction of a new standard for decimal floating-point datatypes in the next revision to 754 IEEE floating-point standard [1]. Decimal format is optimal for financial applications though binary is best for scientific applications due both to its mathematical properties and performance advantage.

Vojin G. Oklobdzija and Ram K. Krishnamurthy (eds.),
High-Performance Energy-Efficient Microprocessor Design, 189–208.

Binary floating-point units are available on every microprocessor and are very common in embedded applications including game systems. Most designs center around a fused multiply-add dataflow due to its simplicity and performance advantage over separate multiply and add pipelines. One technique used for increasing performance is to use Horner's rule for transforming a set of equations into a series of multiply-adds [2]. This numerical analysis technique is very common and takes full advantage of this type of dataflow.

The first processor to contain a fused multiply-add dataflow was the first IBM RS/6000 workstation which was introduced around 1990 [3]. Many of the hardware implementation algorithms of this machine are still popular today. The optimizing compiler was key to enabling C programs to be expanded into a series of fused multiply-adds.

This chapter will detail current design techniques that have become common hardware design practices in creating fused multiply-add implementations. These techniques have not been well documented except in patent filings or are just part of folklore passed on by designers. This chapter will cover an introduction to the IEEE 754 binary floating-point format followed by a detailed look at a typical dataflow. The multiplier counter tree will be developed and the counter tree reduction will be shown. The addend will also be discussed, which is summed with the two partial products from the multiplier. This presents a complication that is not well known due to the sign extension encoding of Booth encoded partial products. Then the end around carry (EAC) adder will be discussed, which results in a positive magnitude result. Also, current techniques in leading zero anticipation (LZA) design will be discussed. Then normalization and rounding will be discussed to complete the design of fused multiply-add dataflow. References will be given for more in-depth study of these topics. This chapter will provide an overview of the complications of floating-point unit design.

2. IEEE 754 Format

In the 1960s and 1970s many proprietary formats were popular, such as IBM S/360 hexadecimal format [4], VAX, and CRAY [5], which were incompatible and did not have the best mathematical properties. In 1985 the "IEEE standard for binary floating-point arithmetic" [6], otherwise known as the IEEE 754 standard, was ratified, and has become the universal format in all microprocessor designs. The standard defines two basic formats: single-precision (32 bits) and double-precision (64 bits), and also provides extended formats. Most manufacturers implement at least single and double. In addition, Intel, as well as clone manufacturers such as AMD, implement a double extended format [7, 8]. Intel and its clones optimize their dataflow for a double extended format while other manufacturers such as IBM and Sun Microsystems optimize for the

Table 1. IEEE 754 standard formats

Type	Sign	Exponent	Fraction	Format width	Bias	Emax	Emin
Single	1	8	23	32	127	+127	−126
Double	1	11	52	64	1023	+1023	−1022
Double Extended	1	15+	63+	79+	$2^{(n-1)}-1$	$2^{(n-1)}-1$	$-(2^{(n-1)}-2)$

double format. Table 1 shows the bitwise breakdown of IEEE 754 standard formats as well as their associated biases and signed exponent ranges.

Normalized numbers can be represented by the following equation:

$$X = (-1)^{Xs} * (1 \cdot Xf) * 2^{(Xe-bias)}$$

where X is the value, Xs is the sign bit, Xf is the fraction of the significand which is augmented by an integer bit equal to one, and Xe is the unsigned binary exponent. If the exponent, Xe, is not equal to the maximum (all ones) or the minimum (all zeros) then the value of the number, X, is specified by this equation. If the exponent is all ones or all zeros a special number is represented. There are four types of special numbers. If the exponent is all ones and the fraction is zero, a positive or negative infinity is symbolized. If the exponent is all ones and the fraction is non-zero, then a not-a-number (called NaN) is represented which is useful for representing non-numeric or non-representable results. If the exponent is all zeros and the fraction is all zeros, a positive or negative zero is symbolized. The fourth special number is a denormalized number, sometimes called a denormal or subnormal number, and is represented by an all-zero exponent and a non-zero fraction. Denormals, unlike the other special numbers, require non-trivial calculations of a result. The following equation specifies the value of a denormalized number:

$$X = (-1)^{Xs} *(0 \cdot Xf) * 2^{(1-bias)}$$

where Xe = 0, and bias is the bias given in Table 1. Note that the implicit bit is removed and there is a gradual loss of precision as the significand gets smaller. Also, the exponent is set to (1− bias) rather than (0− bias) which creates less of a discontinuity from the normalized numbers, but creates complications in the implementation. Denormalized numbers are difficult to implement in hardware, though with some added complexity and small addition of hardware, they can be implemented in a pipelined manner [9, 10].

3. Fused Multiply-add Dataflow

A fused multiply-add operation can be described by the following equation:

$$T = B + A * C$$

"A" is the multiplicand, "C" is the multiplier, "B" is the addend, and "T" is the target of the result. The dataflow is designed to require several pipeline stages of latency but to have a throughput of one instruction per cycle. Thus the dataflow supports the performance of effectively two operations (a multiply and an addition) per clock cycle. The dataflow consists of a multiplier, addend alignment, an incrementer, an adder, a leading zero anticipator, a normalizer, and a rounder as shown in Figure 1. The dataflow of only the significand is shown, since the exponent dataflow is rather simple. Only the critical path in the exponent dataflow is discussed where it is used to calculate the addend alignment and its update for normalization. Each of these blocks will be detailed. Unless otherwise specified, examples will be based on double-precision operands which have 11 bits for the exponent and 53 bits for the significand (52 bits of fraction and an implied bit).

3.1. Addend Alignment

Before the addend, B, can be added, it must be properly aligned to the product. At the point where the addend is combined with the product, it is represented by two partial products called the sum and the carry. Each of these partial products are two times as wide as the input operands or 106 bits. Usually, in a floating-point adder, the operand with the smaller exponent is aligned. But for a multiply-add dataflow it is very costly to swap and align two 106-bit operands. Instead it is much easier to only align and conditionally complement the addend which is one 53-bit operand. Thus, the product is treated as having a fixed radix point and the addend is aligned to this radix point. The range of shifting is from when the addend is 53 bits plus two guard bits greater than the product, to when the addend's most significant bit is less than the product's least significant bit. This is approximately three times the width of the data plus some guard bits. For a double-precision dataflow the shift amount range is around 161 bits. Since 8 bits can represent a shift amount of up to 256, only the bottom 8 bits of the shift amount need to be considered. Any shift which is greater than 256 can be capped to a maximum right or left shift.

The shift amount is based on the exponent difference which is expressed as follows:

$$D = (\mathrm{Be} - \mathrm{bias} + \mathrm{b0}') - ((\mathrm{Ae} - \mathrm{bias} + \mathrm{a0}') + (\mathrm{Ce} - \mathrm{bias} + \mathrm{c0}'))$$

where D is the exponent difference or shift amount, and a0, b0, and c0 are the implied ones of the significands of A, B, and C respectively.

$$\begin{aligned} D &= \mathrm{Be} - \mathrm{Ae} - \mathrm{Ce} + \mathrm{bias} + \mathrm{b0}' - \mathrm{a0}' - \mathrm{c0}' \\ D &= \mathrm{Be} - \mathrm{Ae} - \mathrm{Ce} + \mathrm{bias} + z \quad \text{where } z\ \varepsilon\ \{-2, -1, 0, +1\} \\ D^* &= \mathrm{Be} - \mathrm{Ae} - \mathrm{Ce} + \mathrm{Offset} + z \end{aligned}$$

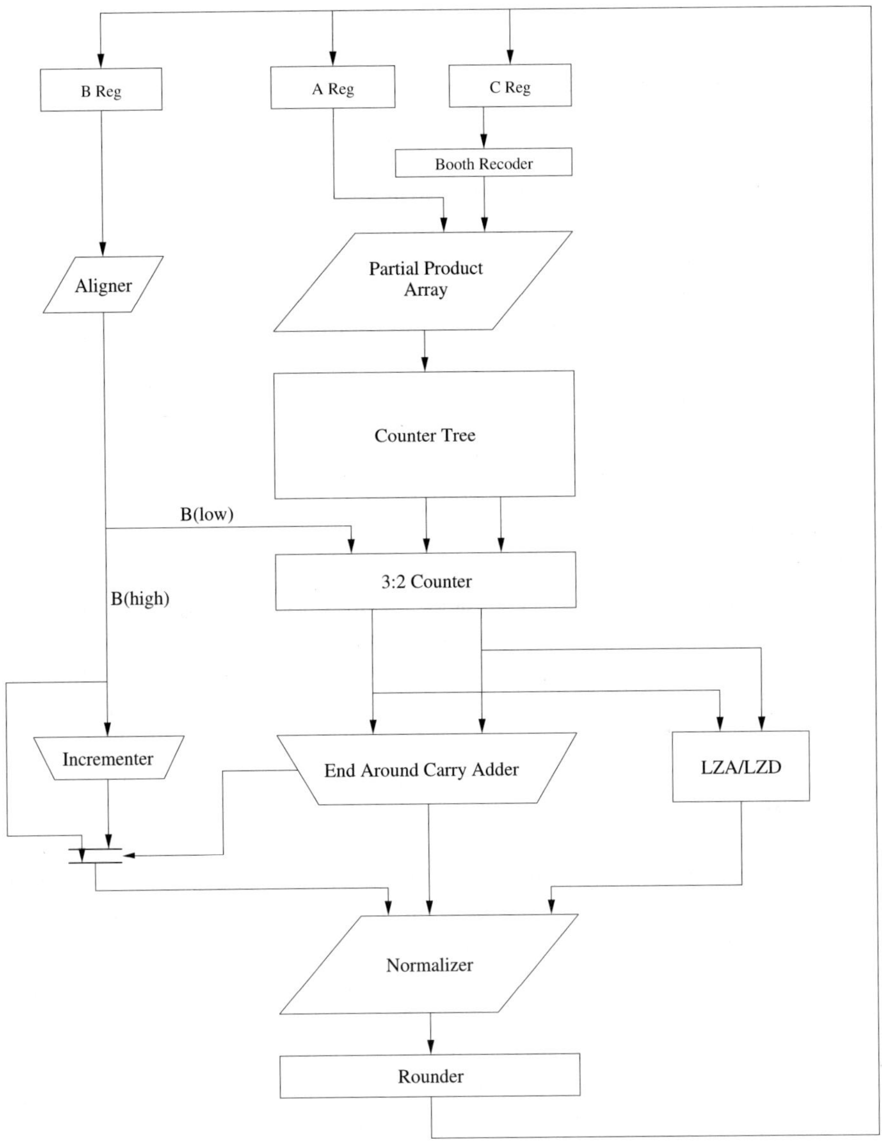

Figure 1. Fused multiply-add dataflow.

The shift amount can be based off any reference point and is usually chosen such that all shift amounts are positive. The D* term represents the exponent difference plus offset due to the change in reference point. The combined offset with bias can be chosen such that the low-order 8 bits are all zeros by just finding the appropriate reference point. This simplifies the calculation to just a three-way add except for the case of denormal operands where z is non-zero. There are several ways to take care of these cases [9, 10]. One simple solution is to design multiple adders that take care of different values of z and choose between them when it has been determined which operands are denormals. Another solution is to first shift the addend by the exponent difference without z, and then have a late stage which shifts by z.

The shift amount calculation is a critical path, so reducing this calculation by choosing the proper offset is essential. After the shift amount is determined, the alignment of the significand is performed. Shifting usually involves multiple levels of multiplexing. The aligner is designed to shift small amounts first, followed by large amounts. This reduces the width of the shifter in early stages. Note that the alignment approximately involves a shift of the addend between 0 and 161 bits to the right. The result needs to be preserved for at least 161 bits and any bits shifted out need to be logically ORed into a sticky bit representing inexactness. The inexact sticky bit will be needed later for rounding. Inexactness can occur wherever there is shifting right or where there is a reduction in the width of the dataflow.

The low-order 106 bits are forwarded to the multiplier counter tree to be reduced with the last two partial products. The high-order 55 bits are forwarded to an incrementer. These bits, if non-zero, could be incremented by a carry out of the adder. Once the carry out of the adder is known, the high addend bits are selected from the incrementer output or from a non-incremented path.

3.2. Multiplier Design

Multiplication involves creating multiples of the multiplicand to form a partial product array followed by their summation to form the final product. A radix of the multiplier is chosen which determines the range of multiples possible. A larger radix involves more work to create and choose the multiples, but less effort to add. In grade school we are taught to multiply using a decimal format, and perform a radix-10 multiplication to create partial products in the range of 0 to 9 times the multiplicand. A partial product array is formed out of digit multiplications and then the partial products are summed to create the final product. In binary, the radix of the multiplier is typically a power of 2, and for double precision, the operands are 53 bits. A radix-2 method is the simplest method since it creates one partial product per bit of the multiplier operand, but requires a huge counter tree to reduce 53 partial products. A radix-4 multiplication method will reduce the number of partial products to approximately 2 bits per partial product, or more exactly $\text{ceil}((n + 1)/2) = 27$. However, there is added complexity to the partial product creation. In a pure non-Booth, radix-4 method, the multiples of (0X, 1X, 2X, and 3X) need to be created. The 3X multiple is non-trivial to form and requires extra delay and area for an adder. In the 1950s, Booth [11] showed a technique used in accounting where digits could be recoded into both negative and positive digits. In binary, this technique is based on the string recoding theorem where a string of ones can be replaced by 1 in the bit position immediately to the left of the most significant bit in the string, and a -1 in the least significant bit of the string as shown by the following equation:

$$\sum_{(i=0 \text{ to } n-1)} 1*2^i = 1*2^n - 1*2^0$$

Example: 0001111100 = 0010000(−1)00

This transformation eliminates two consecutive ones and thus will eliminate the 3X multiple. A Booth radix-4 scanning simplifies the multiples to the set of (−2X, −1X, 0X, +1X, and +2X). 0X, 1X, and 2X are trivial to create as well as their negative counterparts and this will be explained later. Many implementations today use a Booth radix-4 multiplier and our example will assume this type of design.

For IEEE double precision, Figure 2 describes the scanning of the multiplier operand. Multiplier, Y, has bits labeled 0 to 52 where 0 is the most significant and 52 is the least significant. As an example, partial product 10 is formed from scanning bits 19 to 21. There are also two extra bits added at the ends to make sure the recoding completes the scanning of strings. The most significant extra bit, "o", equal to zero also insures the 0th partial product is positive.

The scanning involves examining 3 bits of the multiplier and recoding them as though they were a string. There are 27 partial products which are labeled 0 to 26 where 0 is most significant. The recoding of every 3 bits follows the equations below:

$$\text{sel\_1x} \Leftarrow Y_{i+1} \text{ xor } Y_{i+2}$$
$$\text{sel\_2x} \Leftarrow (Y_i' Y_{i+1} Y_{i+2}) + (Y_i Y_{i+1}' Y_{i+2}')$$
$$\text{sel\_sign} \Leftarrow Y_i$$

where Y_i, Y_{i+1}, and Y_{i+2} are three consecutive bits of the multiplier from more significant to less significant, and Y' indicates the complement of Y, and sel_1x, sel_2x, and sel_sign are selection signals to choose between the 1X multiple, the 2X multiple, and the two's complement of the partial product.

```
Bit Positions of the Multiplier Operand, 0 to 52

           11111111112222222222333333333344444444445555
o01234567890123456789012345678901234567890123456789012o
-------------------------------------------------------
                     1   1   1   1   1   2   2   2   2
-0- -2- -4- -6- -8- -0- -2- -4- -6- -8- -0- -2- -4- -6-
                       1   1   1   1   1   2   2   2
  -1- -3- -5- -7- -9- -1- -3- -5- -7- -9- -1- -3- -5-
Partial Product Numbers and Bits Scanned
```

Figure 2. Radix-4 scanning of multiplier.

```
          1         2         3         4         5         6         7         8         9         10
012345678901234567890123456789012345678901234567890123456789012345678901234567890123456789012345678901234

                                                 S̄SS012345678901234567890123456789012345678901234567890123
                                               _1S̄012345678901234567890123456789012345678901234567890123-H
                                             _1S̄012345678901234567890123456789012345678901234567890123-H
                                           _1S̄012345678901234567890123456789012345678901234567890123-H
                                         _1S̄012345678901234567890123456789012345678901234567890123-H
                                       _1S̄012345678901234567890123456789012345678901234567890123-H
                                     _1S̄012345678901234567890123456789012345678901234567890123-H
                                   _1S̄012345678901234567890123456789012345678901234567890123-H
                                 _1S̄012345678901234567890123456789012345678901234567890123-H
                               _1S̄012345678901234567890123456789012345678901234567890123-H
                             _1S̄012345678901234567890123456789012345678901234567890123-H
                           _1S̄012345678901234567890123456789012345678901234567890123-H
                         _1S̄012345678901234567890123456789012345678901234567890123-H
                       _1S̄012345678901234567890123456789012345678901234567890123-H
                     _1S̄012345678901234567890123456789012345678901234567890123-H
                   _1S̄012345678901234567890123456789012345678901234567890123-H
                 _1S̄012345678901234567890123456789012345678901234567890123-H
               _1S̄012345678901234567890123456789012345678901234567890123-H
             _1S̄012345678901234567890123456789012345678901234567890123-H
           _1S̄012345678901234567890123456789012345678901234567890123-H
         _1S̄012345678901234567890123456789012345678901234567890123-H
       _1S̄012345678901234567890123456789012345678901234567890123-H
     _1S̄012345678901234567890123456789012345678901234567890123-H
   _1S̄012345678901234567890123456789012345678901234567890123-H
 _1S̄012345678901234567890123456789012345678901234567890123-H
1S̄012345678901234567890123456789012345678901234567890123-H
012345678901234567890123456789012345678901234567890123-H
```

Figure 3. Partial product array.

Two's complementing is performed by one's complementing, or inverting the partial product, adding a hot one in the subsequent partial product, and encoding a negative sign extension for this partial product.

The resulting partial product array is shown in Figure 3. Each partial product has 54 bits to represent a multiple of the multiplicand. The multiple is selected by using a 2:1 true/complement multiplexor which chooses either the 1X or 2X multiple, and then outputs either the true or the complement of this multiple. If neither 1X or 2X is chosen, a zero is output. All but the top and bottom partial product have $1S'$ appended to the left side of the partial product. This is done to encode the sign extension. It is equal to "10" for a negative partial product and "11" for a positive partial product. The top partial product is shown with $S'SS$, which equals "011" for a negative partial product and "100" for a positive partial product. Creating a negative partial product involves one's complementing a positive multiple of the multiplicand and adding a hot one, which is represented by "-H", in the row below it. This encoding of the partial product is shown in detail for any radix [12, 13]. The bottom row must be positive since there is no position for adding a hot one to complement it, which is insured by the leading zero (represented with an "o") in Figure 2 as discussed earlier.

The next step in performing the multiplication is to sum the partial products [14, 15]. Many types of counters are available, but 3:2 and 4:2 counters are very popular. 4:2 counters were introduced by Weinberger in 1981 [16] as two cascaded 3:2 counters, and were later implemented as very fast, aggressive pass gate multiplexors [17]. Similar implementations of larger counters are possible, but they have not become popular. The pass gate implementation of a

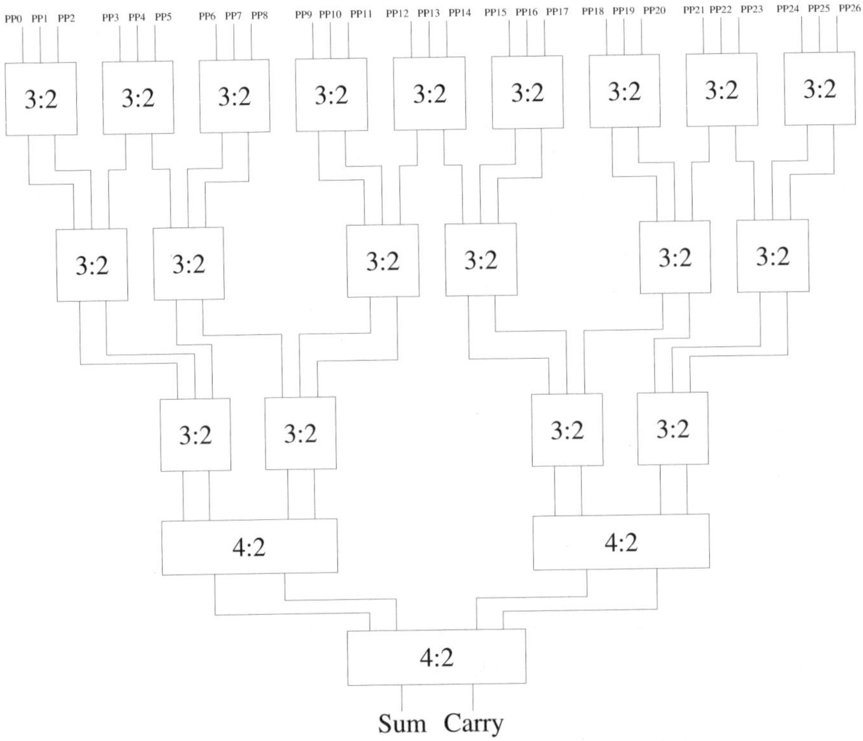

Figure 4. Counter tree of multiply-add dataflow.

4:2 counter takes about 1.5 times the delay of a 3:2 counter versus twice the delay if it were implemented using the cascaded method. For the reduction of 27 partial products a mixture of 4:2 and 3:2 counters works best as is shown in Figure 4.

In the first level of the tree, 3:2 counters reduce 27 partial products to 18. In the second level, 3:2 counters reduce 18 partial products to 12. In the third level, 3:2 counters reduce 12 partial products to 8. At this point it is best to switch to 4:2 counters to reduce to 4 partial products. In the last level, a 4:2 counter is used to reduce the 4 partial products to 2. At this point the addend is ready to be reduced. The aligned addend, along with the 2 partial products from the multiplier, are reduced by a 3:2 counter to 2 partial products. The counter tree takes the equivalent of six 3:2 counter delays to reduce the 27 partial products to 2, and an additional 3:2 counter level to reduce the addend.

3.2.1. *Suppression of sign extension*

There is a problem with combining the addend with the partial products of the multiplier prior to reducing them to one final product. If the final product is formed first, the sign extension encoding has a chance to propagate any carry outs that need to be ignored. The carry out of the sign extension is similar to the carry out of a two's complement subtraction. In a subtraction, if there is a carry out then the sum is positive, and if there is no carry out, the sum is negative. The carry out of an effective subtraction with a positive sum is ignored since it is in a bit position more significant than the most significant bit of either operand being subtracted. The multiplier's sign extension encoding should also cause a carry out which should be ignored.

The sign extension encoding can be thought as a series:

$$\text{Sign Ext} = 1, \text{S}1', 1, \text{S}2', \ldots, 1, \text{S}(\text{n}-2)', 1, \text{S}(\text{n}-1)', 1, (\text{S}(\text{n})' + 1)$$

where Si is the sign bit of the i-th partial product. Given that

$$1 - \text{Si} = \text{Si}'$$

Then,

$$\begin{aligned}\text{Sign Ext} &= 1, (1-\text{S}1), 1, (1-\text{S}2), \ldots, 1, (1-\text{S}(\text{n})+1)\\ \text{Sign Ext} &= 2^0 - \Sigma(\text{i}=1,\text{n})\,\text{Si}\,{}^*4^{-\text{i}}\end{aligned}$$

All other terms can be considered to be positive. If all partial products are positive, then the sum of these sign extension terms is one which is a carry out.

It can be proven that an effective subtract of two's complement numbers that results in a positive sum will always have a carry out. So, for either case of having any negative partial products or having all positive partial products there will be a carry out. And it can be shown that there will always be only one carry out caused by this multiplier sign extension encoding.

This carry out, due to multiplier sign extension, must be detected and separated from carries caused by adding the addend with the product. In other words, we must figure out whether there was a carry out of the sign extension prior to the last 3:2 counter, which combines the product with the addend. To do this, an extra bit in the counter tree is maintained for the carry out position. If the carry out bit position is a one, the carry out has occurred and the carry out bit of the last 3:2 counter is correct. However, if the carry out bit position is a zero, then the carry out bit of the last 3:2 counter is incorrect and must be inverted.

This has been implemented on many fused multiply-add dataflows but never really discussed. It is a simple correction to implement. Designers learn about this anomaly when they first try to simulate the counter tree and figure out why they are getting failures. The multiplier sign extension encoding causes a carry out which must be accounted for especially when combining an addend into the counter tree.

3.3. End Around Carry (EAC) Adders

The adder in a fused multiply-add dataflow has to produce a magnitude for the result, since floating-point is in sign magnitude format. In a fused multiply-add dataflow it is very difficult to determine *a priori* which operand is bigger. The magnitude of the product is unknown in the early stages prior to combination with the addend. Even if it were determined early that the product was bigger, there would be the problem of conditionally complementing two intermediate operands, the carry and sum outputs of the counter tree, rather than just one. Thus, an adder needs to be designed that will always output a positive magnitude result and preferably only need to conditionally complement one operand, the addend. This type of adder is called an "end around carry" adder (EAC) and has been implemented on several microprocessors, though very few details on their formulation and how they work exist in today's literature.

The effective subtraction of operand B from operand P can be formulated as

$$\begin{aligned} P - B &= P + 2^n - B \\ P - B &= (P + B' + 1) \end{aligned} \tag{1}$$

This is useful if the B operand is smaller, but if the P operand is smaller, then a more useful equation is:

$$\begin{aligned} B - P &= -(P - B) = -(P + B' + 1) = -(P + B') - 1 \\ B - P &= (P + B')' + 1 - 1 \\ B - P &= (P + B' + 0)' \end{aligned} \tag{2}$$

Equation (1) is useful if B is the smaller operand, and Equation (2) is useful if P is smaller. The operand which is smaller can be determined by using a comparator. The best design of a greater than comparator is to use the carry out of an adder performing an effective subtraction.

$$P - B = P + 2^n - B = 2^n + (P - B) \tag{3}$$

In Equation (3) the 2^n refers to a carry out of the adder and will be equal to 1 if P is greater than or equal to B. Remember, when doing two's complement subtraction, a positive result always produces a carry out, and a negative result does not. If the carry out is 1, then Equation (1) should be used, but if the carry out is 0 then Equation (2) should be used. Note that the difference between Equations (1) and (2) is whether the carry in gets set to zero or one and whether the final sum is inverted or not. One could think of implementing these equations by taking the carry out of (P - B) and driving that to another adder's carry in. This is how the adder gets its name as an end around carry adder. There have been several microprocessors that have implemented the adder with two carry chains since the designers didn't know these equations could be simplified. There have also been many implementations that have two adders: (P-B) and (B-P) and select between the two. This is common implementation practice even today, especially in the implementation of the exponent difference circuit.

For decades there has been a better way to design an EAC adder. Unfortunately, since it has not been documented, other less elegant designs have been employed. The propagate and generate terms for both the comparator of Equation (3) and the subsequent adder are similar. It would appear that the common terms would combine. Several authors have mentioned this and have actually said the carry should not propagate twice the length but only one length of the adder [18, 19]. This is commonly known, but why and how does one describe the combination of the comparator with the subsequent adder.

First as an example assume that the width of the adder is separated into four groups labeled 0 to 3 where 0 is the most significant. Then the equations of the comparator are derived. The comparator is described by a carry out equation of P-B where the carry in equals 1.

$$\mathrm{Cout} = \mathrm{G0} + \mathrm{P0\,G1} + \mathrm{P0\,P1\,G2} + \mathrm{P0\,P1\,P2\,G3} + \mathrm{P0\,P1\,P2\,P3} \tag{4}$$

The next step is to describe the group carries of the subsequent adder:

$$\mathrm{C0} = \mathrm{G0} + \mathrm{P0\,G1} + \mathrm{P0\,P1\,G2} + \mathrm{P0\,P1\,P2\,G3} + \mathrm{P0\,P1\,P2\,P3\,Cin} \tag{5a}$$
$$\mathrm{C1} = \mathrm{G1} + \mathrm{P1\,G2} + \mathrm{P1\,P2\,G3} + \mathrm{P1\,P2\,P3\,Cin} \tag{5b}$$
$$\mathrm{C2} = \mathrm{G2} + \mathrm{P2\,G3} + \mathrm{P2\,P3\,Cin} \tag{5c}$$
$$\mathrm{C3} = \mathrm{G3} + \mathrm{P3\,Cin} \tag{5d}$$

where Ci is the carry out of the i-th group. If the Cout equation of the comparator is substituted for the Cin of the adder, the following are the reduced equations:

$$\mathrm{C0} = \mathrm{G0} + \mathrm{P0\,G1} + \mathrm{P0\,P1\,G2} + \mathrm{P0\,P1\,P2\,G3} + \mathrm{P0\,P1\,P2\,P3} \tag{6a}$$
$$\mathrm{C1} = \mathrm{G1} + \mathrm{P1\,G2} + \mathrm{P1\,P2\,G3} + \mathrm{P1\,P2\,P3\,G0} + \mathrm{P0\,P1\,P2\,P3} \tag{6b}$$
$$\mathrm{C2} = \mathrm{G2} + \mathrm{P2\,G3} + \mathrm{P2\,P3\,G0} + \mathrm{P2\,P3\,P0\,G1} + \mathrm{P0\,P1\,P2\,P3} \tag{6c}$$
$$\mathrm{C3} = \mathrm{G3} + \mathrm{P3\,G0} + \mathrm{P3\,P0\,G1} + \mathrm{P3\,P0\,P1\,G2} + \mathrm{P0\,P1\,P2\,P3} \tag{6d}$$

This shows the combination of the comparator and adder equations result in a carry chain for every group that is the length of the width of the adder. An EAC adder has equal length carry chains. This wrapping of the carries is needed for effective subtraction but is not correct for addition. To make the adder selectable for addition and subtraction, the P3 term is modified. An extra bit is added to the least significant bit of the adder to signal whether there is an effective subtract operation and the carry chain should be propagated. This is factored into P3, and P3 is equal to zero for an effective addition operation.

Another complication to the design is the inexact sticky bit. The two's complementation of P or B is supposed to be calculated for the full precision operand. If the addend is shifted to the right and has any bits that are aligned to the right of the least significant bit of the product, these bits must be included in the complementation operation and inexact sticky bit generation. If B has any ones shifted out, the two's complementation of B will not cause a carry in to the adder

Assume B is separated into two parts: Bh which is aligned with the product and Bl which is less significant than the product. The subtraction can be described by:

$$\mathrm{P - B = P + Bh' + Bl' + 1}$$

if Bl′ has zeros in it, (Bl′ + 1) will not propagate a carry into Bh′. The truncated result is given by the following:

$$\mathrm{(P - B)tr = (P + Bh' + Bl' + 1)tr = (P + Bh' + 0)}$$

and if B is greater than P, the following holds:

$$\mathrm{(B - P)tr = ((P + Bh' + Bl' + 0)')tr = (P + Bh' + 0)'}$$

This adjustment for an inexact aligned addend, B, can be handled in one of two ways: (1) a Cin term can be added to Equations (6a) to (6d) which is equal to 1 if there is an effective subtract and Bh is exact (no stickyness), or (2) the P3 term can be modified by adding an additional bit to the propagation equation which is equal to the complement of the inexact sticky bit. Method 2 of modifying P3 to have an AND with effective subtract and an AND with not inexact, is straightforward and the preferred method.

With the minor adjustments to the least significant propagate term (P3) and utilizing the EAC from Equations (6a) to (6d), an adder can be designed which always returns a positive magnitude. The carry equation for each group is equal in length and equal to the width of the adder.

Also, note that the last stage of the adder is an exclusive-OR level used to conditionally complement the sum based on the complement of the carry out of the adder for an effective subtraction. This allows the selection of P-B (Equation 1) or B-P (Equation 2).

3.4. Normalization and Leading Zero Anticipation

The next step in the calculation of the fused multiply-add operation is normalization. Rather than waiting for the adder to complete and then performing a leading zero detection (LZD) off the sum, the number of leading zeros is predicted using a leading zero anticipation (LZA) circuit, first implemented on the IBM RS/6000 [20]. Many LZA circuits are discussed in [21]. A common implementation of an LZA is shown.

The equations of an LZA are derived from examining the possible patterns that have to be detected. Assume the equations for a generic adder with inputs A and B, and result, R. Z is used to represent the case when A and B are both zero in a given bit position. H refers to the exclusive OR of A and B which is called the half-sum, and G is the generate of this bit. A case study is shown first starting with the simple case of effective addition followed by three cases for effective subtraction.

3.4.1. Case 1. $A > 0, B > 0, R > 0$

Case 1 is for A, B, and R positive. This case has the following pattern up to the leading one: $Z^+ : Z'$ where + symbolizes 1 or more, * symbolizes zero or more, and A' represents the complement of A. Two examples are given in Tables 2 and 3. The pattern in terms of Z, H, and G terms is shown, as well as the position of the first one in the sum.

By predicting the pattern $Z^+ : Z'$ the result could be off by one bit as shown by the two examples. The leading one will either be at the location of Z' or at the next higher weighted bit position.

3.4.2. Case 2. $A < 0, B < 0, R < 0$

Case 2 is for A and B both negative two's complement numbers. This case will always have the following leading pattern: $G^+ : G'$. An example is given in Table 4. This case may not apply to a multiply-add dataflow since only the

Table 2. Example 1A

A:	0	0	0	0	1	0	1	1	0	0	0	1	1	1	1	0	0	0
B:	0	0	0	0	0	1	0	0	0	0	1	1	1	1	1	0	1	0
Pattern	Z	Z	Z	Z	H	H	H	H	Z	Z	H	G	G	G	G	Z	H	Z
First One					X													

Table 3. Example 1B

A:	0	0	0	0	1	0	1	1	0	0	0	0	1	1	1	0	0	1
B:	0	0	0	0	1	1	0	0	0	0	1	1	1	1	1	1	1	1
Pattern	Z	Z	Z	Z	G	H	H	H	Z	Z	H	H	G	G	G	H	H	G
First One				X														

Table 4. Example 2

A:	1	1	1	1	0	0	0	0	1	1	0	0	0	1	0	1	0	1
B:	1	1	1	1	1	0	0	0	0	1	1	0	1	0	1	1	0	0
Pattern	G	G	G	G	H	Z	Z	Z	H	G	H	Z	H	H	H	G	Z	H
First One					X													

Table 5. Example 3A

A:	0	0	0	0	1	0	0	0	0	1	0	1	1	1	1	0
B:	1	1	1	1	1	0	0	0	1	0	0	1	0	0	0	1
Pattern	H	H	H	H	G	Z	Z	Z	H	H	Z	G	H	H	H	H
First One									X							

Table 6. Example 3B

A:	0	0	0	0	1	0	0	0	0	1	0	1	1	1	1	0
B:	1	1	1	1	1	0	0	0	1	0	1	1	0	0	1	1
Pattern	H	H	H	H	G	Z	Z	Z	H	H	H	G	H	H	G	H
First One								X								

addend is negative, which will probably not cause both the carry and sum to be negative. This case is included to have a complete LZA implementation that could be used for any adder.

3.4.3. *Case 3. A, B have different signs;* R > 0

Case 3 is for one operand positive and the other negative and with a result that is positive. This case will always have the pattern: $H^+ : G : Z^* : Z'$. Note that $H^+ : G = 2^n$. For the result to be positive for an effective subtraction, there must be a carry out which is equal to this term. Examples are given in Tables 5 and 6.

3.4.4. *Case 4. A, B have different signs;* R < 0

Case 4 is for one operand positive and the other negative, with a negative result. This case will always have the form: $H^+ : Z : G^* : G'$. Note $H^+ : Z < 2^n$ implies there is no carry out. This is required to have a negative result. This case

Table 7. Example 4

A:	0	0	0	0	0	1	1	1	1	1	0	0	0	1	0	1
B:	1	1	1	1	0	1	1	1	0	1	1	0	0	1	0	0
Pattern	H	H	H	H	Z	G	G	G	H	G	H	Z	Z	G	Z	H
First One									X							

occurs for the EAC adder design where the sum must be one's complemented. An example of this case is shown in Table 7. The position of the first one could be off by one as with the other three cases. The carry information is not used to fix this one-bit misprediction. Instead, the normalizer checks to see if the most significant bit is a zero after the normalization process. If it is a zero, the result is shifted to the left by one more bit position.

3.4.5. LZA Equations

The four patterns that indicate the leading one are known, and can be utilized to design a predictor. A three-bit pattern is sufficient to indicate when the position of the leading one has been identified. Case 1 suggests a pattern of $Z\colon Z'$ but this won't work because the scenario of Case 4 will fire for $H^+ : Z : G^*$ and not $H^+ : Z : G^*G'$. A pattern is needed that will not fire for $H^+ : Z : G^*$. The sequence $H'_{(i-2)}Z_{(i-1)}Z'_{(i)}$ for three consecutive bits from bit i-2 to bit i, works for Case 1 and does not accidentally fire on Case 4 too early. For Case 2 the same type of prefixing is needed to avoid firing on Case 3, so the sequence: $H'_{(i-2)}G_{(i-1)}G'_{(i)}$ is used. Case 3 can be accounted for by $H_{(i-2)}G_{(i-1)}Z'_{(i)}$ when Z^* occupies zero positions and $H'_{(i-2)}Z_{(i-1)}Z'_{(i)}$ when Z^* occupies one of more positions. Case 4 can be pattern matched by $H_{(i-2)}Z_{(i-1)}G_{(i)}$ when G^* occupies zero positions and by $H'_{(i-2)}G_{(i-1)}G'_{(i)}$ otherwise.

Using the four, three-bit, comparison patterns described above, the following equation can be derived. This equation can be used to generate the vector, which predicts the leading one.

$$\begin{aligned} V_{(i)} = {} & (H'_{(i-2)}Z_{(i-1)}Z'_{(i)}) + (H'_{(i-2)}G_{(i-1)}G'_{(i)}) \\ & +(H_{(i-2)}G_{(i-1)}Z'_{(i)}) + H_{(i-2)}Z_{(i-1)}G'_{(i)}) \end{aligned} \tag{7}$$

The predicted position of the leading one by the LZA vector may be off by one bit. More precisely, an incorrectly predicted position is always one bit to the right of the correct one. Due to the regularity in the incorrect prediction, LZA logic can still be used with minor manipulation. The normalizer can account for this misprediction by checking to see if the most significant bit of the normalized result is a zero. If it is, then the normalizer shifts the result to the left by one more bit.

The LZA vector (V) formed by Equation (7) is formed for each bit in parallel. Typically, another vector (U) is logically ORed to this vector to prevent shifting a denormalized result past the smallest representable exponent. Thus, the shift amount can be capped to not go beyond Emin. After ORing the two vectors together, a LZD is performed to encode the shift amount needed by the normalization [22, 23].

3.4.6. Normalization

The normalizer is formed out of a series of multiplexors that perform shifts of different amounts. In the first level the coarse shift amounts are performed to reduce the width of the dataflow. For instance, there might be a level to choose between an incremented addend's high bits concatenated with the high 53 bits of the product versus selecting all 106 bits of the product. Then the next stage will receive only about 106 bits as input and will perform shifts of multiple of 32 bits such as 0, 32, 64, or 96 bits shifted to the left. The following stage might choose between shifts of 0, 8, 16, and 24 to the left. Futhermore, the last stage shifts between 0, 1, 2, 3, 4, 5, 6, and 7 bits to the left. An extra stage is needed with an inexact LZA as previously suggested. This stage is a 2 to 1 multiplexor that receives as its select line the most significant bit of data from the previous stage, and chooses between a shift of zero or a shift of 1 bit to the left.

3.5. Rounding

The rounder design is fairly simple. At this stage of the dataflow a truncated intermediate result has been computed. Rounding will pick between the two closest machine-representable numbers to the actual value. The two choices are either this truncated result itself or incremented in the least significant bit. The rounder needs to have an incrementer which can increment at different bit positions. For instance, if single and double precision are supported then the upper 24 bits and 53 bits respectively need to be incremented. An incrementer can be developed from the equations of an adder with one operand equal to zero and the carry in set to one. Then all the generate terms are equal to zero and all the propagate terms are an AND string of all the bits of the operand being incremented.

Following the incrementer, is a multiplexor which selects either a truncated or incremented result. The choice between truncated and incremented result is dependent on the result sign, rounding mode, least significant bit (L), guard

Table 8. Rounding table

Rnd mode	Sign	L	G	I	Action
Nearest			0		Truncate
Nearest		0	1	0	Truncate
Nearest		1	1	0	Increment
Nearest			1	1	Increment
Zero					Truncate
+ Infinity	+		0	0	Truncate
+ Infinity	+			1	Increment
+ Infinity	+		1		Increment
+ Infinity	−				Truncate
− Infinity	−		0	0	Truncate
− Infinity	−			1	Increment
− Infinity	−		1		Increment
− Infinity	+				Truncate

bit (G), and inexact sticky bit (I) as shown in Table 8. Either combined with this multiplexor or separate is the choice between formats.

4. Evolving Design

The design of a fused multiply-add dataflow has changed since it was first introduced in 1990. The most efficient adder design is now an EAC adder. The multiplier sign extension encoding is suppressed, and the LZA algorithms have been refined. There have been advancements in minimizing the delay of denormal operands as well as for the cases of underflow and overflow, and the counter tree designs have evolved.

Future designs will need to consider advances in counter tree designs as well as many other factors. Some futuristic designs have been explored that separate the addition path into two paths: a far and near path [24, 25]. These designs reduce the latency by reducing the number of shifts in the critical path, but at the cost of increased area.

Other research possibilities exist in trying to design a fused multiply-add dataflow for decimal floating-point format. The first problem in this area is how to design efficient decimal counters. This design area is fertile since it was last explored in the 1950s and is being reinvestigated due to the new decimal format defined in the IEEE 754R standard.

Hopefully these future advances will be documented and not become tricks that are known only by the implementers. Eventually the folklore will turn into literature, as has been finally accomplished for current binary floating-point unit designs.

Acknowledgements

Some of the algorithms described were learned through very experienced designers and educators such as Martin Schmookler and Stamatis Vassiliadis. The author is indebted to them for having the opportunity to learn and work with them.

References

[1] "IEEE standard for floating-point arithmetic, ANSI/IEEE Std 754R," The Institute of Electrical and Electronic Engineers, Inc., In progress, http://754r.ucbtest.org/ drafts/ 754r.pdf .

[2] Knuth, D. "*The Art of Computer Programming*, Vol. 2: *Seminumerical Algorithms*, 3rd ed." Addison-Wesley, Reading, MA, **1998**, 467–469.

[3] Montoye, R.K.; Hokenek, E.; Runyon, S.L. "Design of the IBM RISC System/6000 floating-point execution unit", *IBM J. Res. Dev.*, **1990**, *34*(1), 59–70.

[4] "Enterprise Systems Architecture/390 Principles of Operation", Order No. SA22-7201-5, available through IBM branch offices, Sept **1998**.

[5] Waser, S.; Flynn, M.J. *Introduction to Arithmetic for Digital Systems Designers,* Holt, Rinehart, & Winston, **1982.**

[6] "IEEE standard for binary floating-point arithmetic, ANSI/IEEE Std 754-1985," Institute of Electrical and Electronic Engineers, Inc., New York, Aug. **1985.**

[7] Intel Corporation, "Intel Itanium Architecture Sofware Developer's Manual, Volume 1 Application Architecture," ftp://download.intel.com/design/Itanium/Downloads/ 24531703s.pdf, Dec. **2001.**

[8] Intel Corporation, "IA-32 Intel Architecture Sofware Developer's Manual, Volume 1: Basic Architecture," ftp://download.intel.com/design/Pentium4/manuals/ 24547008.pdf, **1997**.

[9] Schwarz, E.; Schmookler, M.; Dao Trong, S. "FPU implementations with denormalized numbers", *IEEE Trans. Computers*, **2005**, *54*(7), 825–836.

[10] Schwarz, E.; Schmookler, M.; Dao Trong, S. "Hardware Implementations of Denormalized Number Handling", *Proc. 16th IEEE Symp. on Computer Arith. Metic*, June **2003**, 70–78.

[11] Booth, A.D. "A signed multiplication technique", *Q. J. Mech. Appl. Math.*, **1951**, *4*(2), 236–240.

[12] Vassiliadis, S.; Schwarz, E.; Hanrahan, D. "A general proof for overlapped multi-bit scanning multiplications", *IEEE Trans. Computers*, **1998**, *38*(2), 172–183.

[13] Vassiliadis, S.; Schwarz, E.; Sung, B. "Hard-wired multipliers with encoded partial products," *IEEE Trans. Computers*, **1991**, *40*(11), 1181–1197.

[14] Wallace, C.S. "A suggestion for parallel multipliers", *IEEE Trans. Electron. Comput.*, **1964**, *EC-13*, 14–17.

[15] Dadda, L. "Some schemes for parallel multipliers", *Alta Frequenza*, **1965**, *34*, 349–356.

[16] Weinberger, A. "4:2 carry-save adder module", *IBM Technical Disclosure Bull.*, **1981**, *23*, 3811–3814.

[17] Ohkubo N.; *et al.* "A 4.4 ns CMOS 54 $\times$ 54-b multiplier using pass-transistor multiplexer", *IEEE J. Solid-State Circuits*, **1995**, *30*(3), 251–257.

[18] Richards, R.K. *Arithmetic operations in digital computers*, D. Van Nostrand Co., Inc., New York, 120, **1955**, 120.

[19] Beaumont-Smith, A.; Lim, C. "Parallel prefix adder design", *Proc. 15th IEEE Symp. Comp. Arith.*, Vail, June **2001**, 218–225.

[20] Hokenek, E.; Montoye, R.K. "Leading-zero anticipator (LZA) in the IBM RISC System/6000 floating-point execution unit", *IBM J. Res. Dev.*, **1990**, *34*(1), 71–77.

[21] Schmookler, M.S.; Nowka, K.J. "Leading zero anticipation and detection – a comparison of methods", *Proc. 15th IEEE Symp Computer Arithmetic*, Vail, 11–13 June, **2001.**

[22] Oklobdzija, V. "An implementation algorithm and design of a novel leading zero detector circuit", *Proc. 26th Asilomar Conf. on Signals, Systems, and Computers*, **1992**, 391–395.

[23] Oklobdzija, V. "An algorithmic and novel design of a leading zero detector circuit: comparison with logic synthesis", *IEEE Trans. on VLSI Systems*, **1993** *2*(1), 124–128.

[24] Seidel, P.M. "Multiple path IEEE floating-point fused multiply-add", *Proc. 46th Int. IEEE Midwest Symp. Circuits and Systems (MWS-CAS)*, **2003.**

[25] Bruguera, J.D.; Lang, T. "Floating-point fused mulipy-add: reduced latency for floating-point addition", *Proc. 17th IEEE Symp. Computer Arithmetic*, Hyannis, 27–29 June, **2005.**

Chapter 9

MICROPROCESSOR ARCHITECTURE FOR YIELD ENHANCEMENT AND RELIABLE OPERATION

Hisashige Ando
Fujitsu Ltd., 4-1-1 Kamikodanaka, Nakahara-ku, Kawasaki, Japan 211-8588

Abstract: With the advance of semiconductor scaling, smaller devices become more vulnerable to an SEU (single event upset, i.e. neutron hit etc.) and operating margin of the circuits is reduced both due to reduced operating voltage and larger process variations. Robust circuit design alone cannot solve these problems. Micro-architectural techniques for avoiding defects and error detection and correction microarchitecure can significantly reduce the probability of failure and enhance the yield and the reliable operation of a microprocessor. The failure mechanisms of nanometer class semiconductor VLSI circuits are described as a background. Then concept and methods of error detection and correction are described, followed by microarchitecture/logic design error detection and recovery techniques. Commercial microprocessors using error detection and recovery techniques are also presented.

Key words: redundancy; error detection; error correction; checkpoint; microprocessor.

1. Introduction

Microprocessors have been taking advantage of continuing semiconductor scaling, from Intel 4004 microprocessors introduced in 1971 containing only 2300 transistors to recent processors integrating more than 1 billion transistors. Semiconductor scaling reduced the minimum feature size of mass-production integrated circuits down to 90 nm in 2004 and the scaling trend is expected to continue down to 22 nm in 2016.

Vojin G. Oklobdzija and Ram K. Krishnamurthy (eds.),
High-Performance Energy-Efficient Microprocessor Design, 209–233.

However, CMOS semiconductor scaling becomes harder and harder as it nears the end of scaling due to fundamental physical limits. As the feature size becomes smaller, a smaller dust particle can cause fatal defects, which lowers the yield of a chip. Also smaller features are harder to control and the electrical parameter variation in the chips becomes larger. This also lowers the yield of the chips.

Smaller transistors and reduced supply voltage reduce the operating margin of the circuits as the noise margin is mostly proportional to the supply voltage and the parameter variation of the smaller devices is larger. Electrostatic charge stored in a circuit node decreases rapidly as device sizes and voltage are reduced. This makes circuits more susceptible to an alpha particle and a neutron hit as the charge generated by a hit is constant and does not scale with the shrinking geometry.

Circuits must be designed to accommodate larger device parameter variations, but it is inefficient to design circuits robust enough to accommodate very large parameter variations or to withstand neutron hits.

Microarchitectural approaches, for repairing fixed errors and recovering from transient errors, are becoming important means to enable microprocessors and/or other VLSI systems to be manufactured economically and to operate reliably.

2. Semiconductor Scaling Issues

2.1. Device Parameter Variation

In a photolithographic process, the intensity of the light, focus, sensitivity and thickness of photo-resist etc. affect the size of the patterns exposed. In other processes, variations in the mixture and flow rate of gases, process temperature and so on, affect the thickness of a film or diffusion of impurities. Although equipment manufacturers and semiconductor engineers are working hard to reduce these variations, it is a tough battle to keep device parameter variation small in line with the continuing scaling.

As the feature size becomes smaller than the wavelength of light, the interference of light affects the intensity of exposure and distorts the pattern to be exposed. OPC (optical proximity correction) is used to compensate this distortion, but the residual distortion becomes the source of device parameter variation [1–3].

The light used for the photolithographic process is a beam of photons. Near the edge of a line some roughness cannot be avoided as the photons hit the photoresist like bullets. As the line width (which defines gate length) becomes narrower, rough edges occupy an increasing percentage of the line width [4]. As the length of the transistor channel becomes a few tens of nanometers or less, the number of impurity atoms in a transistor channel becomes smaller and causes a larger statistical variation of the threshold voltage [5]. These statistical parameter variations cannot be avoided and make relative device parameter variations larger as the scaling progresses.

2.2. Internal Noise

The power density of microprocessors is increasing rapidly [6, 7] as the scaling increases transistor and wire capacitance per unit area roughly inversely proportional to the feature size shrink. The clock frequency of a microprocessor increases as the smaller transistor switches faster. The improvements in circuit design, logic design and microarchitecture design further help to increase clock frequency.

The power supply current of a microprocessor chip increases faster than the power density increase as the supply voltage decreases with the semiconductor scaling. This increase in supply current results in a larger IR drop and even larger L*di/dt noise as dt decreases with the faster switching transistors [8, 9].

Capacitive coupling noise may increase due to the thick wires needed to reduce wire resistance.

On the other hand, the noise margin of a static complementary CMOS circuit decreases with the reduced supply voltage and the larger device variations as the noise margin is expressed as $(\mathrm{Vdd} - V_t)/2$ where V_t is a threshold voltage variation. Analog circuits are more sensitive to device parameter variation and they also have lesser noise margins than complementary CMOS logic circuits.

In addition to these noise sources, chip temperature varies as circuit activity varies. This causes local temperature variation on a chip [10]. Since the temperature affects the transistor parameters and resistivity of a metal wire, the temperature variation further reduces the noise margin [11, 12].

2.3. Single Event Upset

When an energetic atomic or subatomic particle hits a silicon substrate and collides with a silicon atom, electron and hole pairs are generated [13, 14]. The generated electrons are attracted to a positively biased nearby N+ drain when the hit occurs in an NMOS region. On the other hand, the holes are attracted to a negatively biased P+ drain when it hits a PMOS region. This charge injection occurs in a short period of time, usually less than 1 ns, and is equivalent to a current pulse injected into the drain nodes. This pulse causes the noise spike proportional to the amount of charge injected and inversely proportional to the parasitic capacitance of the node.

If the noise exceeds the threshold voltage of the receiving gate by a certain amount or more, it will be amplified, and it can flip the state of a memory cell or a latch [15]. The minimum amount of charge causing the flip is called a critical charge. The frequency of failure is dependent on the design of the circuit and semiconductor process used.

With the improvement in removal of radioactive isotopes from the materials used for chip and package fabrication, the alpha particle hit became less significant and the cosmic neutron hit became the dominant cause of failure.

The semiconductor scaling makes each RAM cell/latch become more sensitive to a more abundant lower-energy particle hit as the operating voltage and parasitic capacitance are reduced. But the scaling reduces the probability of hit as the target size becomes smaller. References 16 and 17 report that the frequency of a single bit failure is relatively constant with the scaling as these two factors mostly cancel out. However, the frequency of a failure per chip increases with the scaling as the number of bits per chip increases.

When the noise is large enough it can also affect logic gates. However, propagation of the noise through a logic gate is affected by the state of the other inputs of the gate. The arrival of the noise to a receiving flip-flop must also be within a narrow latching window of the receiving circuit to cause an error. Due to these masking effects, a radiation particle hit to a combinatorial logic gate is far less likely to manifest as an error than the hit to a storage element [18].

3. Redundancy and Repair

3.1. Yield Improvement with Redundancy

Yield (fraction of good chips among the total chips manufactured) is a function of defect density and the size of a chip. Assuming the distribution of the defects is random, the yield of a chip is expressed as follows.

$$Y_c = e^{-(D_0 \times A)} \tag{1}$$

where D_0 is the defect density and A is the size of the chip. With a clustered defect model used in ITRS2003 roadmap [19], the yield is expressed as follows:

$$Y_c = \left(1 + \frac{D_0 \times A}{\alpha}\right)^{-\alpha} \tag{2}$$

where α is a cluster factor and $\alpha = 2$ is used in ITRS2003 roadmap.

If we build $N + 1$ set of identical units and allow one unit to be defective, by eliminating one bad unit by some means after fabrication of the chip, the yield of the chip can be improved.

The yield of each unit is expressed as follows:

$$Y_u = \left(1 + \frac{D_0 \times A/N}{\alpha}\right)^{-\alpha} \tag{3}$$

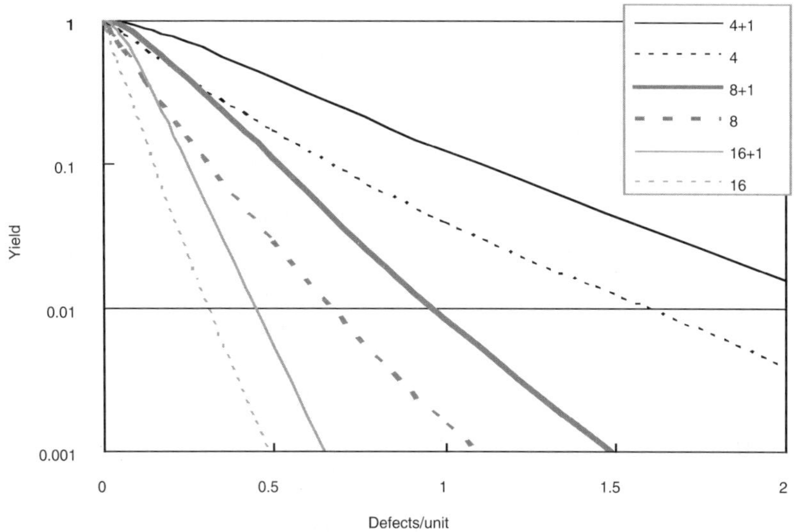

Figure 1. Yield with and without redundancy.

The yield of the chip with one redundant unit is expressed as follows:

$$Y_c = Y_u^{(N+1)} + (N+1) \times Y_u^N \times (1 - Y_u) \quad (4)$$

$$= Y_u^N \times (1 + N \times (1 - Y_u)) \quad (5)$$

$$= \left(1 + \frac{D_0 \times A/N}{\alpha}\right)^{-\alpha N} \times \left(1 + N \times \left(1 - \left(1 + \frac{D_0 \times A/N}{\alpha}\right)^{-\alpha}\right)\right) \quad (6)$$

The first term in the equation 6 is Y_u^N and is the yield without redundancy. Hence the yield gain of the $N + 1$ redundancy is $1 + N(1 - Y_u)$. When Y_u is very close to 1.0, the gain is small. When Y_u is low and N is large, the yield improvement is significant as shown in Figure 1. However, when Y_u is too low or N is too large, Y_u^N becomes extremely small and the resulting yield Y_c is uninteresting even with the large improvement. It is important to pick the correct combination of Y_u and N for $N + 1$ redundancy to be beneficial.

3.2. RAM

Since random access memories have a regular array of storage cells, they are a natural candidate for the use of redundancy-based yield enhancement. A column redundancy [20] is the most widely used redundancy technique for

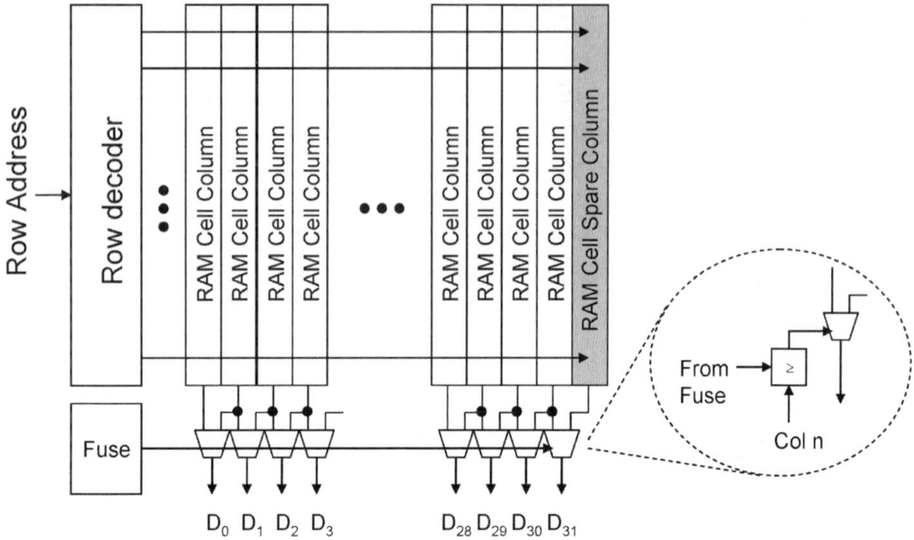

Figure 2. RAM column redundancy scheme.

RAM. As shown in Figure 2, one spare column for every 32 columns and a multiplexer is added to the output of each column to select either the read signal from its own column or the one from the adjacent column on the right. The multiplexer selects the right-hand input when the column number input is greater than or equal to the fuse input. Assuming column i has a defective memory cell, the multiplexers in column 0 to $i-1$ select their own column read signal. However, the multiplexers in column i to 32 select the right-hand side adjacent column and avoid using the defective column i.

Identifying a defective column and substituting a good spare is done when the manufacturing test is performed. The information on which column to be avoided is usually stored in the programmable (by laser or electronically) fuses on the same chip.

3.3. Logic Blocks

The application of a redundancy to logic blocks is more difficult than RAM since it is usually not the repetition of the same blocks. Even with the repeated logic blocks, multiplexer switch circuits which remove a defective logic block may become significant in size compared to the logic block itself when it has large number of interface signals. In this case the yield of the multiplexer reduces the yield gain obtained from the redundancy. Also the testing for each logic block is a challenge when it has a large number of signals and the function is complex.

For this reason large logic blocks with a well-defined compact interface, e.g. processor cores, are a promising candidate for a block redundancy scheme.

As the power dissipation issue of a microprocessor is becoming the limit of an increase in processor core size, a chip integrating multiple smaller processor cores is becoming more desirable [21, 22]. It is practical to apply a redundancy scheme for yield improvement of a chip multiprocessor having many cores.

The IBM z900 mainframe processor MCM (multi-chip module) contains 10 dual-core processor chips which totals 20 processor cores, but uses only up to 16 of them for data processing and keeps four of them for redundancy [23]. A game console using a cell processor integrating eight synergistic processor elements [24] is going to use seven of them for processing, and keeps one as a redundant spare for yield enhancement.

The Intel Itanium 2 processor uses block redundancy for its third level on-chip cache [25]. It uses 32 SRAM blocks for the data storage, two blocks for the ECC (error correcting code) and one block for the spare. The defects in decoders and sense amplifiers can be tolerated with use of this block-based redundancy scheme whereas the column-based scheme described in Section 3.2 can only repair cell or column-related defects.

4. Error Detection and Correction

As the noise margin of circuits decreases and SEU due to a neutron hit become more frequent, a malfunction needs to be detected for preventing the output of an incorrect result without any warning. The correction upon the detection of an error is more desirable than the detection as it allows the system to run continuously.

4.1. Redundancy for Error Detection and Correction

To detect an error, redundancy is required. There are many ways to add redundancy and they are categorized as follows:

- Information redundancy
- Hardware redundancy
- Time redundancy

4.1.1. Information redundancy

For a block with multi-bit output signals, redundancy can be added to make all correct outputs different from the wrong outputs. The most frequently used information redundancy is a parity check. Odd parity check adds one redundant check bit to a collection of data bits and sets the value of the check bit to make the total number of the "1" bits odd for correct outputs, whereas the wrong outputs have an even number of "1" bits.

A 2 out of ($_5C_2$) code for decimal numbers is another example. Any correct output has two "1"s whereas the number of "1"s in wrong outputs can be any number but two. Since four bits are sufficient to represent a decimal number, it has redundancy in information; but there is no specific redundant bit like a parity bit.

The history of error detection and correction codes is long and many codes have been invented. Some of them are described in Sections 4.2 and 4.3; also see ref. 28 for more codes.

4.1.2. Hardware redundancy

A duplicate and compare scheme uses two identical logic blocks, one for master and the other for checker, and comparing both outputs for error detection. For a triple modular redundancy (TMR) error correction, three identical blocks are used and a majority voter at the output guarantees correct output unless more than one block produce the same error, which is an extremely low probability event.

Figure 3(a) shows a duplicate and compare scheme. This configuration can detect errors in one module, but, cannot correct an error. Figure 3(b) shows a configuration using two duplicate compare blocks and a switch controlled by the error indication signals from both units to select which block to drive the output with. This configuration can mask the occurrence of an error and continuously produces correct results. Figure 4(a) shows a TMR configuration. The outputs from three identical modules are fed into a majority voter. Assuming at most

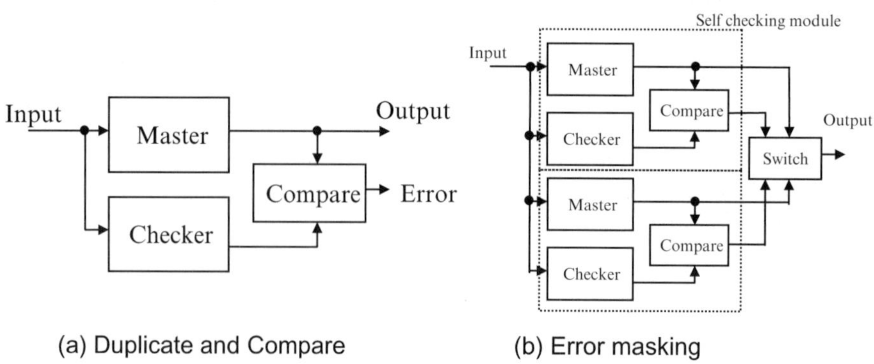

Figure 3. Duplicate and compare.

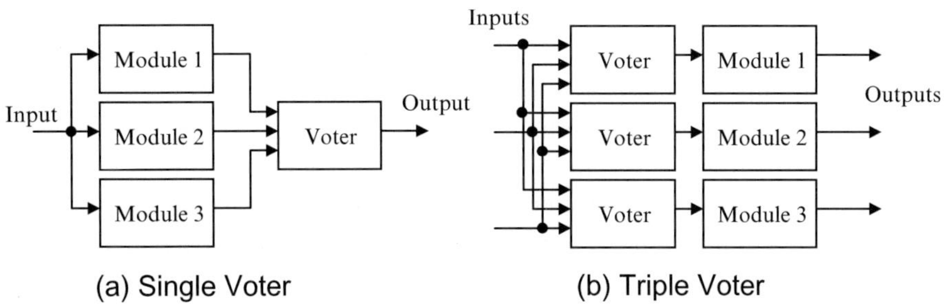

Figure 4. Triple modular redundancy.

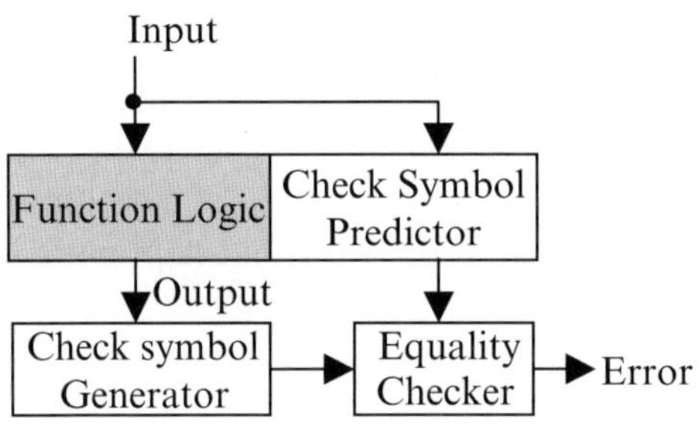

Figure 5. Algorithmic check.

one block generates an erroneous output the triple input majority voter always produces correct output; but this configuration is vulnerable for faults at the common inputs of the modules and the outputs of the voter. Figure 4(b) shows a more reliable configuration with three voters. This configuration is safe for any single point of failure [26].

These hardware redundancy schemes are simple and straightforward, but the overhead is high as they require twice to four times the number of modules and the additional compare or voter circuits.

Algorithmic check is a way to reduce high overheads of duplication. It compresses the output of a target logic block with the use of some algorithm and having a logic block which calculates the same compressed output independently from the target logic block.

An example shown in Figure 5 is the one having a check symbol generator for the output compression from a target function block and having a check symbol predictor which calculates the expected check symbol of the function block output independently from the function logic itself. A detection capability of an algorithmic check depends on the scheme used. For example, when the parity check is used for the check symbol generation, a class of errors with even number of output bits flip cannot be detected. On the other hand, a modulo-3 check can detect any single point of error in a partial sum adder in a multiplier.

4.1.3. Time redundancy

Assuming the nature of the fault is transient, such as SEU, performing the same calculation again and comparing both results will detect an error if it exists. If they differ, the system will do the same calculation one more time to get the correct result. These are equivalent to duplicate and compare or TMR, but, the replication is done in the time domain instead of hardware.

Simple time redundancy cannot detect fixed faults as they give the same wrong results for the two comparing calculations. This problem can be avoided by using two processors and making the first and second calculation in the different processors. When all the calculations are duplicated, it is similar to the duplicate and compare hardware redundancy.

Time redundancy is often implemented in a software layer. It generates a pair of threads or processes for the code sections which require high reliability and an operating system assigns one thread each to a processor. For the rest of the code sections two processors can be used as a multiprocessor system with each processor doing different processing and improving the total throughput of the system.

4.2. Error Detecting Codes

Various error detecting codes were designed for different purposes, for example, random bidirectional error detecting codes, unidirectional error detecting codes, burst error detecting codes and so on. For the error detection in a microprocessor, bidirectional error detecting codes are the most frequently used as both "0" to "1" and "1" to "0" errors can occur. For some circumstances unidirectional error detecting codes may be useful as those codes can detect any number of bits of unidirectional error with a relatively small number of check bits compared to bidirectional error detecting codes.

4.2.1. Parity check

The Hamming distance between the two N-bit binary numbers is defined as the number of differing bit positions. Mathematically, this is the Manhattan distance between the two points corresponding to those binary numbers in N-dimensional binary space. Since an m bit error moves a position of a code word by distance m from its original position, an m bit error cannot change one code word into another code word with distance $m + 1$ or more. So any error with m or less bit flips can be detected if we can choose a set of code words with minimum distance $m + 1$ between any two code words.

Parity check adds one check bit to data to make the total number of "1"s to even (even parity) or odd (odd parity). With even parity, any code word (data + parity bit) has an even number of "1"s. Since any one bit change makes the number of "1"s to odd, which is not a code word, any single bit error can be detected. Parity check is a simple way to make the minimum distance between any two code words to two.

4.2.2. All unidirectional error detection code

The Berger code [27] is an all unidirectional error detection (AUED) code. It adds a count of the number of "0"s in data bits as the check bits. For example, when the data bits are 0011001, the number of "0"s is four, hence the check bits are 100. Any error with a direction of "0" to "1" occurring in data bits reduces the "0" count, whereas any error with the same direction in check bits increases the binary representation of the count. Hence they do not match. Any error in reverse direction can also be detected.

Constant weight code, for example ${}_5C_2$ code has two "1"s in a 5bit code word. This code also is an AUED code since any unidirectional error can increase or decrease the number of "1"s from the original two.

4.3. Error Correcting Codes

The minimum distance between any two code words must be $2m + 1$ for m bit error correction. Since m bit error moves one code word by distance m, a data word with an error is contained in a (hyper) sphere with radius m centered at the original code word. Any two spheres with radius m do not overlap since the minimum distance between any two code words is $2m + 1$. So any data word with less than or equal to m bit error can be identified with the code it originated from.

Hamming code is a minimum distance 3 code generated as follows. As shown in Figure 6, each column of the Hamming code H matrix is a binary representation of the column number. The columns with single "1" are check bits (C_0 to C_2) and remaining columns are data bits (D_0 to D_3). The first row of the H matrix contains "1" in C_2 and D_1 through D_3. This means that the EOR of these four bits must be "0", or $C_2 = D_1 \oplus D_2 \oplus D_3$. The second and third rows define C_1 and C_0 respectively.

When an error occurs at bit position i, for example bit position 6, the EOR for some H matrix row(s) will not be "0". Since the row 1 and row 2 contain "1" in column 5, row-wise EOR will be 110. This indicates that bit position 6 is flipped and this makes the correction possible. A set of these row-wise EOR results is called a syndrome.

$$
\begin{array}{ccccccccc}
 & & C_0 & C_1 & D_0 & C_2 & D_1 & D_2 & D_3 \\
 & & \downarrow & \downarrow & & \downarrow & & & \\
 & & 1 & 2 & 3 & 4 & 5 & 6 & 7
\end{array}
$$

$$
H = \begin{bmatrix} 0 & 0 & 0 & 1 & 1 & 1 & 1 \\ 0 & 1 & 1 & 0 & 0 & 1 & 1 \\ 1 & 0 & 1 & 0 & 1 & 0 & 1 \end{bmatrix}
$$

Figure 6. Hamming code H matrix.

Table 1. SECDED code length

	Total length	Check bits	Data bits
General relationship	$\leq 2^N$	$N + 1$	$<2^N - N$
Common use	72	8	64
	137	9	128

Two bit error moves a code word by distance two and can make it within distance one from some other code word which is distance three from the original code word. In this case, Hamming code mistakenly corrects it to the other code word without giving any indication of a mistake. Since this is highly undesirable, a code which can correct single bit errors and also detect double bit errors is commonly used. This type of code is called a SECDED (single error correction double error detection) code.

Hamming code can be extended to make it detect double bit errors by adding an all "1"s row and a column with all "0"s except the added row to the original H matrix. The all "1" row is an even parity for all code word bits. If this parity check for the received word fails, the syndrome should indicate the failed bit position as described above. When the parity check of all received word is correct and the syndrome is non-zero, it is an indication of an even number of bits flipped.

Although the construction of the Hamming code is clever but conceptually simple, finding a set of code words with minimum distance $2m + 1$ by trial and error is almost impossible. The modern SEC (single error correction) and SECDED codes are mostly constructed using abstract algebra [28].

There are many SEC or SECDED codes with the same correction capability, but some of them, for example the Hsiao code [29], which contains a smaller number of "1"s in H matrix, the hardware required for encoding and error detection is simpler than the Hamming code.

Table 1 shows the total length, number of check bits and usable data bits for optimal SECDED codes. For the 64 bit data, 8 check bits are required and the resulting total code length is 72 bits.

For chips with multiple output bits a single physical fault can flip multiple bits. For example, a failure in power supply connection makes all outputs go to ground level. To cope with this situation the codes which can detect and correct an error with any number of bits as long as all these errors are within one block of bits (called byte regardless of the number of bits) were developed. These codes are called SbEC (single byte error correction) or SbECDbED (single byte error correction and double byte error detection) codes. The Kaneda–Fujiwara [30] code can make a S8ECD8ED code for 64 bit data with 14 check bits. Both the Kaneda–Fujiwara code and the Reed–Solomon code [31] can make an S8EC code for 128 bit data with 16 check bits. These codes are used for chip-kill error correction for 8 bit output DRAM chips.

5. Error Detection and Correction in Microprocessors

5.1. Error Detection in Data Path

Figure 7 shows a simplified diagram of a microprocessor data path. The data is read from a memory and stored into a register file entry through the cache and the load store unit. Both the value and the representation of the data are not altered in this path.

Parity check is an efficient way to detect errors in the data path without data alteration. A parity generator is placed at the point where new data values are generated; at the output of the function unit in Figure 7. As the other blocks store and/or transfer the data without alteration, only a parity checker is placed at the input of each block.

DRAMs and caches contain a large number of storage elements and the failure rate of these blocks is high compared to the rest of the microprocessor

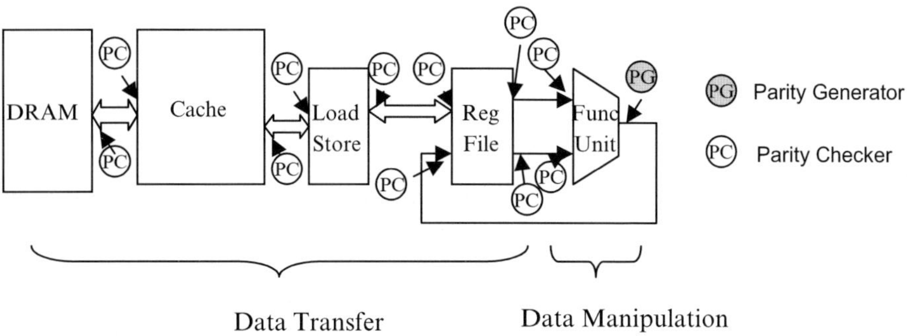

Figure 7. Parity check for the data path.

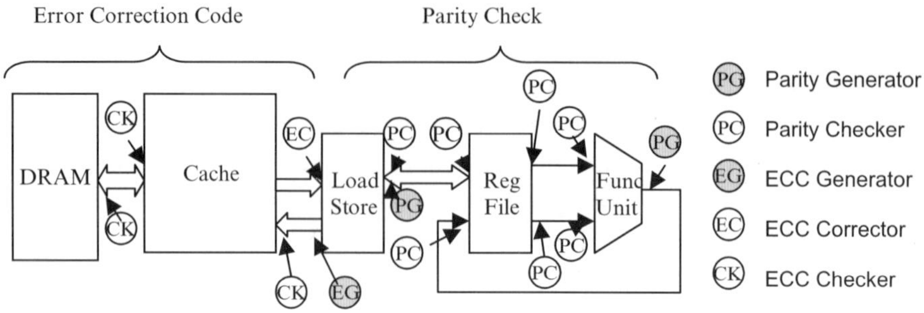

Figure 8. Microprocessor data path with memory ECC.

system. For this reason the use of an error correction code, like the SECDED code described in Section 4.3, is common. As shown in Figure 8, ECC check bits are generated at the output of the store unit and stored into the cache and the DRAMs. For the read, error correction is done at the input of the load unit. The rest of the processor data paths are checked by the parity check as the timing in the processor core is tight and it is difficult to fit an error correction stage in the pipeline. The output of a load unit is a generator of the new data and a parity generator is added.

An ECC checker is added to the inputs of the cache and DRAMs for detecting data corruption. When an uncorrectable double bit error is detected at the DRAM and/or cache interface checkers, the data is changed to a special pattern without raising an error interrupt. The special data pattern indicates where the error is detected and the check bits are arranged to give a special syndrome as an error flag. An error is raised when the flagged data is read into the load unit for use. This technique, called data poisoning, is effective in avoiding unnecessary error detection and interrupt when the corrupted data is not used by the processor core.

5.2. Error Detection for Arithmetic Units

As the number of ones changes from an input of an arithmetic unit to an output, a simple parity check or ECC cannot be used to check an occurrence of an error in the arithmetic unit.

The duplicate and compare gives high detection coverage, but it requires a checker unit and a compare circuit which more than doubles the amount of required hardware. For reducing the amount of hardware required, an algorithmic check can be used.

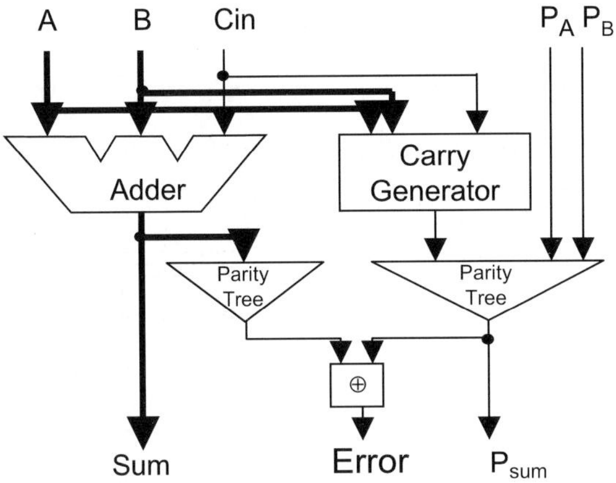

Figure 9. Adder with parity check.

5.2.1. Adder with parity check

The sum of each bit is expressed as $S_i = A_i \oplus B_i \oplus C_i$, where A_i and B_i are the ith input bits and the C_i is the carry into the ith bit. The parity of the whole n-bit sum is expressed as follows:

$$
\begin{aligned}
P_{\text{sum}} &= S_0 \oplus S_1 \oplus \ldots \oplus S_n \\
&= A_0 \oplus B_0 \oplus C_0 \oplus A_1 \oplus B_1 \oplus C_1 \oplus \ldots \oplus A_n \oplus B_n \oplus C_n \\
&= A_0 \oplus A_1 \oplus \ldots \oplus A_n \oplus B_0 \oplus B_1 \oplus \ldots \oplus B_n \\
&\quad \times \oplus C_0 \oplus C_1 \oplus \ldots \oplus C_n \\
&= P_A \oplus P_B \oplus C_0 \oplus C_1 \oplus \ldots \oplus C_n
\end{aligned} \tag{7}
$$

This equation (7) means that the parity of the sum, P_{sum} is calculated with the binary sum of the parity of the input A, P_A and input B, P_B and all the carry bits. These carry bits can be calculated with a carry generator, like the ones used in a carry look-ahead adder. The predicted parity P_{sum} and the parity calculated from the sum itself are compared as shown in Figure 9 for the error detection.

This error check is equivalent to checking an error at the output of the adder and cannot detect a class of faults in the adder which causes an even number of bit flips in the sum. The detection capability can be improved by choosing a smaller parity check unit, for example, 8 bit instead of whole 64 bit.

There are other more complete self-checking adder designs, but, they require more hardware [32, 33].

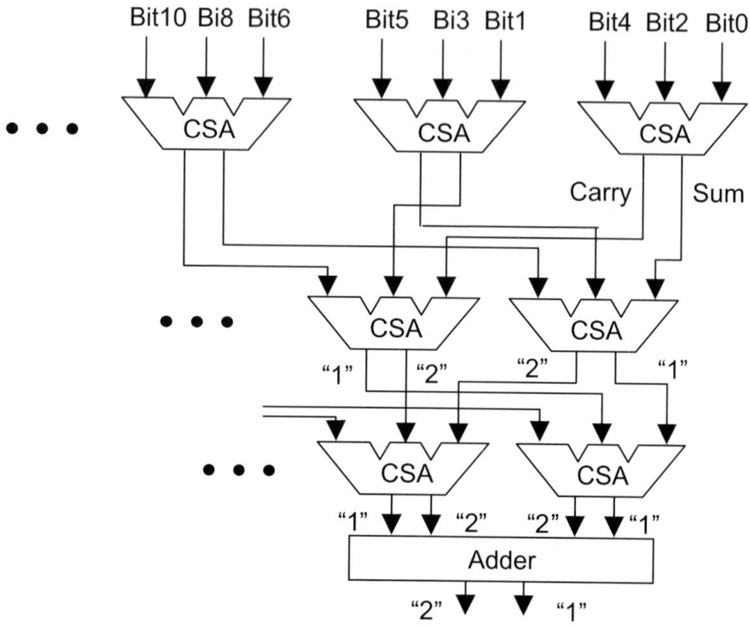

Figure 10. CSA tree for mod_3 calculation.

5.2.2. *Multiplier with residue check*

With the following arithmetic relation a multiplier can be checked with the calculation of residue:

$$\mathrm{Mod}_p(A \times B) = \mathrm{Mod}_p(\mathrm{Mod}_p(A) \times \mathrm{Mod}_p(B)) \tag{8}$$

Mod_p is a Modulo operation with a prime number p. Since Mod_p is the residue with the division by p, this method is called a residue check.

Mod_p can be calculated as follows without using a divider. When the bit is set, mod_3 of the least significant bit (bit 0) is 1, mod_3 of the next bit 1 is 2, mod_3 of the bit 2 is 1 and mod_3 of the bit 3 is 2 and so on. Mod_3 of a binary bit is an alternating sequence of 1 and 2. Mod_7 of a binary bit is a repeating sequence of 1, 2 and 4. Using this relationship, mod_3 can be calculated with a tree of carry save adders as shown in Figure 10.

An error in a multiplier can be detected by implementing Equation (8) as shown in Figure 11.

This error detection is equivalent to the residue check error detection for the output of the multiplier and it cannot detect any faults resulting in the same mod_3 output values. This is possible in cases when the input A is zero with mod_3, and all the stuck- at faults at the input B (or the ones equivalent to) cannot be detected since the mod_3 of the product is zero regardless of the

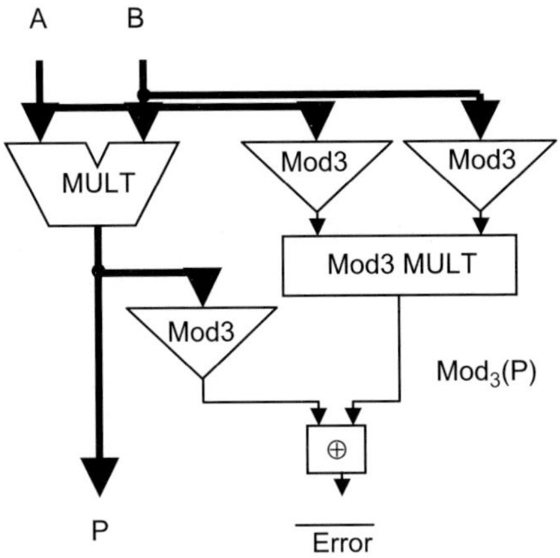

Figure 11. Multiplier with the residue check.

B input value. The same is true for the A input. This shortfall is corrected by adding a parity check to the inputs and the check for the booth encoder in the multiplier. This complete checker configuration is described in ref. 34.

Both the last stage adder in Figure 10 and the mod_3 multiplier in Figure 11 are simple combinatorial circuits with four inputs and two outputs.

5.3. Error Detection for Control Circuits

Duplicate and compare is often expensive for control circuits since they need a large number of compare circuits for the outputs whereas the amount of control logic is relatively small.

Control circuits can be checked with an assertion as shown in Figure 12. With the introduction of information redundancy in a state assignment and the state transition logic which makes transition from one of the correct states into a correct state, an error can be detected as a manifestation of a wrong state [35]. Although this scheme does not cover the decode logic block, it is effective as the neutron hit failure rate of latches is higher than combinatorial circuits.

5.4. Other Error Detection Methods

Based on knowledge of the microarchitecture, some apparently anomalous condition can be detected as an error by assertion. A watchdog timer is an error detection mechanism in this category. A timer is reset at the issue of any instruction and the timer is counted up with the processor clock. If the issue of the next instruction occurs within the predetermined number of cycles, the timer is reset again and will not fire. But if something goes wrong and the next

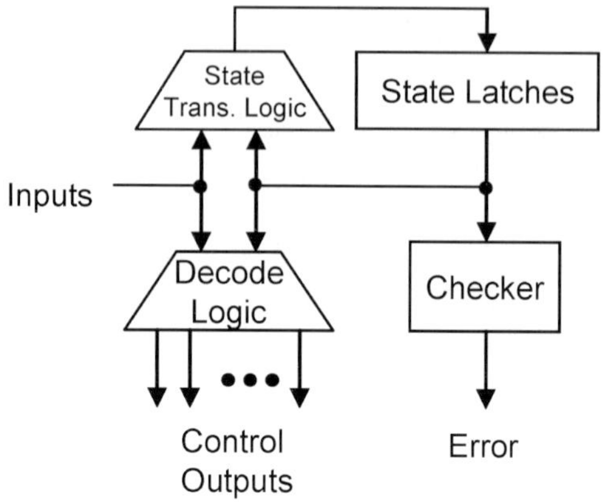

Figure 12. Error detection with redundancy check.

issue did not happen within a certain number of clock cycles, the watchdog timer fires and reports an error.

A special logic can be designed to check the correctness of the state of a microprocessor based on the knowledge of microarchitecture and its correct operation, like the watchdog timer; these checkers in general are called assertions. Assertion can also be implemented in software [36]. Strategically placed assertion checkers are useful for efficient error detection.

5.5. Fixed Fault Detection vs. Transient Fault Detection

Any error detection circuit can detect both transient and fixed faults, but the detection of a fixed fault is much easier because of its persistent nature, allowing it to be detected in multiple processor cycles.

For example, the simple adder parity check described in Section 5.2.1 cannot detect any fault generating an even number of bit flips. But with a different input combination, the same physical fault can cause an odd number of bit flips and the existence of the fault can be detected. Assuming an error detection circuit can detect 60% of the total possible faults, a chance of detecting a transient fault is only 60% . But with the randomly changing inputs, a probability of successful detection of a fixed fault is $1 - (1 - 0.6)^n$, where n is the number of cycles tested. With only five cycles of testing with varying inputs the probability of detection increases to 99% .

The detection after multiple cycles may be too late for error correction with the checkpoint recovery described in the next section. But it is certainly useful for the prevention of the most damaging unnoticed erroneous output case. Also, it gives vital information for the failed component replacement for later repair.

6. Checkpoint Recovery

6.1. Basic Concept

More often than not it is desirable to recover from a state of error to resume normal operation. The error detection techniques described so far allow for efficient detection but the erred process must be stopped for recovery. Although TMR can mask the occurrence of an error, it is a very expensive solution.

When the known correct state is kept as a checkpoint, a microprocessor or a system can go back to the checkpoint and redo the processing after the checkpoint is made when an error is detected. Assuming the nature of the error is transient, it will not happen again and the system recovers from the error.

Cache-based recovery [37, 38] is the method to keep a checkpointed state in the memory. It keeps state changes due to the instruction execution, after the checkpoint is made, within CPU registers and cache memories. When a cache line needs to be written back into the memory, all modified cache lines are also written back in addition to CPU registers to make a consistent checkpoint. There are other memory based checkpoint schemes [39] as well.

Multiple generations of IBM mainframe processors [42, 43] have a special storage unit called an R-unit (recovery unit) for storing checkpoints for the recovery. The R-unit gathers processor state changes to make checkpoints.

Since it takes some time to do the recovery, an error correction by means of checkpoint recovery may not be appropriate for time-critical real-time systems, like fly-by-wire controls.

6.2. Checkpoint Recovery for the Microprocessor with Speculative Execution

Modern high-end microprocessors often do speculative execution, speculating an outcome of a conditional branch before the condition calculation is completed, and execute instructions in the predicted path. This method improves the performance of the microprocessor as it can proceed without waiting for the resolution of the branch condition. With modern branch prediction mechanisms, the speculated branch direction is correct for over 90% of the time.

However, when the speculation turns out to be wrong, the processor needs to go back to the mis-predicted conditional branch and re-start from that instruction. This is done by making a checkpoint before the conditional branch.

The Fujitsu SPARC64 V processor [46, 47] used this speculative execution checkpoint-recovery for transient hardware error recovery [40, 41]. The processor implemented an error detection network similar to the one described in Section 5. The design guarantees that the error detection report arrives at the commit stage earlier than the commit time of the instruction with the detected error.

This implementation is efficient as it uses an already existing checkpoint-recovery mechanism for the speculative execution and the additional hardware required is for error detection only.

7. Microprocessors Examples

High reliability and continuous non-stop operation is very important for large computer systems handling social, governmental and/or enterprise infrastructure information processing. To meet this demand, some commercial microprocessors, especially designed for large multiprocessor servers, have implemented error detection and correction as well as a checkpoint-recovery system.

7.1. Intel Itanium 2

The Intel Itanium 2 microprocessor implemented block redundancy for its large level 3 on-chip cache for yield enhancement. The first Itanium 2 processor [44] has 135 24 KB SRAM blocks, and each of them produces a 2 bits output. These SRAM blocks are divided into two groups including one spare block each. The selection of the outputs from the SRAM blocks in each group is done in the same manner as the column redundancy described in Section 3.2.

The newer dual core Itanium 2 processor with 24 MB level-3 cache [45] added one redundant block for each 34 data+ECC blocks. Although this level of redundancy is about the same as having one spare column for each 32 columns, the major advantage of this design is its capability to tolerate failures other than column failure.

It also implements an SECDED code for error correction for the on-chip caches which hold the original data. The level-1 instruction cache which is a read-only cache is protected by the parity check. The level-1 data cache is also parity protected since it is a store-through cache and the same content is written into the SECDED code protected level-2 data cache. The dual core Itanium 2 chip with 24MB level-3 cache added parity check to the integer register file.

It also uses ECC for the data bits and parity check for the address bits of the chip I/O bus.

It can correct an error that occurs in the ECC protected data. It also can recover from an error in level-1 caches by invalidating the entry where the error is detected and reading the error-free data from the ECC protected storage; but it does not have a checkpoint-based error recovery mechanism.

Table 2. Error detection and recovery scheme

Hardware unit	Error detection	Recovery
L1I$ data	Parity check	Invalidate & Retry
Translation lookaside buffer	Parity check	Invalidate & Retry
Branch target address Buffer	Parity check	Miss prediction recovery
L1I$ /L1D$ tag	Parity check + Duplication	Pick correct side and rewrite
L1D$ data	SECDED code	Hardware ECC
L2$ data and tag	SECDED code	Hardware ECC
Registers	Parity check	Checkpoint recovery
ALU, shifter, VIS	Improved parity prediction	Checkpoint recovery
Multiplier/divider	Residue check and improved parity prediction	Checkpoint recovery

7.2. Fujitsu SPARC64 V

The error detection and recovery scheme for the Fujitsu SPARC64 V processor [46, 47] is summarized in Table 2.

Level-1 instruction cache (L1I$) data is protected by parity check as it is read-only and the same information is in the SECDED code protected level-2 cache (L2$) or the memory. The Translation lookaside buffer is also a copy of the memory map stored in the memory. The Branch target address buffer (BTAB) gives a branch direction and a target address, but the whole information is a hint for performance improvement. Even if it is wrong it only makes a wrong speculative branch, which will be corrected later. An error in BTAB only slows down the execution, but it does not affect the correctness of the processing.

The level-1 data cache (L1D$) tag contains information which is not stored in other places. This array is designed to survive from any single bit error. The L1D$ tag is protected with the duplication with parity check. When an error is detected in one set, the output from the other set is used and the error-free data is rewritten into the other set for transient error correction. Although the L1I$ tag does not require error correction, the same structure as the L1D$ tag is duplicated. The L1D$ data array is protected with the SECDED code. When an error is detected, error correction and the re-write for correcting array content is performed by hardware. The level-2 cache tag and data are also using the same SECDED scheme.

Logic circuits are protected with the parity check for the data transfer portion as shown in Figure 13. The data manipulation portion is covered by the various algorithmic checks described in Section 5.2.

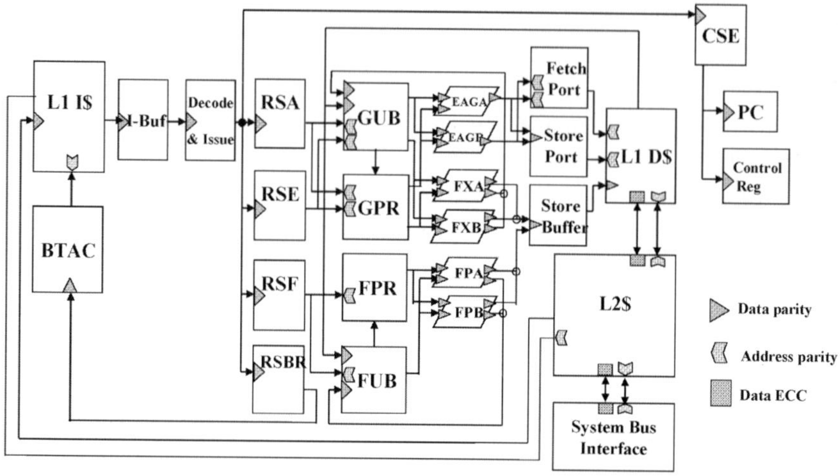

Figure 13. Error checker placement in SPARC64 V processor.

When an error is detected in the registers and the arithmetic units, checkpoint recovery is performed.

7.3. IBM z990

The IBM z990 processor [48] duplicates instruction and execution units and compares both outputs. Although the duplicate and compare more than doubles the amount of hardware, it achieves an excellent error-detection capability. Processor states are continuously stored into the R-unit which checkpoints the processor state after every instruction execution. When a hardware error is detected, an appropriate checkpoint is recovered from the R-unit and the recovery is done.

An additional benefit of keeping the checkpoint in the R-unit is that the contents of the R-unit can easily be extracted and transferred to the other z990 processor and continues the execution. This capability gives the z990 the ability to survive from a fixed hardware failure in addition to a transient failure.

Level-1 caches, TLB and R-unit are protected by SECDED code or other appropriate means to be able to recover from a single bit error.

8. Summary

Semiconductor scaling makes device parameter variations larger and negatively affects the yield of VLSI chips. The mechanisms showing why the semiconductor scaling makes VLSI devices having larger device parameter variations and why yield enhancement with microarchitecture is beneficial are presented.

Scaling also reduces the noise margin of the circuits and the devices are more susceptible to an error caused by a cosmic neutron hit. Various means for adding redundancy for error detection and correction are presented. The codes for error detection and error correction are also described.

In Section 7 three microprocessors examples for high-reliability large-scale servers are explained as regards their error detection and correction features.

Error detection and recovery by microarchitecture is practically the only way to assure reliable operation of the scaled VLSI devices.

Microarchitectural techniques described in this chapter will be more widely used for yield enhancement and reliable operation as semiconductor scaling continues.

References

[1] Li, X.-Y. *et al.*, "An effective method of characterization poly gate CD variation and Its impact on product performance and yield", *Proc. Int. Symp. Semicond. Manuf.*, **2003**, 259–262.

[2] Nagase, M.; Tokashiki, K. "Advanced gate etching for accurate CD control for 130-nm node ASIC manufacturing", **2004** *17*(3), 281–285.

[3] Schellenberg, F. "A little light magic [optical lithography]", *Spectrum IEEE*, **2003**, *40*(9), 34–39.

[4] Asenov, A.; Kaya, S.; Brown, A.R. "Intrinsic parameter fluctuations in decananometer MOSFETs introduced by gate line edge roughness", *Electron Devices*, **2003** *50*(5), 1254–1260.

[5] Asenov, A. "Random dopant induced threshold voltage lowering and fluctuations in sub 0.1 micron MOSFETs: A 3D 'atomistic' simulation study," *IEEE Trans. Electron Dev.*, **1998**, *45*, 2505–2513.

[6] Pollack, F. "New microarchitecture challenges in the coming generations of CMOS", **1999**. MICRO-32. Keynote, 32nd Annual International Symposium on Microarchitecture.

[7] Gelsinger, P.P. "Microprocessors for the new millennium: challenges, opportunities, and new frontiers," *IEEE Int. Solid-State Circuits Conf.*, **2001**, *XLIV*, 22–25.

[8] Chen, H.H.; Ling, D.D. "Power supply noise analysis methodology for deep-submicron Vlsi chip design", *Proc. 34th Design Automation Conf.*, **1997**, 638–643.

[9] Zhao, S.; Roy, K.; Koh, C.-K. "Estimation of inductive and resistive switching noise on power supply network in deep sub-micron CMOS circuits", *Proceedings, Int. Conf. Computer Design*, **2000**, 65–72.

[10] Prakash, M. "Cooling challenges for silicon integrated circuits", *Proc. 9th Int. Conf. on Thermal and Thermomechanical Phenomena in Electronic Systems*, **2004**. (ITHERM '04) 2, 705–706

[11] Bota, S.A. *et al.*, "Within die thermal gradient impact on clock-skew: a new type of delay-fault mechanism", Proc., Int. Test Conf., **2004**, 1276–1283

[12] Banerjee, K.; Mehrotra, A. "Global (interconnect) warming", *Circuits and Devices IEEE*, *17*(5), 16–32

[13] Ziegler J.F. *et.al.*, "IBM experiments in soft fails in computer electronics (1978–1994) ", *IBM J. Res. Dev.* **1996**, *40*(1), 3–18
[14] Ziegler, J.F. "Terrestrial cosmic rays", *IBM J. Res. Dev.*, **1996**, *40*(1), 19–40.
[15] Freeman, L.B. "Critical charge calculations for a bipolar SRAM array", *IBM J. Res. Dev.* **1996**, *40*(1), 119–130.
[16] Baumann, R. "The impact of technology scaling on soft error rate performance and limits to the efficacy of error correction," *IEDM Tech. Dig.*, December **2002**, 329–332.
[17] Tosaka, Y.; Ehara, H.; Igeta, M. *et al.*, "Comprehensive study of soft errors in advanced CMOS circuits with 90/130 nm technology," *IEDM Tech. Dig.*, December **2004**, 941–944.
[18] Liden, P. *et al.*, "On latching probability of particle induced transient in combinational networks", *Int. Symp. on Fault Tolerant Computing FTCS24*, June **1994**, 340–349.
[19] http://public.itrs.net/Files/2003ITRS/Home2003.htm
[20] Smith, R. *et al.*, "32K and 16K MOS RAMs using laser redundancy techniques", Solid-State Circuits Conference. *Digest of Technical Papers*, **1982**, *XXV*, 252–253.
[21] Horowitz, M.; Dally, W. "How scaling will change processor architecture", *ISSCC Dig Tech Pap.* **2004**, 132–133.
[22] Ando, H.; Tzartzanis N.; Walker, W. "A case study: power and performance improvement of a chip multi-processor for transaction processing", *IEEE Trans. VLSI Syst.*, July **2005** (To be published).
[23] Alves, L.C. *et al.*, "RAS design for the IBM eServer z900", *IBM J. Res. Dev.*, **2002**, *46*(4/5), 503–522.
[24] Pham, D. *et al.*, "The design and implementation of a first-generation CELL Processor", *Tech. Digest, Int. Solid-State Ciruit Conf.*, February **2005**, 184–185.
[25] Wuu, J. *et al.*, "The asynchronous 24MB on-chip level-3 cache for a dual-core Itanium-family processor", *Tech. Digest, Int. Solid-State Ciruit Conf.*, February **2005**, 488–489.
[26] Nelson, V.P. "Fault-tolerant computing: fundamental concepts", *IEEE Computer*, **1990**, *23*(7), 19–25.
[27] Berger, J. "A note on error detecting codes for asymmetric channel", *Info Control*, **1961**, 68–73
[28] Fujiwara, E.; Pradhan, D.K. "Error-control coding in computers", *IEEE Computer*, **1990**, *23*(7), 63–72.
[29] Hsiao, M.Y. "A class of optimal minimum odd-weight-column SECDED codes", *IBM J. Res. Dev.*, **1970**, *14*, 395–401.
[30] Kaneda, S.; Fujiwara, E. "Single byte error correcting–double byte error Detecting codes for memory systems", *IEEE Trans. Computers*, July **1982**, C31(7), 737–739.
[31] Reed I.S.; Solomon, G. "Polynomial codes over certain finite fields", *J. Soc. Indust. Appl. Math.*" **1960**, *8*, 300–304.
[32] Gorshe, S.; Bose, B. "A self-checking ALU design with efficient codes", *14th VLSI Test Symp.* April **1996**, 157–161.
[33] Nocolaidis, M. "Efficient implementations of self-checking adders and ALUs", *Int. Symp. on Fault Tolerant Computing FTCS23*, June **1993**, 586–595
[34] Sparmann, U.; Reddy, S. "On the effectiveness of residue checking for parallel two's complement multipliers", *IEEE Trans VLSI Systems*, **1996**, *4*(2), 219–228.
[35] Matrosova, A.; Ostanin, S. "Self-checking FSM design with observing only FSM outputs", *Proc. 6th Int. On-Line Testing Workshop*, **2000**. 153–154.
[36] Goloubeva, O. *et al.*, "Soft-error detection using control flow assertions", *Int. Symp. on Defect and Fault Tolerance in VLSI Systems*, November **2003**, 581–588.
[37] Hint, D.B.; Marinos, P.N. "A general purpose cache-aided rollback error recovery (CARER) technique", *17th Symp. on Fault Tolerant Computing*, **1987**, 170–175

[38] Ahmed, R.E.; Frazier R.C.; Marinos, P.N. "Cache-aided rollback recovery (CARER) algorithms for shared-memory multiprocessor systems", Digest of Papers, *20th Int. Symp. on Fault-tolerant Computing*, **1990**, 82–88.

[39] Bowen, N.S.; Pradhan, D.K. "Processor- and memory-based checkpoint and rollback recovery", *IEEE Computer*, **1993**, *26*(2), 22–31.

[40] Ando, H.; Kitamura, T.; Shebanow, M.; Butler, M. US Patent 6,519,730 "Computer and error recovery method for the same"; February 11, **2003**, Filed on March 16, 2000,

[41] Sato, T. "Exploiting instruction redundancy for transient fault tolerance", *Int. Symp. on Defect and Fault Tolerance in VLSI Systems*, November **2003**, 547–554.

[42] Webb, C.F.; Liptay, J.S. "A high-frequency custom CMOS S/390 microprocessor", *IBM J. Res. Dev.* **1997**, *41*(4/5), 463–474.

[43] Schwarz, E.M. *et al.*, "The microarchitecture of the IBM eServer z900 processor", *IBM J. Res. Dev.* **2002**, *46*(4/5), 381–396.

[44] Rusu, S. *et al.*, "A 1.5-GHz 130-nm Itanium 2 processor with 6-MB on-die L3 cache", *IEEE J. Solid-State Circuits*, **2003**, *38*(11), 1887–1895.

[45] Naffziger, S.; Stackhouse, B.; Grutkowski, T. "The implementation of a 2-core multi-threaded Itanium®-family processor", *ISSCC Dig. Tech. Pap.*, **2005**, 182–183.

[46] Inoue, A. "Fujitsu's new SPARC64 V for mission-critical servers", *Microprocessor Forum* **2002**.

[47] Ando, H. *et al.*, "A 1.3GHz fifth-generation SPARC64 microprocessor", *IEEE J. Solid-State Circuits*, **2003**, *11*, 1896–1903.

[48] Slegel, T.J.; Pfeffer E.; Magee, A. "The IBM eServer z990 microprocessor", *IBM J. Res. Dev.* **2004**, *48*(3/4), 295–310.

Chapter 10

HOW IS BANDWIDTH USED IN COMPUTERS?

Why Bandwidth is the Next Major Hurdle in Computer Systems Evolution and What Technologies Will Emerge to Address the Bandwidth Problem

Phil Emma
IBM Corp., USA

Abstract: This chapter will explore the issue of bandwidth: how it is used, and how that effects the performance of an individual processor in a system, as well as how it impacts the overall system. It will examine the current trends in system evolution, and explain what those trends imply in light of bandwidth. Finally, the chapter will explore the technologies that are likely to emerge to satisfy those trends.

Key words: bandwidth; bus width; cache; cache line; cache miss; central electronic complex; CEC; computer system; DRAM; embedded DRAM; finite cache effect; line size; miss penalty; miss rate; miss ratio; multicore; multithread; optical interconnect; optics; prefetching; queuing; trailing edge; trailing edge effect; virtualization; wavelength division multiplexing; WDM.

1. Definition of Bandwidth

Figure 1 shows two entities, A and B. The *bandwidth* from A to B is equal to the product of the number of channels connecting A to B and the data rate per channel:

$$\text{Bandwidth} = (\#\ \text{channels}) \times (\text{data rate per channel})$$

In computer systems, channels are aggregated into byte-oriented busses, where a byte is 8 bits (or 9 bits with parity, or more bits for more powerful error-control

Vojin G. Oklobdzija and Ram K. Krishnamurthy (eds.),
High-Performance Energy-Efficient Microprocessor Design, 235–287.

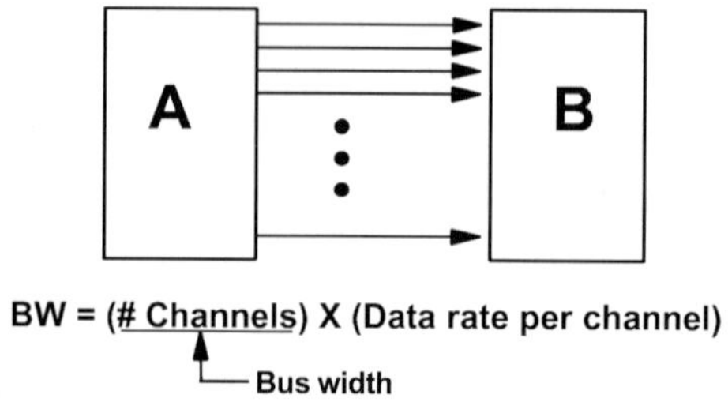

Figure 1. Definition of bandwidth.

codes). That is, a bus will generally be some number of bytes wide, where that number is a power of 2, and a byte comprises 8 (or 9) channels.

Common on-chip bus widths in modern systems are in the 16 byte–32 byte range, and run at frequencies equal to (or perhaps half as much as) the processor frequency. Therefore, on-chip busses today move 16–32 bytes per processor cycle (or on every-other processor cycle) [1, 2].

Off-chip busses tend to be narrower (by a factor of 2 or so) and slower (also by a factor of 2 or so) [3]. This is because off-chip drivers are more complicated and take more power than on-chip drivers, and the packaging infrastructure required to support wide fast busses is much more expensive than for on-chip busses.

Similarly, when busses leave each successively higher level of packaging (board, backplane, and frame), they get progressively more narrow, and generally slower. Again, this is because the connector density, cost, and electrical performance become less and less favorable as the level of integration becomes cruder, and the physical dimensions grow. Still, within the structure of the central electronic complex (CEC) that comprises a monolithic computer system, the busses remain byte-oriented. Again, a byte requires 8 (or 9) channels.

How is it that, as connectivity goes from intrachip to interchip to interboard to interframe, the bandwidth need not remain constant? In fact, how can it be allowed to drop by factors of 2, 4, and 8 at each level, without hampering performance?

The first answer is that the bandwidth leaving an entity can be mitigated by providing more on-entity storage. I will refer to the amount of on-entity storage as the *content*. Typically, content takes the form of cache memory. The first and most important message of this chapter is that *bandwidth and content are mutually fungible*. That is, each can manifest the properties of the other (excess bandwidth can have the same effect as excess content, and vice-versa), hence they can be exchanged. The bandwidth within a CEC is primarily used to service cache misses, which is basically why content and bandwidth are manifestations of each other. I will explain this in detail later.

The second answer to the question of reducing bandwidth capability is that as you reduce the bandwidth leaving an entity, you certainly *can* hamper performance. Further, there are nonlinearities in how this occurs. If you are unaware of what the nonlinearities are, you can be in for some unhappy surprises when you put a system together. I will explore these effects in this chapter.

When data traffic leaves the CEC (or the room, or building, or city), the physical infrastructure changes – usually to bit-serial optics. This is transport technology (telecommunications or data communications), which is different technology from what is currently used within the CEC. I will not address the transport area in this chapter.

At the end of this chapter I will explain the directions in which the intra-CEC signaling technology is likely to evolve. While much of this technology will certainly become optical, it will be fundamentally different technology than what is currently used in high-speed serial communications.

2. Bandwidth Scaling

From first principles it is useful to ask how bandwidth needs to scale over time, and whether the technology is on track to provide that scaling. In the case of a uniprocessor on a chip, the purpose of the off-chip bandwidth is to service the misses from the highest level of on-chip cache.

It has been known for several decades that the cache miss rate (in misses per instruction) is proportional to the reciprocal of some root (typically, the square root) of the cache capacity. Since, for a fixed capacity, the miss rate is independent of the speed of the processor, the bandwidth must scale with the speed of the processor, but is palliated by the cache capacity. If we give the speed of the processor in millions of instructions per second (MIPS), the relationship is:

$$\text{Bandwidth} = \text{MIPS/SQRT(Cache size)}$$

Until fairly recently the SIA roadmap projected that uniprocessor MIPS would double every 18 months. At the same time, Moore's Law has the cache capacity doubling every 3 years. If these projections hold, then to keep the bandwidth on track requires that bandwidth increases by 2SQRT(2) every 3 years.

Recall that bandwidth is equal to the number of channels times the data rate per channel. Therefore, scaling bandwidth at this rate either requires that the bus width (which I will use from here on as a proxy for the number of channels) increases, or that the bus frequency (which I will use from here on as a proxy for data rate) increases, or both.

In the case of off-chip busses, increasing the bus width requires more advanced packaging to accommodate a denser I/O pitch, and more layers of metal in the package to escape the larger number of signals from the I/O pitch to whatever wiring pitch is used within the package. Historically, the number of I/Os per chip has not evolved nearly at this rate. Roughly speaking, it has increased less than 3 orders of magnitude (from the order of 10 to the order of 1000) in about 50 years. Do not plan on achieving sufficient scaling from I/Os.

On the other hand, frequency has scaled much more quickly, although not quite quickly enough. In the past two decades bus frequency has remained on the same scale as processor frequency, although it has usually been a little slower – by a factor of 2 or 4. Since frequency scaling has been the primary factor driving the SIA projection of MIPS growth, we might say that frequency scaling should give us a factor of $2\times$, but not of $2\mathrm{SQRT}(2)\times$.

Therefore, it appears that bandwidth cannot quite scale to keep pace with the SIA roadmap, even when taking Moore's Law into account. On the other hand, the trend as we came into the new millennium has become much more focused on power-efficiency [4], and is now de-emphasizing frequency scaling. Power and power density are seen as the next major limitations on system performance. For a fixed amount of power, a system can deliver more total performance if its (growing number of) constituent parts use power more efficiently.

Since continuing to scale frequency is not power efficient, there has been a general industry consensus that processor frequency will not grow much beyond 5 Gigahertz in the next few CMOS generations [5]. This would appear to offer relief. On the other hand, there are three emerging trends in system design and processor design that could put us on a trajectory that is even worse than that of frequency scaling.

First, all major processor manufacturers are putting multiple processors on the same chip [2]. So while today's available off-chip bandwidth may be sufficient to supply the needs of a single 5 Gigahertz processor, putting 2, or 4, or 8 processors on a single chip will eventually stress this bandwidth. This is because the bandwidth must now be used to satisfy the misses of 2, or 4, or 8 processors. If there is no sharing of data, the aggregate miss rate will scale linearly with the number of processors.

As the number of processors on a chip grows, the on-chip systems will be run as symmetric multiprocessors (SMP) that share data – perhaps by sharing a large on-chip cache – with a capacity that we will say (for now) scales with Moore's Law. If some fraction of the working set is shared by all processors, then the aggregate bandwidth is mitigated by this fraction, but the remaining bandwidth must still scale linearly with the number of processors.

Second, the processors are becoming increasingly multithreaded. Essentially, this means that each processor appears to be running as if it is multiple processors [6]. This requires that the on-chip caches be sufficiently large to hold the working sets of all threads. For a fixed cache size, this will strain the bandwidth; again, linearly with the degree of multithreading.

Third, systems are becoming virtualized, so that multiple logical partitions can be co-resident on the hardware [7]. Since the logical partitions are independent, so are the working sets. Again, this places a linear strain on the on-chip caches, and on the off-chip bandwidth.

For example, we could have 4 processors on a chip, and each processor could be 4-way multithreaded. The chip then appears to have 16 processors. The hypervisor may then bring up 4 independent (virtual) systems on that chip. Thus, the 4 processor chip has to be have like it is four independent 16-way SMPs, i.e., the chip must logically behave as 64 processors. This will put a huge strain on the on-chip caches, and on the off-chip bandwidth.

3. Components of Uniprocessor Performance

Before understanding and evaluating bandwidth quantitatively, it is helpful to review processor performance from the systems-design point of view [8]. Figure 2 shows a conceptualization of a uniprocessor system. The processor (or "core") comprises: (1) an instruction unit (I unit) which fetches, decodes, and stages instructions for execution; (2) an execution unit (E unit) which executes the instructions; and (3) a first-level cache (L1 cache), which is a small local memory system that holds the instructions to be executed, and the data to be operated on.

In many modern machines the L1 cache is actually two separate caches: one to hold the instructions, and one to hold the data. The I unit, E unit, and L1 cache are designed to be a strongly coupled aggregation. The fetching and execution of instructions and data are pipelined in assembly-line fashion so that multiple instructions are in flight simultaneously. This enables a processor

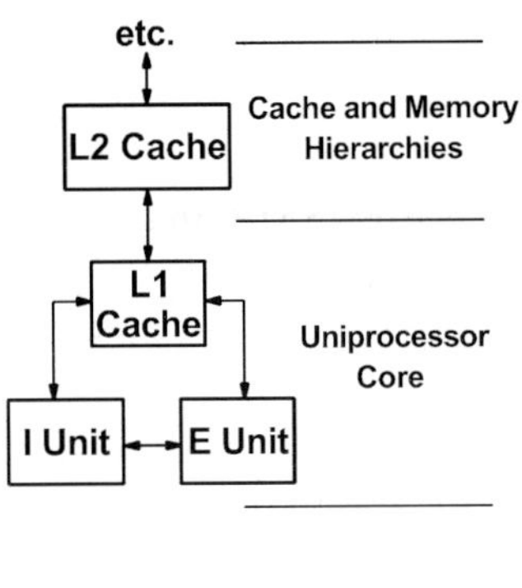

Figure 2. Conceptualization of a uniprocessor.

cycle time to be much faster than the time it would take to process any single instruction, and it allows multiple instructions to be processed at the same time ("in parallel").

Since the L1 cache is integral to the processor pipeline, its capacity is determined by the cycle time. That is, if the processor is to run at a given frequency, and a fetch is specified to take a given number of cycles in the pipeline flow (usually a small number if the pipeline flow is to be smooth), then these two constraints determine the maximum size of the (integral) L1 cache.

The cache is a buffer that temporarily holds instructions and data that are brought in from the memory subsystem when they are needed. As long as the data in the cache continues to be referenced by the processor, it will remain there. When it stops being referenced, it will eventually be replaced by newly referenced data.

Outside the processor core is a hierarchy of progressively larger (ergo slower) caches, which eventually reaches the main storage of the CEC. Main storage in a server is many Gigabytes of commodity dynamic random access memories (DRAMs) packaged on hundreds of dual in-line memory modules (DIMMs). Main memory is connected to magnetic storage (usually disks) via channel processors.

In Figure 2 the first level of cache outside the processor has been labeled "L2." In principle, the choice of labels is arbitrary; but in some circles the label "L1" specifically connotes the cache that is integral to the processor pipeline, and the label "L2" specifically connotes the point of coherency in a multiprocessor. I will not explain coherency here [9, 10]. I mention it only because if the first level beyond the L1 is not the coherency point, then a label such as "L1.5" would be used in some lexicons.

If the processor finds all of the data and instructions that it needs in the L1, it is said to run at its *infinite cache performance* (which is a limit). When the processor references instructions or data that are not in the L1, these events are called cache *misses*. When a miss occurs, the referenced data must be brought in from the cache hierarchy, and the delay associated with fetching data from the slower levels of cache will generally cause the pipeline to stall. The resulting slowdown is called the *finite cache effect* (FCE).

Figure 3 shows how the components of Figure 2 map into a conceptualization of performance. The performance metric that I will use in this chapter is *cycles per instruction* (CPI). This is the average number of cycles required to process an instruction. Note that the rate of processing instructions, MIPS, is proportional to the processor frequency *divided by* CPI. That is, a small CPI is desirable, and a large CPI is bad.

In Figure 3, CPI has been plotted as a function of the cache miss rate. When there are no misses, the miss rate is zero, and the CPI line intersects the *y*-axis at the point corresponding to the infinite cache performance. The infinite cache performance comprises a piece called "E Busy" (which accounts

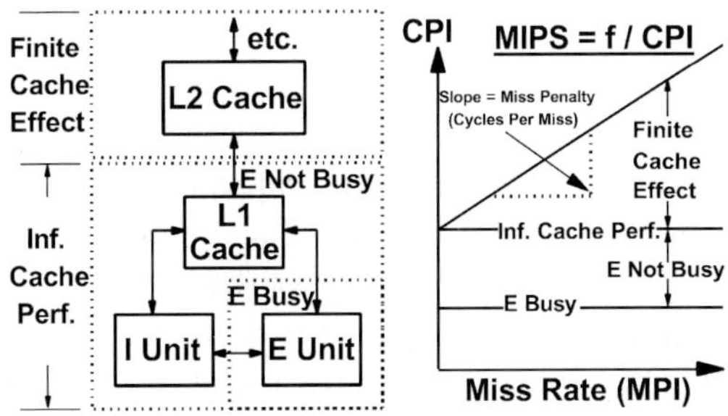

Figure 3. Conceptualization of uniprocessor performance.

for the inherent work done by the E unit), and a piece called "E Not Busy" (which accounts for pipeline effects – those things that inhibit the processor from running at the inherent E Busy limit).

The finite cache effect is determined by the slope of the CPI line, and the miss rate. The miss rate (the x coordinate) is measured in *misses per instruction* (MPI), and the slope of the CPI line is the average miss penalty, measured in *cycles per miss* (CPM). Determining the slope of this line – the CPM – can be done by measuring a running machine to get a cluster-plot of CPI versus MPI, or it can be done by running a simulation using different L1 cache sizes so as to get a range of miss rates. Or, if you have lots of experience working with the particular microarchitecture in question, you can probably intimate its value from various parameters (logic paths, access times, etc.).

Bandwidth affects MPI in several different ways. These are all explored in this chapter. While CPI is shown as a straight line in Figure 3, it is not really straight; instead it has three linear regimes. As I will explain later, miss events are random in time; as such, they can cluster. For a typical operating L1 miss rate (the middle regime), the clustering of misses causes some of the penalty associated with them to be overlapped. Thus, the average penalty of clustered misses may be less than the penalty that would have accrued to the same misses were they taken in isolation.

If the miss rate is extremely low (as it approaches zero – the leftmost regime), clustering will not happen to the same degree. This causes the average miss penalty to be somewhat worse than when the miss rate is higher. Thus, as the CPI "line" approaches the y-axis (from the right), it will actually curve upwards a little bit.

At the other end of the spectrum (the rightmost regime), if the miss rate is extremely high, the cache hierarchy (and the bus to it) becomes saturated. The systems performance has very little to do with the processor when this happens. Instead, the CPI is limited by the cache hierarchy. This is another linear regime, but it has a (much) steeper slope than for a typical steady-state miss rate (the middle regime).

4. Cache Miss Rate

It has been known for several decades that the cache miss rate is proportional to the reciprocal of some root of the cache capacity [11]. The exponent (corresponding to the root) varies with workload, and can change if the capacity is changed drastically. Figure 4 is a log–log plot (base 2 is used on both axes so as to make the slope readily apparent) of the miss rate versus capacity for a typical business workload (TPC). This is shown over a range of 9 successive doublings (512×) of cache size for three different line sizes (to be discussed later).

As is apparent in the figure, there are two linear regimes that predominate over the 9 doublings, with a fairly clean break on a particular doubling. The clean break occurs because some significant aspect of the working set becomes subsumed at a particular capacity (in this case, 2^{18} kbytes). Obviously (although not shown) after the doubling that causes the entire workload to be subsumed (which is off the right end of the graph), the miss rate will drop precipitously toward zero.

The first regime in Figure 4 has a slope of $-\frac{1}{2}$ (two doublings of the capacity cuts the miss rate in half), which indicates that the miss rate is proportional to the reciprocal of the square root of the capacity. The second regime has a slope of $-\frac{3}{4}$ (four doublings of the capacity cuts the miss rate by three doublings), so the miss rate is proportional to the reciprocal of the $\frac{3}{4}$ root of the capacity.

This curve is shown for a uniprocessor. In a uniprocessor there are only two reasons that a miss can occur within any set [12]. When the processor first references a datum that it has *never* referenced before, there is no way that the datum can be in the processor's cache, hence the reference *must* be a miss. These misses are called *compulsory misses.*

If the processor had referenced a particular datum in the past, but misses again when re-referencing the same datum, then it must be that the datum was

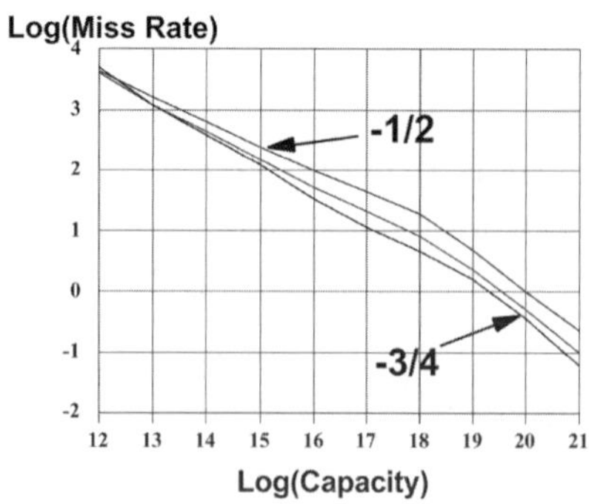

Figure 4. Log (miss rate) versus log (capacity) over nine doublings in capacity for a TPC workload.

replaced by other (more recently referenced) data because the cache was not large enough to hold the first datum while also accommodating more recently referenced data. These misses (re-references to data previously referenced) are called *capacity misses*. They occur because capacity of the cache is finite.

Now consider a multiprocessor system, as shown in Figure 5. In a multiprocessor there is another mechanism for missing. If processes running on different processors share (and modify) data, then for the processes to be *coherent* (that is, for them to have the same view of the state stored in memory), a processor modifying data must first ensure that the data is removed from all remote caches. By doing so, it ensures that when any remote processors reference the data, the remote references will miss, and will obtain the new (modified) data when the misses are satisfied.

Misses that occur because remotely running processes cause the data to be removed from the local cache (when the cache is large enough to have retained the data) are called *intervention misses*. They occur because of intervention events (invalidation requests) coming from other processors.

Figure 6 shows how the three kinds of misses interact as a function of capacity in a multiprocessor system. In accordance with their definition, compulsory

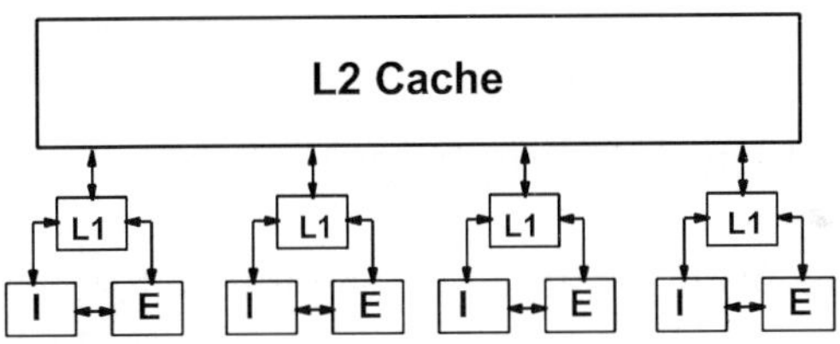

Figure 5. A multiprocessor system with a shared L2 cache.

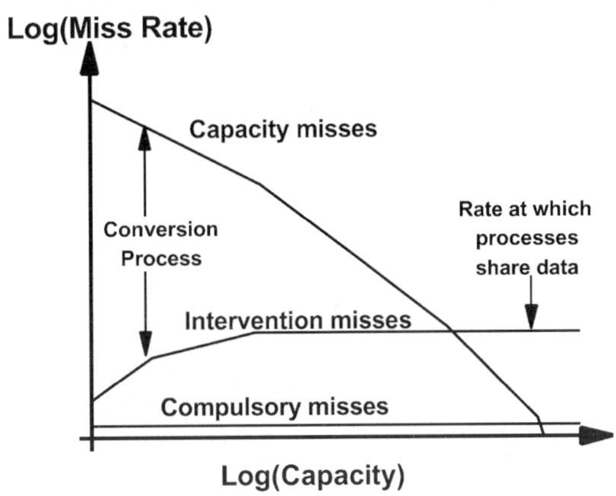

Figure 6. Three kinds of misses in a multiprocessor, and their interaction as a function of capacity.

misses are independent of the cache size. They are also statistically negligible over the long run, and are shown as greatly exaggerated in the figure just for the sake of acknowledging their existence. I make no further reference to compulsory misses.

The interaction of interest is between the capacity misses and the intervention misses. If the cache size is inadequately small (near the *y* axis), the capacity miss rate is very high, and data in the cache is very short-lived. It is unlikely that data will reside in the cache long enough to be invalidated by a remote process, hence the intervention miss rate is very low. As the cache capacity increases, data in it are resident longer, and the likelihood of intervention events increases too. Thus (at least initially), capacity can be thought of as a "converter" that converts (a fraction of the) capacity misses into intervention misses. Once the cache gets sufficiently large (so as to cause all modified data to generate intervention events), the intervention misses level off. The asymptote for intervention misses is determined by the rate at which running processes share data. The intervention miss rate is independent of cache size past this point. On the other hand, the capacity miss rate will continue to fall as the cache size is increased further. As a general rule, bigger is always better (insofar as miss rate is concerned).

5. Misses, and the Trailing Edge

Most of the bandwidth within a computer is used for moving cache lines so as to satisfy the misses in the system. There is other traffic too, depending on the coherency protocols, the desired recoverability, and other factors. The other traffic is primarily single-store traffic, and some control signaling. This other traffic will not be addressed in this chapter, and is (generally) not the primary performance concern. In the interest of focusing on the main issue, I consider only cache misses.

Figure 7 is a temporal depiction of a cache miss, with a timeline going from left to right. The impinging arrows above the timeline depict events occurring at the processor, and the impinging arrows below the timeline depict bus events.

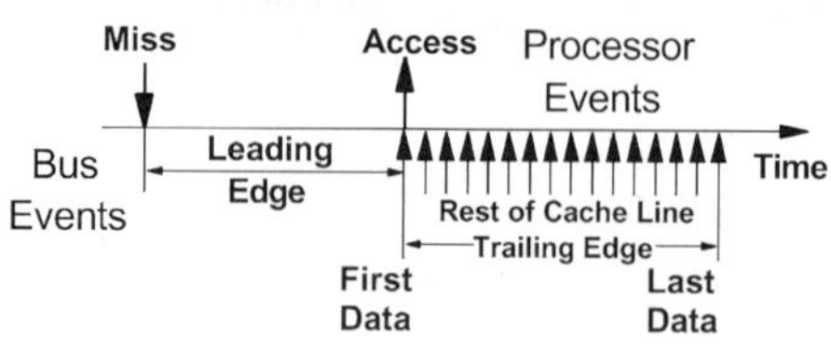

Figure 7. Temporal depiction of a cache miss.

The first event is that the processor references a datum that is not in its cache, and this causes a miss (shown as the first impinging arrow).

As a result of the miss event at the processor, numerous events happen within the system (not shown in the figure) to fetch the desired data from the cache hierarchy and deliver the referenced datum to the processor. This flow would generally be: (1) generate ECC code for the miss address and request the address bus to the L2; (2) on the granting of the bus request, transmit the miss address to the L2; (3) on receipt of the miss address by the L2, check the ECC, and request an L2 access; (4) on the granting of the access request, look up the address in the L2 directory, and fetch the data from the L2 (assuming that it's there); (5) check the ECC that was stored with the data in the L2 and request the data bus returning to the processor; (6) on the granting of the data bus, return the requested datum to the processor, followed by the other data within the cache line being returned; and (7) check the ECC on the arriving data at the processor. The data is now ready for use, and its access (to the first data) is shown as the next event in Figure 7.

This flow contains a number of logic stages (mostly ECC generation and checking, and prioritization and multiplexing), some requesting and granting of shared resources (the busses, and the L2) with opportunities to queue (wait) at these resources, wire delays from the processor to the L2 and back, and the nominal access time of the L2 directory and arrays. This flow of events – from the miss event to the receipt of the desired datum – is what is nominally thought of as the "miss penalty."

But this is not the entire miss penalty. We refer to this first set of events (up until the delivery of the datum that caused the miss – the first data back) as the *leading edge* (LE) of the miss. Data are not stored in cache levels as individual words. Rather, data are stored in contiguous blocks, each block containing multiple data. Originally, these blocks were called *cache blocks*; now they are usually referred to as *lines*. In modern processors, line sizes are at least 32 bytes. In large servers, the line sizes are as large as 128 or even 256 bytes. Since the busses in the system are not nearly this wide, the cache line is delivered to the processor as a sequence of packets, each packet being the width of the bus.

For example, if a 256 byte line is transmitted over a 16 byte bus, the number of packets required is $256/16 = 16$. When a line is returned to the processor, modern processors start by returning the packet containing the datum that was actually requested. By returning this packet first, the processor is able to start running (if the miss caused it to stall) as soon as the first packet arrives, i.e., immediately after the leading edge of the miss.

The remaining packets are transmitted to the processor at the bus frequency. The transmission of the remaining packets is called the *trailing edge* (TE) of the miss. The trailing edge is the number of processor cycles required to move the line. Since the packets are moved at the bus frequency, if the bus frequency is slower than the processor frequency, the trailing edge is equal to the number of packets times the ratio of the two frequencies:

$$\text{TE} = (\text{Line size/Bus width}) \times (\text{Processor frequency/Bus frequency})$$

In the example above in which there are 16 packets, if the bus runs at half the speed of the processor, the trailing edge is $16 \times 2 = 32$ cycles. If the bus runs at one-third the speed of the processor, the trailing edge is $16 \times 3 = 48$ cycles. The complete miss penalty is then equal to the sum of the leading edge, plus whatever effects are caused by the trailing edge:

$$\text{Miss penalty} = \text{Leading edge} + \text{Effects[Trailing edge]}$$

The trailing edge effects are bandwidth effects. For the time being we will assume that the trailing edge effects are proportional to the size of the trailing edge. This is not really true, but it can be used as a good approximation if the miss rate is held relatively constant. Since the trailing edge effects are bandwidth effects, their significance depends to a large extent – as I will show – on the bus utilization, which depends on the size of the trailing edge, and on the miss rate as follows:

$$\text{Bus utilization} = (\text{Miss rate/CPI}) \times \text{TE}$$

Note that the miss rate (in misses per instruction) divided by CPI is a temporal miss rate (in misses per cycle), and since TE is in cycles per miss, the bus utilization is dimensionless. Physically, utilization is a probability (the probability that the bus is busy), so it can only have physically meaningful values between 0 and 1. Also, note that the reciprocal of the temporal miss rate is the average intermiss distance (in cycles per miss) which is the average number of cycles between misses. An equivalent formula is then:

$$\text{Bus utilization} = \text{TE/Intermiss distance}$$

Since the utilization cannot exceed 1, TE cannot be greater than the intermiss distance.

6. Choosing Cache Line Size

If trailing edge effects hurt performance, why do we store data as lines – which cause the trailing edge effects in the first place – instead of as individual words?

From a cost perspective, storing data as individual words would be prohibitively expensive. This is because, in a cache structure, each item that is stored must have a directory entry that identifies (by address) what the item is.

Depending on the format of the directory entry and the granularity of information stored, a directory entry will be on the order of a doubleword (8 bytes) wide. If we kept a doubleword of directory information for every word of data, the directory would be larger (and slower) than the cache. This would be a terrible use of space. If a directory entry is a doubleword, it is desirable to have the item to which the directory entry pertains be much larger than this.

Even if cost were not the issue, making lines large will help performance (up to a point), despite the fact that doing so causes trailing edge effects. The choice of line size is an optimization problem.

There are two (orthogonal) heuristically observable characteristics of reference patterns, called *spatial locality of reference*, and *temporal locality of reference*. If a particular datum is referenced, then the property of *spatial locality* means that it is very likely that other data that are spatially close to that datum (by address) will also be referenced. Because reference patterns have spatial locality, it helps to fetch that locality (as a cache line) when a miss occurs, i.e., spatial locality is the rationale for cache lines. The property of *temporal locality* is that if a particular datum is referenced, then it is very likely that the same datum will be re-referenced in the near future. Temporal locality is the rationale for caching data in the first place. If reference patterns were not temporal, the cache would not be useful.

No reference pattern is purely spatial or purely temporal; all reference patterns contain both elements. Thus the choice of line size is a choice that trades temporal locality against spatial locality as illustrated in Figure 8. This figure shows two caches, each having the same total capacity. The leftmost cache is labeled "spatial cache," and it is divided into a small number of large lines. To the extent that spatial locality predominates a reference pattern, it will be captured by the large lines. The rightmost cache is labeled "temporal cache," and it is divided into a large number of small lines. To the extent that temporal locality predominates a reference pattern, it is better captured by this cache because there are many more lines with which to capture it.

Note that filling the temporal cache will require incurring many more misses than will be incurred by the spatial cache, because there are many more lines

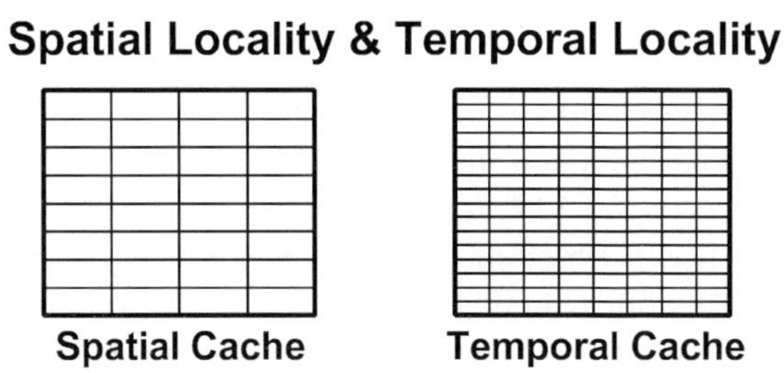

Figure 8. Choosing cache line size to capture spatial context or to capture temporal context.

to bring in. However, even though the spatial cache will incur fewer misses than the temporal cache, each spatial miss costs more because it has a longer trailing edge.

The optimization (first for miss rate) can be done by first defining a parameter called the *occupancy* of the line, denoted P, as the probability that both halves of a line are used while the line is in the cache [13]. While P will depend on the line size, and should not be extrapolated through many doublings, it can be treated as a constant for a given line size. It will also depend (to a small extent) on the cache miss rate. We will ignore this – we are doing an optimization that is well above this level of accuracy.

To do the optimization we ask the question: "For what occupancy, P, does the spatial cache have the same miss rate as the temporal cache?" We will take the line size of the spatial cache to be twice the line size of the temporal cache. Let the spatial cache be denoted C(N, 2L), meaning that it has N lines of size 2L. Let its miss rate be denoted M(N, 2L). Similarly, let the temporal cache be denoted C(2N, L), meaning that it has 2N lines of size L. Note that the two caches are the same size, 2NL. We want to know what value of P will make M(N, 2L) and M(2N, L) equal.

If the occupancy of a line of length 2L is P, then by definition, spatial cache C(N,2L) has the same useful contents as the temporal cache denoted by $C((1+P)N, L)$, i.e., $1-P$ of the long (2L) lines are not fully used. But $C((1+P)N, L)$ has to take $1+P$ misses for every miss taken by C(N, 2L) to fetch that useful content, because it has $1+P$ times as many lines. That is:

$$(1+P)M(N,2L) = M((1+P)N,L)$$

But from the inverse root law (as was shown in Figure 4), we know that if we make the cache Z times larger while holding the line size constant (by adding more lines), the miss rate decreases by $Z{*}{*}(-\alpha)$ for some fraction α. Then:

$$\begin{aligned}(1+P)M(N,2L) &= M((1+P)N,L) = ((1+P)**(-\alpha))M(N,L)\\ &= ((1+P)**(-\alpha))(2**\alpha)M(2N,L)\end{aligned}$$

Dividing both sides by $(1+P)$ M(2N, L) reveals that the two caches – C(2N, L) and C(N, 2L) – have the same miss rates when:

$$P = (2**(\alpha/(1+\alpha))) - 1$$

This allows us to slice the plain into two pieces as shown in Figure 9. In this figure the value of occupancy, P, that causes the two caches to have the same miss rate is plotted as a function of the root exponent, α. For given set of coordinates α and P (as measured from a workload), look to see whether (α, P) lies above or below this curve. If it lies above, then doubling the line size (to capture more spatial context) will result in a lower miss rate. If it lies below, then doubling the line size will result in a higher miss rate.

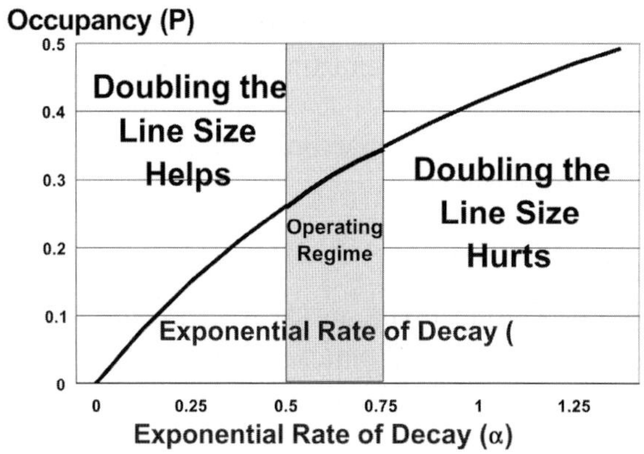

Figure 9. The value of Occupancy that makes the miss rates M(2N, L) = M(N, 2L) as a function of α.

This analysis only considers miss rate; it does not take the trailing edge effect into consideration. Recall that we had said that the trailing edge effect is proportional to TE. Let β denote the constant of proportionality. That is, we assume

$$\text{Effects[TE]} = \beta\text{TE}$$

We had said earlier that this is not really true, and that β depends on the miss rate. We will show this relationship later, and proceed with the optimization assuming that this is a good enough approximation for now. Since LE should be independent of the line size (which is another approximation, as we will see later), the penalty for a miss in C(2N, L) is LE + βTE, and the penalty for a miss in C(N, 2L) is LE + 2βTE. Choosing the breakpoint (at which doubling the line size is a break-even proposition) is matter of setting the finite cache effects of the two caches equal. That is, set:

$$\text{M(2N, L) X (LE} + \beta\text{TE)} = \text{M(N, 2L) X (LE} + 2\beta\text{TE)}$$

From our previous optimization (for miss rate alone), we found the ratio between M(N, 2L) and M(2N, L) as a function of β and P. Substituting this solution into the equation above yields:

$$(\text{LE} + \beta\text{TE})/(\text{LE} + 2\beta\text{TE}) = (2{**}\alpha)\text{X}((1 + \text{P}){**}(-(1 + \alpha)))$$

This equality contains five parameters. To make this more tractable, we define a new parameter called the *significance* of the trailing edge, denoted γ, as the ratio of the trailing edge effect to the leading edge:

$$\gamma = \beta\text{TE/LE}$$

Quite literally, γ captures the significance of the trailing edge effect relative to the leading edge. This allows us to make the substitutions $\mathrm{LE} + \beta\mathrm{TE} = (1 + \gamma)\mathrm{LE}$ and $\mathrm{LE} + 2\beta\mathrm{TE} = (1 + 2\gamma)\mathrm{LE}$. Substituting these into the equation above, and doing some manipulation, results in a break-even point for P expressed as a function of α and γ:

$$\mathrm{P} = (((1 + \gamma)/((2{*}{*}\alpha)\,\mathrm{X}\,(1 + 2\gamma))){*}{*}(-(1/(1 + \alpha))) - 1$$

Figure 10 shows a plot of this function. The break-even point for occupancy, P, is plotted as a function of the significance of the trailing edge, γ, for two different values of α, specifically, for $\alpha = 0.5$ and for $\alpha = 0.75$. For a given α, the curve again splits the plane into two pieces: one for which doubling the line size helps, and the other for which doubling the line size hurts.

Note that the two curves, $\alpha = 0.5$ and $\alpha = 0.75$, intersect the y-axis at the points $\mathrm{P} = 0.25$ and $\mathrm{P} = 0.35$, respectively. On the y-axis $\gamma = 0$, which means that there is no trailing edge. This corresponds to the previous optimization in which the trailing edge was ignored, and we optimized only for miss rate. This corroborates the curve obtained in Figure 9, because the curve in that figure intersected the vertical lines at $\alpha = 0.5$ and $\alpha = 0.75$ at the points $\mathrm{P} = 0.25$ and $\mathrm{P} = 0.35$, respectively.

Figure 10 clearly illustrates how important the trailing edge effect is. If we optimized purely on miss rate, a root value of $\alpha = 0.5$ would lead to the conclusion that it is a good idea to double the line size for occupancies greater than 0.25. But when we take the trailing edge into account, a significance of 0.5 would require an occupancy of nearly 0.55 to reach the same conclusion. And with values of TE nearing 32 or 48 (as they are), a significance of 0.5 could be a realistic number.

In this optimization, I assumed that the trailing edge effect was β TE for some constant β. As I had pointed out, it's more complicated than this. There are multiple effects of moving a line and, by definition, all of them are trailing edge effects. They can be separated into *direct* and *indirect* effects. The

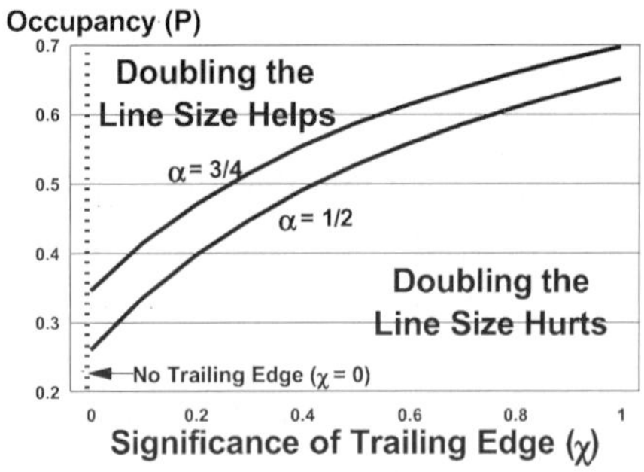

Figure 10. The value of Occupancy that makes the FCEs of C(2N, L) and C(N, 2L) equal as a function of the significance of the TE, γ, shown for $\alpha = 0.5$, and $\alpha = 0.75$.

direct effects are the immediate consequences within a processor that occur independently of what else is going on within the system, i.e., they impact performance regardless. The indirect effects are queuing effects that depend both on the previous trailing edges generated by the local processor, as well as on the other traffic in the system.

7. Direct Trailing Edge Effects

There are three direct effects of the trailing edge. The first is a local bandwidth problem at the L1. The second is a matter of accounting for spatial references, and the last concerns the utilization of the local miss facilities.

For all processors having *non-blocking caches* (meaning that the processor continues to run – if it logically can – when there are misses in progress), the infinite cache portion of the CPI is predicated on having a certain amount of L1 bandwidth available for use by the processor. As is apparent from Figure 7, during the trailing edge portion of a miss, the incoming line uses L1 cache bandwidth. Therefore, the available bandwidth as seen by the processor is diminished, and the processor runs a little slower (on the average) than it otherwise would.

Of course, buffering can be put into the cache interface so that the L1 need not be impacted by the arrival of every packet. The L1 put-away width can be made wider than the packet width so as to reduce the L1 utilization associated with storing the incoming line. For example, a double packet can be stored after the arrival of every other packet, or a quadruple packet can be stored after the arrival of every fourth packet. While input buffering can greatly reduce this effect, it is present nonetheless.

Next, I have said that reference patterns are spatial and, in fact, this is the rationale for having long cache lines. This means that after the datum that generated the miss arrives (in the first packet), the processor can continue to run if the miss had logically blocked its progress. Because the reference pattern is spatial, it is likely that the processor will (very soon) reference another datum that is in the incoming line but that has not arrived yet, i.e., a datum from another packet in the line. This is called an *upstream reference*.

Had the line size been much shorter, this upstream reference would likely have been another miss (assuming that the short line size did not contain both of the referenced data). Since the line size was long enough to capture the spatial context (of the second reference), the second reference does not cause a miss. Instead, the second reference appears as a trailing edge effect. The processor may have to stall to await the arrival of a particular packet within the trailing edge.

Finally, consider the hardware that is needed at the L1-to-bus interface to administer a miss. A miss has an address associated with it (the miss address which is sent to the L2), it may have an instruction tag associated with it (so that it can revive a stream after the leading edge), a stream tag associated with the instruction, a miss sequence number and perhaps other information so that the returning packets can be matched up with the correct misses.

A hardware *miss facility* is needed for each outstanding miss. A miss facility must contain registers for all of the accounting information mentioned above, as well as some buffering to handle the incoming line (if the incoming line is buffered before impinging the L1 as mentioned above).

In order to have multiple outstanding misses or prefeteches, there must be a hardware miss facility for each outstanding miss or prefetch. In addition, there must be arbitration and selection logic to handle the returning misses appropriately. Each time the processor wants to issue a prefetch, and each time the processor has to issue a miss, it must find an available miss facility, and it must allocate the facility to the new prefetch or miss. That miss facility will remain in use until the prefetch or miss assigned to it is completed in its entirety – including its trailing edge.

Once all of the miss facilities are in use, no new misses can be issued until one of the outstanding misses completes and releases its miss facility. The most common problem with real prefetching is that it pre-allocates the miss facilities. This can cause major delays if exigent misses occur that were not anticipated by the prefetch mechanism. The exigent miss may wind up queued behind a number of prefetches which (even if they are correct prefetches) are not needed urgently. This results in a loss of performance even when the prefetches are correct.

If all miss facilities are in use and another miss occurs, the processor must stop. It must wait for a miss facility to become available. The processor remains stopped until the trailing edge that is currently in progress completes. This is the last direct trailing edge effect.

8. Indirect Trailing Edge Effects

Indirect trailing edge effects are caused by bottlenecks that are external to the processor. Specifically, they are bus effects. Miss traffic from a processor will collide with – and be delayed by – the trailing edges of its own misses (which is self interference), as well as with the traffic generated by other processors (which is general queuing delay). In this section I discuss both of these effects.

8.1. Self Interference

Figure 11 shows a hypothetical sequence of four references made by a processor. Let's assume that if there are no misses (i.e., in an infinite-cache run) the distances between the references are 22 cycles, 4 cycles, and 40 cycles as shown. Figure 12 shows this same sequence in a finite-cache run in which all four references are misses. In this figure I have assumed a leading edge of

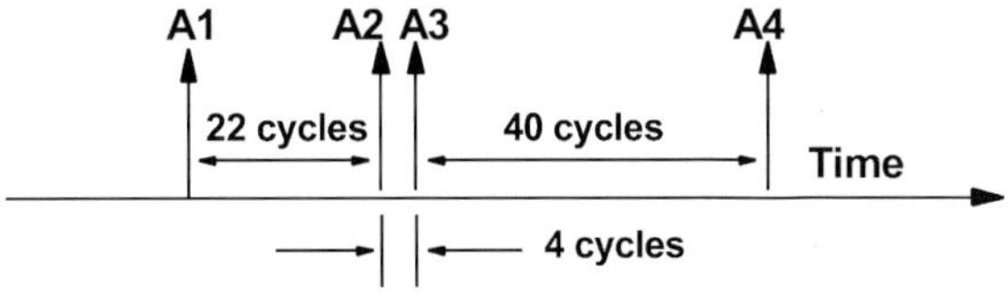

Figure 11. An infinite-cache temporal sequence of four references.

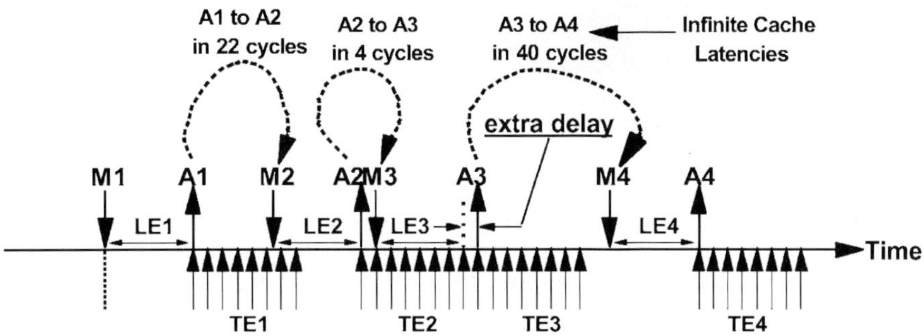

$$E(TE_k) = (TE - (LE + (A_{k-1} \rightarrow A_k)))^+$$

Figure 12. A finite-cache temporal sequence of the same four references in which all of them are misses.

24 cycles, and a trailing edge of 32 cycles. Each "tick" in the trailing edges depicted represents 4 cycles.

Note that the first and second misses are far enough apart so as not to collide. So are the third and fourth misses. But the second and third misses (having a nominal infinite-cache distance of 4 cycles) are clustered together, and there is a trailing edge effect. Specifically, following the leading edge of the second miss (M2), the second access occurs (A2). Four cycles later, the reference to the third datum is attempted, which results in the third miss (M3). The leading edge of M3 takes 24 cycles, so it completes $4 + 24 = 28$ cycles after A2. But the trailing edge of M2 is 32 cycles, starting from A2. So although the leading edge of M3 is completed, the processor is unable to access the datum, because there are $32 - 28 = 4$ cycles of trailing edge left on the bus from the second miss. Thus, the access to the third datum A3 is delayed by this difference.

The general rule is that if the distance between two miss accesses is less than the trailing edge, there is a trailing edge penalty equal to this difference. If the infinite cache distance between two miss references is denoted $A_k \rightarrow A_{k+1}$, then the nominal temporal distance between them is their infinite cache distance plus the leading edge of the second miss. If this nominal temporal distance is

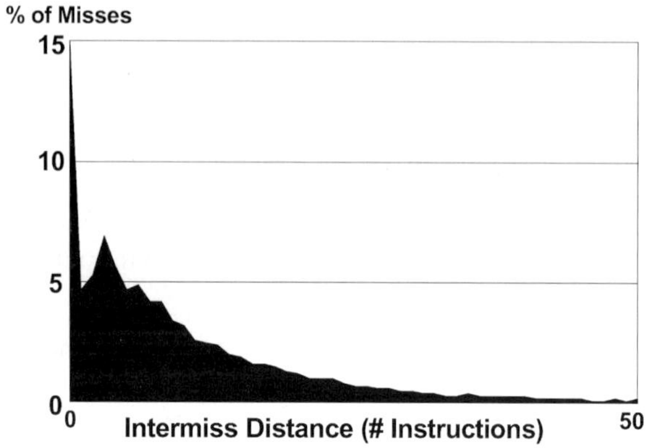

Figure 13. The intermiss-distance density function measured from a typical commercial workload.

less than the trailing edge of the first miss, the second access will be delayed until the trailing edge completes. The trailing edge penalty is then the difference between the two.

$$\text{Effect}[\text{TE}_k] = (\text{TE}_k - (\text{LE}_{k+1} + (\text{A}_k \rightarrow \text{A}_{k+1})))^+$$

The superscript "+" at the end of the rightmost expression denotes that there is only a delay if the expression is positive. For misses that are spread far apart, the rightmost expression will be negative, and there is no trailing edge effect. Note that for the trailing edge to manifest in this particular way, it must be larger than the leading edge.

The next obvious question is: "To what extent do the misses cluster like this, and what is the average trailing edge effect?" Figure 13 shows a real intermiss distance density function. The data used to create this figure were measured from a real processor running a typical business workload (TPC). While, strictly speaking, deeper analysis has shown that this is not actually an exponential function, it is fairly close to one. Hence, for the sake of obtaining a tractable understanding of what's happening, we can treat the miss process as a Poisson process.

Let τ be the trailing edge: $\tau = \text{TE}$. Let r be the temporal miss rate, and let ρ be the average intermiss distance: $\rho = 1/\text{r}$. The trailing edge effect due to self interference is then the integral of miss arrivals at t for t less than τ of $(\tau - \text{t})$ re**(−rt) dt. The solution to this integral is

$$\text{Effect}[\tau] = \tau - \rho(1 - \text{e} **(-\tau/\rho))$$

This says that the trailing edge effect is equal to the trailing edge minus a scaled function of the intermiss distance. The scaling function is an exponential function of τ/ρ, which is the bus utilization, U.

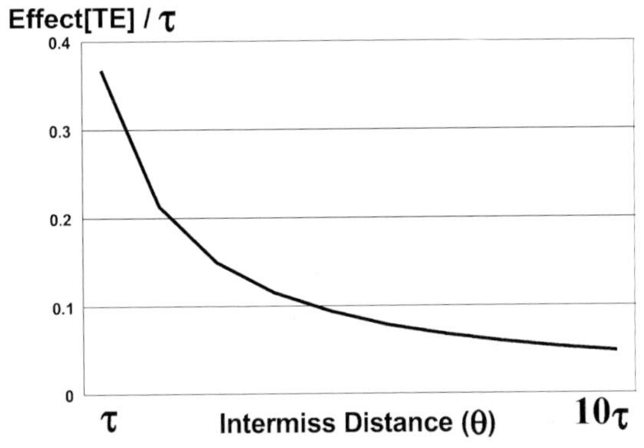

Figure 14. The trailing-edge effect normalized to the trailing edge as a function of the average intermiss distance.

Figure 14 shows the trailing edge effect normalized to (meaning divided by) the trailing edge. As is evident from the equation above, this is 1 – ((1 – e**(–U))/U). This is the fraction of the trailing edge that manifests as delay. It is exactly the fraction β that I used in the line size optimization previously.

Since ρ must be at least as large as τ, the curve has been plotted over the range of τ to 10τ. Note that at the point $\rho = \tau$, the utilization is 100%, and the normalized trailing edge effect is 1/e, which is about 0.37. This means that $\beta = 0.37$ when the bus is completely saturated. Of course, it would not make sense to run the utilization at 100%, as will be seen later. Caches should be made large enough so that ρ is at least several times larger than τ. For example, ρ should be large enough to make the approximation that "β is a constant" reasonable. If a cache is too small this is a very poor approximation, but the system performance in this case will also be very poor; so much so that the accuracy of this approximation is academic.

Having understood the self-interference aspect of trailing edge, we can now revisit an earlier simplifying assumption that CPI was linear in miss rate; shown previously in Figure 3. Figure 15 shows the actual situation. There is a large linear region in the center of the curve (where most machines operate) where there is some clustering of misses. In this region the average miss penalty is equal to a portion of the leading edge, αLE, because the clustering effect hides part of the leading edge, plus a portion of the trailing edge, βTE, because the clustering also causes self-interference.

For very low miss rates (at the left end of the curve), there is very little clustering, so the full leading edge is exposed, but there is no self-interference. The miss penalty in this region is just LE (which is how most people conceive of miss penalty). For very high miss rates the performance is dominated by the miss process, and the miss process is limited by whichever of {LE,TE} is the largest.

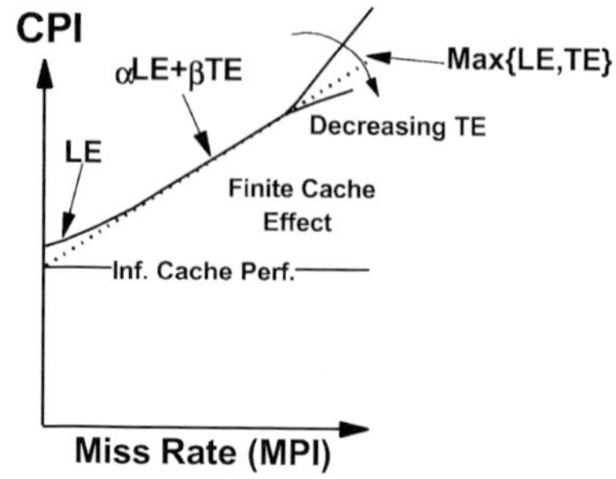

Figure 15. Three linear regimes: a revised view of uniprocessor performance (in CPI) as a function of the miss rate.

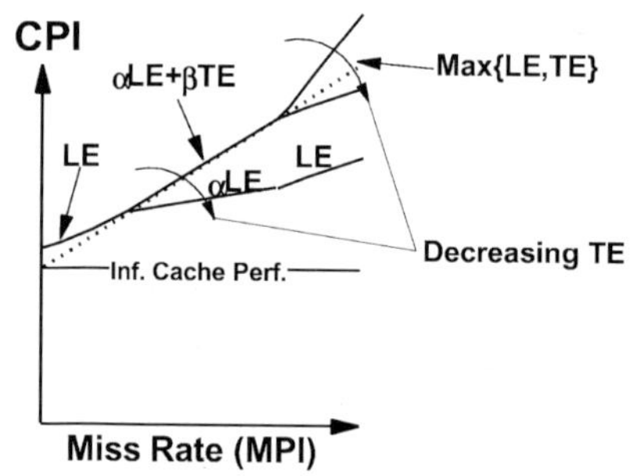

Figure 16. Three linear regimes: showing how a reduction in trailing edge changes the revised view of uniprocessor performance.

Figure 16 shows that if there were a way of creating enough bandwidth to make the trailing edge very small (e.g., moving an entire cache line in a cycle), the curve of the previous figure would degenerate into three regions. The outer two regions (very high and very low miss rates) would have a miss penalties of LE. The center region, where most machines operate, would have a smaller miss penalty, αLE. Excessive bandwidth is a good thing.

8.2. Queuing

Figure 17 shows a single bus drawn as an open queuing system. The single bus is the "server" in the system, and it has a fixed service time, τ. There is an infinite queue in front of the bus, and there are n "producers" (processors) that generate miss requests with Poisson arrival rates of r/n. The aggregate arrival rate is then r.

The question is: "How long are misses delayed because they have to wait in the queue?" In other words, what is the expected waiting time, E[wait]? This is an M/D/1 queue; the M denotes a Poisson arrival process, the D denotes a

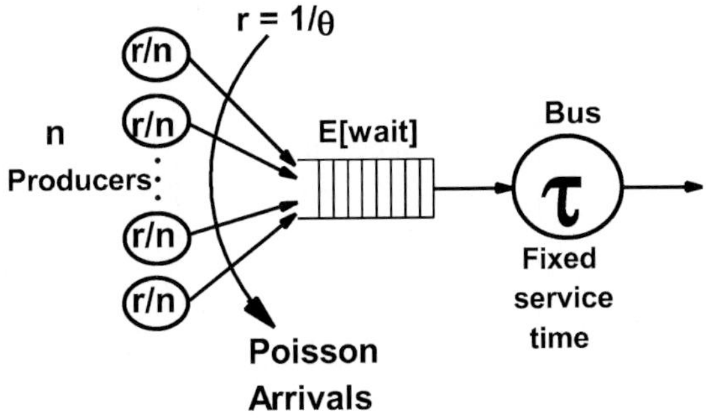

Figure 17. An open M/D/1 queuing model for a single bus.

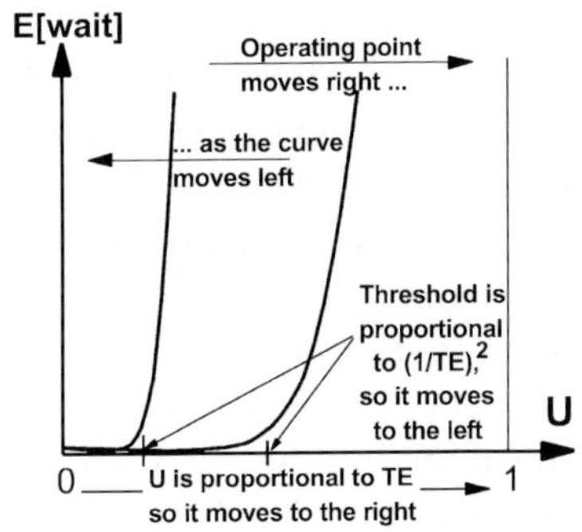

Figure 18. Queuing delay (E[wait]) in the open M/D/1 bus model as a function of utilization, depicted abstractly to show the effects of trailing edge.

deterministic (constant) service time, and the 1 indicates that there is a single server (bus). This model is *open* because there is no feedback: when misses are serviced, they leave the system; the fact that they were delayed (by E[wait]) has no effect on the arrival rate. We will revise this later. For an M/D/1 queue [14]:

$$\text{E[wait]} = \tfrac{1}{2}\,\text{U}\tau/(1 - \text{U})$$

Figure 18 shows E[wait] drawn as a function of U in cartoon style to illustrate a point. (Note that E[wait] and U are both functions of τ, so plotting this is not straightforward.) What this shows is that if the utilization is kept low, E[wait] is ignorable. At some "threshold" (in quotes, because the term is a little too strong), E[wait] "explodes." There is a serious nonlinearity in this relationship that is compounded by two things as τ increases. First, the threshold is determined by a series of terms that is dominated by a term in $(1/\tau)$**2. This means that as τ increases, the threshold point moves to the left. But since U is proportional to τ, as τ increases, the operating point moves to the right!

This is a "double whammy." What it means is that many systems designers (because of satisfactory bus speeds and smaller line sizes) are accustomed to

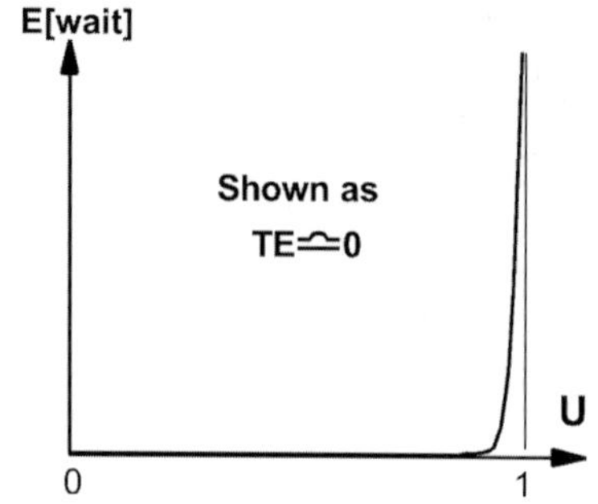

Figure 19. Queuing delay (E[wait]) in the open M/D/1 bus model as a function of utilization, depicted abstractly for very high bandwidth (i.e. as the trailing edge approaches 0).

having a low enough utilization so that there are no queuing effects. It is off their radar screens. But in a single doubling of the trailing edge (either by scaling the processor speed without being able to scale the bus speed, or by doubling the line size), queuing delay can cause a major problem. Repeating the point: *increasing the trailing edge moves the operating point out (to the right) while pulling the threshold point in (to the left)*.

This is not good. It is a direct statement about the importance of bandwidth, which can be used to reduce the trailing edge directly. Figure 19 shows the potential benefit of having excessive bandwidth. If very high bandwidth is used to remove trailing edge (say, by allowing an entire cache line to be moved in a single cycle), the "threshold" point moves far to the right. This would allow us to drive the bus traffic very hard, without seeing queuing delay. This could be used to prefetch very aggressively, with little concern for prefetching accuracy, and would allow the processor to run down multiple speculative paths. This is discussed in the next section.

Figure 19 does not depict reality. The bus utilization cannot approach 1 without causing queuing delay. This graph is an artifact of solving the system as an *open* queuing network, i.e., we did not feed back the delay to slow down the requesting processors. Figure 20 shows the corresponding closed queuing system. In this system, feedback is used to modulate the request rate as $r' = f(r, E[wait]))$, where r is the original (unmodulated) rate.

This can be solved by taking r out of the time domain, and then putting it back into the time domain. Recall that r is the *temporal* miss rate (in misses per cycle). Define the *time-independent* miss rate, r_I, as the number of misses per instruction. The assumption is that r_I is independent of how fast the processor runs. This cannot be extrapolated too far because of high-MIPS effects [8], but is a reasonable assumption for a processor that will not run at a drastically different speed. Now define a base performance, CPI_0, as the performance in the absence of queuing delay, and the actual (modulated) performance as CPI. The next three equations follow from things already known, and these lead

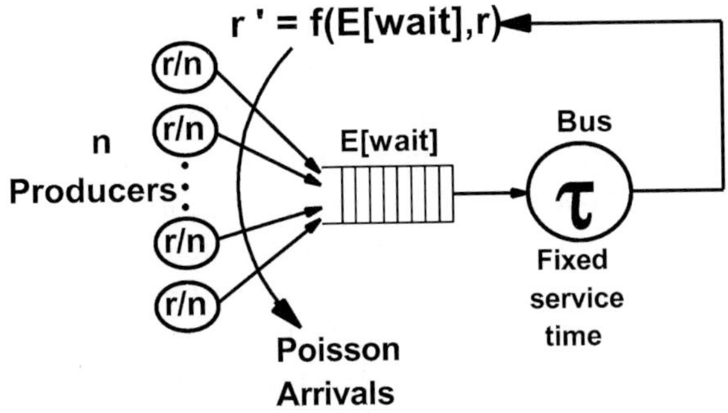

Figure 20. A closed M/D/1 queuing model for a single bus.

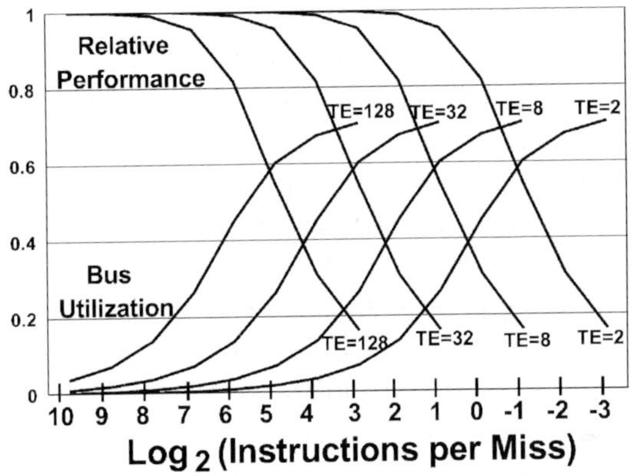

Figure 21. Bus utilization and relative system performance as a function of the log of the miss rate, shown for four values (three quadruplings) of trailing edge.

to the fourth equation, which (when substituting in the appropriate expressions from the first three equations) gives E[wait] as a function of itself, and the other known parameters. The fifth equation is the solution.

$$\begin{aligned}
\text{CPI} &= \text{CPI}_0 + \text{r}_\text{I}\text{E[wait]} \\
\rho &= \text{CPI}/\text{r}_\text{I} = (\text{CPI}_0/\text{r}_\text{I}) + \text{E[wait]} \\
\text{U} &= \tau/\rho = \text{r}_\text{I}\tau/(\text{CPI}_0 + \text{r}_\text{I}\text{E[wait]}) \\
\text{E[wait]} &= \tfrac{1}{2}\,\text{U}\tau/(1 - \text{U}) \\
\text{E[wait]} &= \tfrac{1}{2}\,(\tau - \theta + \text{SQRT}(\theta(\theta - 2\tau) + 3(\tau{**}2)))
\end{aligned}$$

where $\theta = \text{CPI}_0/\text{r}_I$. Based on this queuing delay, we recalculated CPI and U, and used these to create the plots in Figure 21.

Figure 21 shows four identical pairs of curves. Each pair of curves was constructed for a particular value of the trailing edge, these being 128, 32, 8, and 2 cycles per trailing edge, respectively as we go from left to right; each pair having a factor of 4 less trailing edge than the previous pair. For each pair of curves, the lower curve (which curves upward) is the bus utilization, and the upper curve (which curves downward) is the relative performance. Relative performance is CPI_0/CPI. (Recall that performance in MIPS is proportional to the reciprocal of CPI. That is, as CPI increases, performance decreases.)

Relative performance is plotted because it has the same range as utilization. That is, as utilization starts at 0 and climbs upward to something less than 1, relative performance starts at 1 (recall that CPI_0 is defined as the value of CPI when there is no queuing delay), and declines as queuing delay grows with utilization. The x-axis is the log (base 2) of the miss rate. Note that when the trailing edge is cut by a factor of 4 (going from one pair of curves to the next pair), the next pair is identical to the first pair, but is shifted to the right by 2 places. Since it is a log (base 2) scale, shifting to the right by 2 places represents quadrupling the miss rate.

That is, if we cut the trailing edge by a factor of 4, and increase the miss rate by a factor of 4, the utilization stays the same. At any fixed value of utilization (in any pair of curves), the relative performance also has a fixed value. This means that relative performance is purely a function of utilization. It doesn't matter whether the utilization is achieved by having lots of short events (a high miss rate and a low trailing edge) or by having fewer long events (a larger trailing edge with a smaller miss rate). The only thing that matters is the utilization itself. The bus doesn't care why the utilization is what it is. This is why every pair of curves is identical.

So how does system performance vary with bus utilization? Figure 21 shows that, as soon as the bus utilization exceeds 15%, system performance begins to drop off very slowly. At a bus utilization of 30%, system performance is degraded by more than 10%. And at a bus utilization of 40%, system performance has fallen off by nearly 20%. It is unacceptable to be losing more than 10% of the entire system performance to bus queuing. For this reason it is a good rule of thumb that bus utilization should not be driven much higher than 30% if the miss process resembles a Poisson process. (Note that this is not true for scientific workloads, for which the busses can be driven much harder.)

Therefore, if you can calculate how much data needs to be moved for a given workload (given the cache size – which will determine the miss rate, and assuming a nominal CPI – so that the miss rate can have a temporal interpretation), you should plan on providing at least $3\times$ the bandwidth required to move this much data. Without at least $3\times$ this bandwidth, the system performance will be unacceptable. When it comes to bandwidth, more is always better.

9. Bandwidth and Prefetching

The principal method that has been proposed by many to reduce the finite cache effect is *prefetching* [15, 16]. The idea is that if data are prefetched far ahead of their actual reference points, the misses that would normally occur will not occur. This requires that the reference pattern be predictable sufficiently far ahead in time so that misses are anticipated and avoided. In fact, this is the *only* way to avoid misses, excepting some hypothetical (but as yet, unspecified) new ways of programming that do not need to touch the same data in the same general order.

That is, misses are inherent to a program that ruminates on a working set at a certain rate, and that runs on a machine having a cache that is not large enough to hold that working set. There are four parameters that are used to characterize the efficacy of prefetching:

1. *Timeliness* is the degree to which correct prefetches are initiated early enough so that their associated misses do not occur.
2. *Coverage* is the percentage of misses that are correctly anticipated.
3. *Accuracy* is the percentage of prefetches that bring in data that is actually used.
4. *Bandwidth* will limit the ability to prefetch in a number of ways to be discussed below.

Note that, while these four parameters give lots of insight into the efficacy of a prefetching mechanism, they are not sufficient for accurately predicting performance. For example, coverage is the percentage of (original) misses avoided, but the miss ratio is not necessarily reduced by this amount exactly. The prefetching mechanism will cause other lines in the cache to be replaced, and some of that replacement activity will result in new misses that were not part of the original set of misses.

Note also that timeliness, coverage, and accuracy all work against each other. To achieve excellent timeliness, prefetches must be done well ahead of time – before there is much certainty as to the path being followed by the program. Thus, an algorithm that is very timely might not achieve good coverage (because the wrong path was anticipated), and may similarly have poor accuracy (because it brings in the wrong lines). Similarly, to get high coverage, a prefetching mechanism cannot be conservative about deciding whether to prefetch things. High coverage usually has the side-effect of poor accuracy. To get high accuracy requires that the prefetching mechanism be very certain about what is prefetched, so the coverage will be lower.

Since the accuracy cannot be 100%, and since prefetching (correct or not) will cause new misses because of replacements done by the prefetched data, and since correct prefetching should enable the processor to run further down speculative paths and generate new misses, the amount of traffic will increase when any type of prefetching is done. This can, and usually does, put a strain on the bus bandwidth.

In modern machines, real prefetching mechanisms seldom improve performance even when scoring high marks on the first three metrics: timeliness, coverage, and accuracy. This is because machines today do not have a huge surplus of excess bandwidth available. When the prefetching mechanism makes any "mistake," the limited bus bandwidth makes the system very unforgiving. The cost for mistakes can more than negate any gains made by correct prefetching.

The definition of a "mistake" can be very loose. A mistake could mean bringing in the right line at an inopportune time so that it gets in the way of a demand miss – because there is insufficient bandwidth to handle both. Or it could mean bringing the right things in, except doing it in the wrong order. A demand miss sequence may allow the machine to run faster than it could with prefetching if the prefetching fetches the right things in the wrong order.

For example, Figure 22 shows four different flows for the same sequence of four misses. The first flow (on the top line) is exactly the flow that was described in Figure 12. It is a nominal sequence of misses, M1, M2, M3, and M4, with no prefetching.

Suppose that a prefetching mechanism knew that these four misses were coming in this order. The second line of the figure shows the flow starting with a sequence of four prefetches P1, P2, P3, and P4, for the four corresponding misses M1, M2, M3, and M4. Note that all prefetches are issued far enough ahead of time (in terms of the leading edges) to eliminate all four misses. The first miss is eliminated: A1 appears as the first event following P4; there is no miss M1. However, the trailing edge of the first miss prevents the second miss from disappearing, despite the fact that prefetch P2 had been issued well enough in advance of when miss M2 was to occur. So miss M2 is not removed, although its penalty is reduced somewhat. Similarly, the trailing edge of M2 prevents prefetch P3 from eliminating miss M3. The fourth miss, M4, is eliminated.

If we put the event A4 in correspondence between the first two lines, we can see the amount of delay that was removed by "perfect" prefetching. By "perfect," we mean that:

1. All of the misses were predicted (100% coverage).
2. All of the predicted misses were predicted in the correct order.
3. All of the prefetched lines were used (100% accuracy).
4. All prefetches were issued far enough ahead of time so that they should have eliminated their corresponding misses (adequate timeliness).

Despite having perfect prefetching, we only eliminated half of the misses. This shortcoming is *entirely* because of the bandwidth limitation.

Real prefetching mechanisms will not be perfect. On the third line of Figure 22 the same flow is shown with the prefetching being only slightly less "perfect." In this flow I slightly permuted the order of the prefetching as follows: P2, P1, P4, and P3, i.e., I swapped the first two prefetches, and swapped the last two prefetches. Otherwise, all other aspects of the prefetching are perfect. In this flow only one miss, M2, is eliminated. If we put the event A4 in correspondence between the second and third lines, we can see

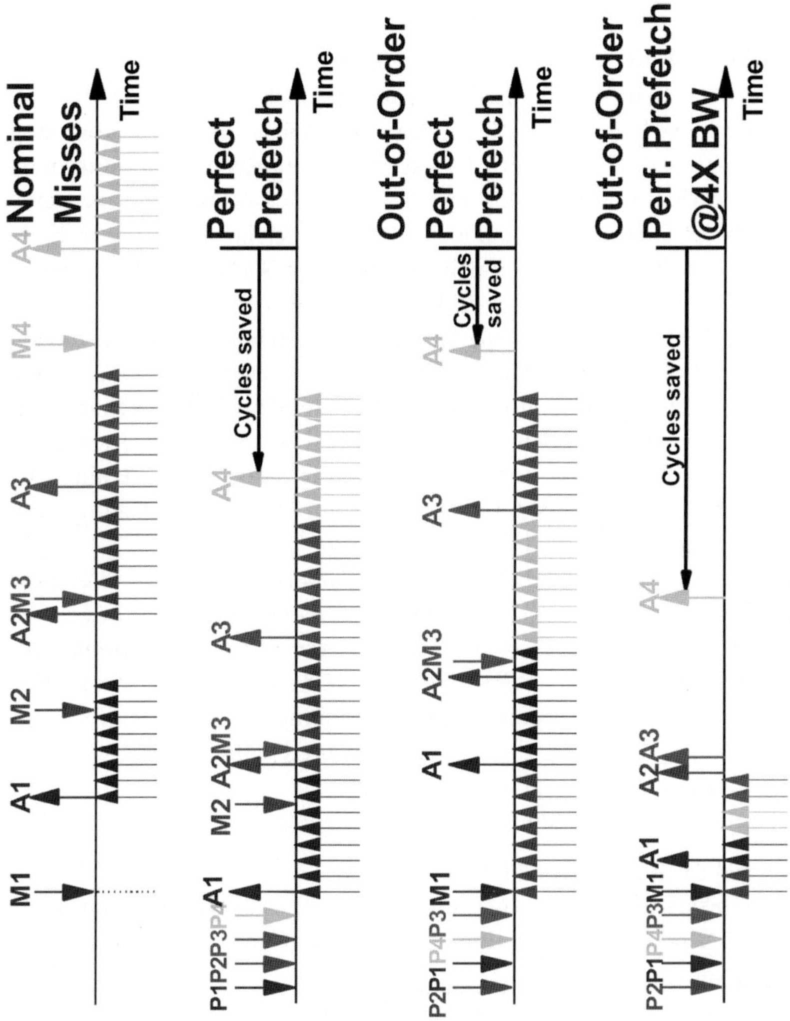

Figure 22. Four flows showing the same sequence of four misses using: no prefetching; perfect prefetching; out-of-order perfect prefetching; and out-of-order perfect prefetching using 4× more bandwidth.

that, by making this slight mistake, we have lost about half of the gain made by "perfect" prefetching.

The fourth line of the figure shows the same flow as the third line, except the bandwidth has been quadrupled. This cuts the trailing edges by a factor of 4. In this flow three of the misses have disappeared, and the performance gain is much more than what was achieved with "perfect" prefetching on the second line. This shows that bandwidth has a profound effect on the efficacy of prefetching.

In fact, no real prefetching mechanism is as good as the mechanism depicted on the third and fourth lines of Figure 22. As explained, this mechanism achieves perfect coverage and accuracy, with adequate timeliness. To take this one step closer to reality (but still unrealistic), let's see what happens if we insert a single bad prefetch into the middle of the prefetch sequence. This would keep the coverage at 100%, and drop the accuracy to 80% – which is unrealistically high.

Figure 23 shows this flow. The top line in the figure is a copy of the third line in Figure 22. It is repeated here for easy comparison with the second line. As we saw, permuting the order of the perfect prefetch sequence caused us to give back roughly half of the performance achieved by the same sequence issued in the correct order. In the bottom line of Figure 23, an incorrect prefetch, Px, has been inserted into the stream. The result of this is that so many cycles are lost so as to make this flow worse than the original flow without prefetching (the first line of Figure 22). Access A4 appears to the right of the finishing point shown for no prefetching.

Remember that, while this sequence is hypothetical, no real prefetching mechanism is quite this good. *In real processors, whenever prefetching is used, despite achieving fair levels of coverage and accuracy, bandwidth limitations almost always cause the prefetching mechanism to hurt performance*. This statement pertains to business computing. Prefetching can be used effectively in scientific computing, where there are regular predictable strides, very predictable branches, and few surprises.

I have said that real prefetching mechanisms are less than perfect because:

1. They fail to anticipate all of the misses.
2. Of the misses anticipated, their ordering changes as the algorithm starts to work, hence as the algorithm adapts, it issues many of the prefetches in the wrong order.
3. In addition to prefetching useful data, they prefetch data that is not useful.
4. Even when prefetching correct data, they prefetch at inopportune times, and sometimes get in the way of exigent misses that were unanticipated.

In addition to these obvious drawbacks of imperfect prefetching, there is another major effect that is intrinsic to how I have defined the leading edge. Recall that I had said that, when a miss occurs, the first packet returned through the hierarchy is the packet containing the specific datum that was requested

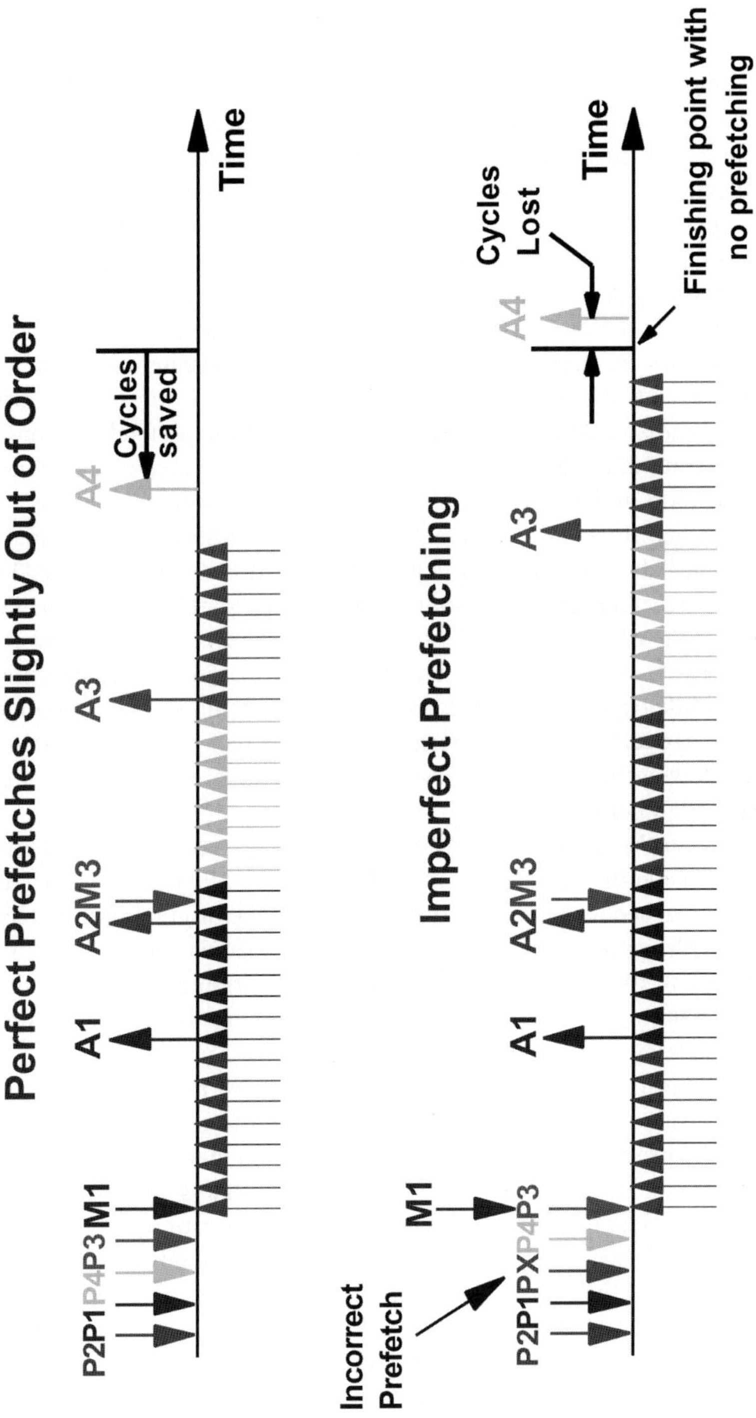

Figure 23. Two flows showing the same sequence of four misses using: out-of-order perfect prefetching; and imperfect prefetching.

by the processor, i.e., the packet containing the datum that generated the miss. A processor-generated miss stream contains specific (byte or word) addresses, since these are the actual data being requested.

Many prefetching algorithms work only at the granularity of cache lines, and make no attempt to predict what packet within the line will be used first. For example, a commonly used *next sequential prefetching* algorithm simply predicts that if line x is referenced, then line (x + 1) will also be referenced [17]. (This is a simple use of spatial locality; if this algorithm works, doubling the line size achieves much of the same benefit.) But the algorithm does not anticipate which packet in line (x + 1) will be the first packet referenced.

When a prefetching mechanism merely issues line addresses to the miss-handling apparatus, the prefetched line will be returned starting with packet 0, and continuing with the remaining packets of the line in order. If the first referenced packet is packet y, then the leading edge of the prefetch subsumes part of the trailing edge. Specifically, the leading edge becomes lengthened by y times the ratio of the processor and bus frequencies, minus 1.

Because of this effect, if the prefetching mechanism is not extremely timely, it can actually be better to wait for the demand miss to identify the correct packet than to let the prefetching mechanism request the line starting with packet 0. For example, if the prefetching mechanism does a correct prefetch to (packet 0 of) a line 10 cycles before the demand miss happens, but the demand miss goes to the middle packet in a line having a 32-cycle trailing edge, the correct prefetch done 10 cycles early will return the requested data about 6 cycles later than it would have been received had we suppressed the prefetch, and allowed the miss to happen normally.

But remember that excessive bandwidth makes the system very forgiving of "mistakes" made in prefetching. To underscore this, Figure 24 shows the same hypothetical sequence of Figure 23, both with the original trailing edge (which causes the prefetching to lose performance) shown on the first line, and one-quarter of the trailing edge (achieved by quadrupling the bandwidth) shown on the bottom line. On the bottom line a big savings is achieved despite the mistake made by the prefetching mechanism.

In fact, there is a genuine synergy between prefetching and excessive bandwidth. In combination they achieve far more than would be expected by studying either one by itself. Our laboratory did an experiment in which we annotated all of the misses on an execution trace (identified by running the trace once in a finite-cache simulation). These annotations were then post-processed to find the string of branch instructions (up to 20) preceding each miss. We then did 20 more runs to find the optimal point (at one of the 20 branches) to issue a prefetch for each miss [15].

We reran the annotated trace, issuing prefetches for the misses at their optimal timeliness points. We repeated the experiment multiple times preselecting fixed percentages of the misses to prefetch (to allow us to artificially

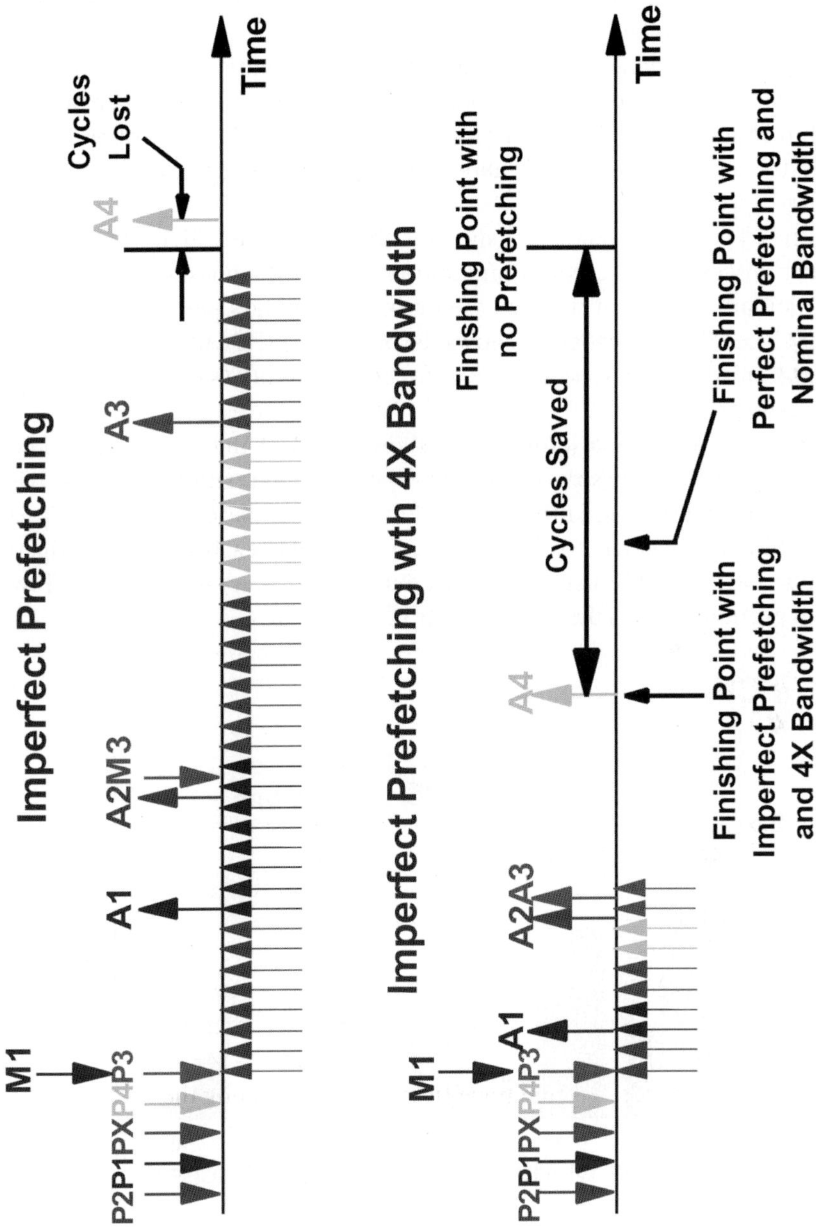

Figure 24. Two flows showing the same sequence of four misses using: imperfect prefetching; and imperfect prefetching using 4× more bandwidth.

specify the prefetching coverage), and also issuing a preselected percentage of prefetches to known wrong addresses (to allow us to artificially specify the prefetching accuracy). We also had the flexibility of specifying other (than the optimal) timeliness points in terms of how many branches ahead of the miss each prefetch was to be done.

This gave us a framework to independently specify the timeliness, coverage, and accuracy of a hypothetical prefetching algorithm – even when no real algorithm could achieve these numbers – so as to explore the space of prefetching, and to study the amount of bandwidth required to support any particular algorithm.

Figure 25 is a set of bar charts showing the finite cache effect (FCE) for a particular processor and cache configuration generated with this technique. In this particular experiment we used the optimal timeliness points, and perfect accuracy. The leftmost set of bars shows the base case FCE with no prefetching. The next four sets of bars show the FCE for coverages of 50%, 67%, 75%, and 100%, respectively.

Each set of bars has a leftmost bar showing all of the misses, a middle bar showing just the misses on the data (operand) side, and a rightmost bar showing just the instruction misses. In addition, each of these bars has a striped portion, which subsumes a solid portion. The height reached by the striped portion denotes the FCE using a 16 byte bus. The (lower) height reached by the solid portion denotes the FCE using a 128 byte bus.

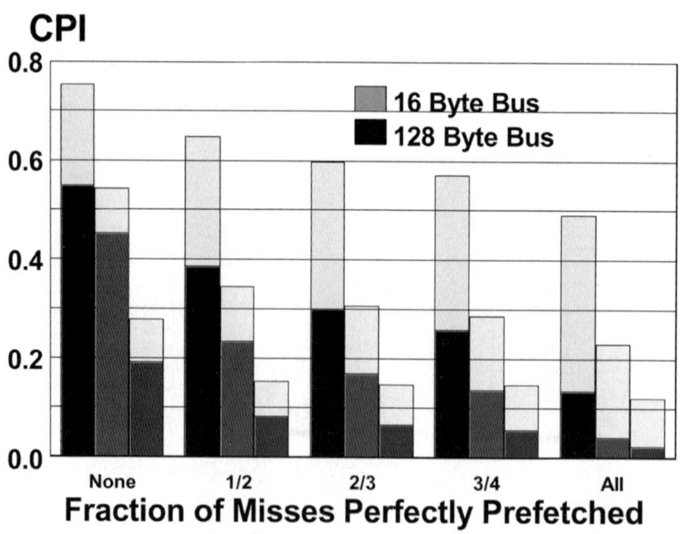

Figure 25. The finite cache effect (in CPI) for a 128-byte line size as a function of the fraction of misses prefetched (perfectly). At each fraction the three bars show: showing all misses, D misses only, and I misses only; and for each bar the effect is shown using a 16 byte bus and using a 128 byte bus.

The line size is 128 bytes, so the solid portions of the bars have no trailing edge effects. It is particularly useful to plot the bars this way, because the striped portion of any bar can then be directly interpreted as the trailing edge effect for that bar.

Looking only at the leftmost bars in the first and last sets, it can be seen that (from the heights of the striped sections) by prefetching all of the misses, we took the FCE down from 0.75 CPI to 0.5 CPI. That is, prefetching with optimal timeliness and with perfect coverage and accuracy only removes 1/3 of the FCE because of bandwidth effects.

Looking just at the first bar in the first set (all misses with no prefetching) it can be seen that by eliminating all bandwidth effects (removing the trailing edge) we took the FCE down from 0.75 CPI to 0.55 CPI. In and of itself, this is not quite as good as perfect prefetching; but it shows that, without prefetching, bandwidth effects account for over 25% of the FCE.

However, if we do perfect prefetching *and* remove the bandwidth effects (the solid portion of the leftmost bar of the last set), we take the FCE down to about 0.12 CPI. This is more than a 6× reduction in FCE, and is quite dramatic. It obviously shows that there is a very strong synergy between prefetching and increasing the bandwidth.

Figure 26 shows this even more directly. All that was done here is that each of the bars in Figure 25 was normalized to itself. All of them are 1 unit high. This allows us to see what portion of each bar (the striped portion) represents bandwidth effects. Note that the heavier the prefetching (as we move from left to right), the bigger the striped portion becomes. This demonstrates that it is difficult to leverage prefetching without an oversupply of bandwidth.

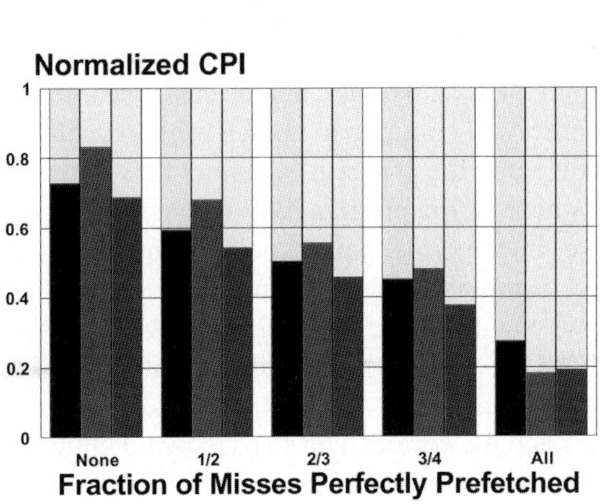

Figure 26. The bar chart of Figure 25, in which each pair of bars (for a 16 byte bus and for a 128 byte bus) is normalized to itself to show the trailing-edge effects (dotted bars) in relation to the leading edges (solid bars).

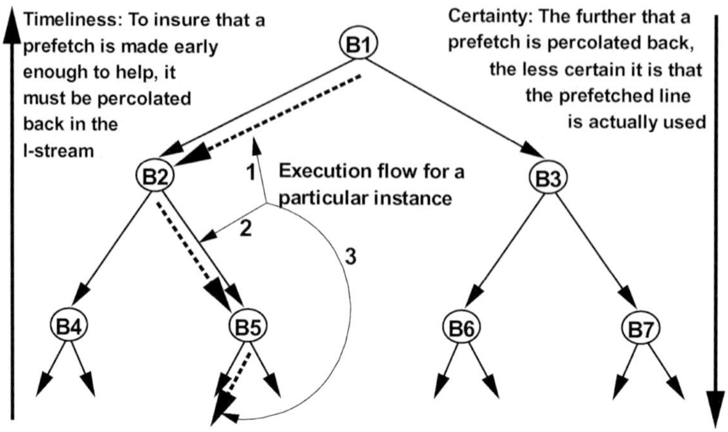

Figure 27. A static program conceived as a directed acyclic graph (a binary tree) in which each node denotes a conditional branch instruction, and in which a dynamic instance of the program is represented by a particular flow through the nodes from the root to a leaf node.

Finally, work is being done in compiler-directed prefetching. The basic idea is that if a compiler can anticipate the path to be taken through a program, and if it can anticipate the data that will be touched along this path, then "touch instructions" (which do non-architected fetches to data) can be put into the code far enough ahead of the actual use of the data to prefetch the data and eliminate misses.

Predicting the path taken through a program is a matter of predicting the branches in it. Figure 27 shows a static conceptualization of a program drawn as a directed acyclic graph; in this case a binary tree. Each node in the graph represents a branch that will be encountered along a permitted dynamic flow through the program. (The same static branch can appear in many of the nodes, e.g., a looping branch will continue to branch back to itself.) The figure shows that, at each branch point, the program can either flow to the left or to the right. Any particular instance of the program will follow exactly one path down the tree to a leaf node. In principle, paths to all leaf nodes are allowed, and will occur for some set of inputs to the program.

Figure 27 shows one particular flow being followed: at branch B1 flow goes to the left; at branch B2 flow goes to the right; and at branch B5 flow goes to the left. Suppose that there is a miss that will occur down the left path following branch B5. To prefetch the correct data so as to avoid taking this miss requires knowing what that data is, and it requires prefetching it early enough in the flow so as to avoid any delay when referencing the relevant data.

For example, let's assume that to be timely enough to eliminate penalty for this miss, the data must be prefetched before the flow reaches branch B1. We percolate the prefetch (for the miss occurring along the left path after branch B5) up the tree past branches B5, B2, and B1. If we have predicted all three of these branches correctly (when we perform the percolation), then the prefetch is a good prefetch. If we predicted any of the branches wrongly, then the prefetch may be a useless prefetch that puts a strain on the available bandwidth, and that replaces other valuable data in the cache.

Therefore, to do a good job of eliminating penalty for misses (assuming that they can be anticipated correctly) requires that the prefetches (or "touch instructions") be percolated up past a number of branch points. The further up the tree we can percolate the prefetch, the better the timeliness. But the further up the tree we percolate the prefetch, the less certain we are that the prefetch is a good prefetch, because the less certain we become that we can predict the entire string of branch instructions correctly. This figure directly shows how timeliness works against accuracy and coverage.

To give a rough idea as to path certainty as a function of percolation distance, Figure 28 shows the approximate relationship. To plot this figure I assumed a fixed predictive accuracy for branches (which is not really the case). Curves are drawn for predictive accuracies of 80%, 85%, 90%, 95%, 98%, and 99%. The x-axis is the percolation distance – the number of branches predicted along a percolation path. Each branch prediction is treated as a Bernoulli trial with the prescribed probability of being correct. The y-axis is the path certainty – the probability that we are still on the correct path following $\times$ branch predictions at the prescribed probability.

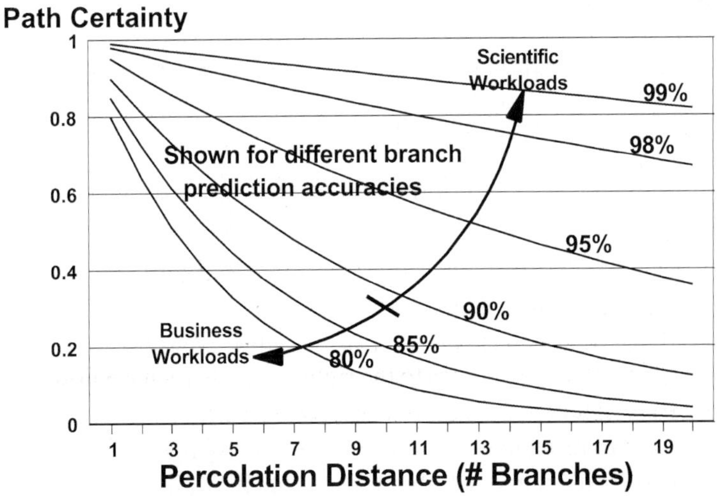

Figure 28. Path certainty (the probability of being on the correct path) as a function of percolation distance (the number of branches through which a prefetch is percolated up the tree) for a range of branch prediction accuracies from 80% to 99%.

Typical commercial business programs have branch prediction accuracies in the 80% range, unless very exotic prediction algorithms are used. At an 80% predictive accuracy, the figure shows that, after 3 branches, there is only a 50% chance that we are on the right path. After 5 branches there is only a 30% chance of being on the right path, and after 10 branches there is only a 10% chance. This shows that, with a predictive accuracy of 80%, it is practically hopeless to do timely prefetching via touch instructions.

At 98–99% predictive accuracy, touch instructions are viable. Again, scientific workloads have very high predictive accuracies, because they are dominated by looping branches. But this is not realistic to do for typical commercial business workloads given an adequate, but not excessive, amount of bandwidth.

However, suppose we wanted to use a percolation distance of three branches, and we were given 8× as much bandwidth to do prefetching. Looking again at Figure 27, a brute force approach would be to prefetch down all 8 (leaf) paths in the figure as soon as we knew we were going to encounter branch B1. With slightly more finesse, we could prune off some of these paths as being unlikely, and could increase the percolation distance, perhaps dramatically.

Again this illustrates that, given huge amounts of excess bandwidth, we would enable very rich prefetching, and push back the "memory wall." But with bandwidth that is merely adequate (adequate for a running program that does not prefetch) the limited bandwidth inhibits the prefetching from working very well.

10. Lessons Learned

There are four main principles that pervade all of the preceding discussions:

1. What seems to be a sufficient amount of bandwidth might not actually be enough.
2. An insufficiency of bandwidth will manifest itself as latency.
3. Bandwidth and content (cache capacity) are exchangeable.
4. Prefetching and high bandwidth are mutually synergistic.

Let's review each of these points.

I showed that trailing edge effects can cause lots of problems, both directly and indirectly. Directly, trailing edge will diminish the L1 bandwidth available to the processor, it will cause the processor to stop when the miss facilities are full, and upstream references to an incoming line will be delayed. Indirectly, clustered misses will incur delays by running into the trailing edges of previous misses, and there are general queuing effects within the system due to both local and remote traffic.

I showed that queuing effects on a bus become unacceptable (10% or more loss of system performance) when the bus utilization is driven beyond 30% if the miss process is Poisson. This means that if you know the *average bandwidth* (how much data must be moved in a certain amount of time) required by a particular application, you should make sure that you have more than 3× this bandwidth available.

When the bandwidth is barely sufficient, perturbations in the miss stream will cause delays that add to the basic leading edge delay of a miss. In addition to the nominal logic, wire, and array flow in the leading edge, there are numerous opportunities for queuing at shared resources. When this happens the leading edge is increased by queuing delays if there is not a large surplus of bandwidth.

Since the principal role of the bandwidth in a processor system is to service cache misses, and since the miss rate depends on the capacity of the cache (and the line size), bandwidth and content are exchangeable. Making caches larger allows them to function well with less bandwidth. Conversely, providing very high bandwidth to a cache allows it to be smaller.

Bandwidth that has been provided to do an adequate job of servicing misses is probably insufficient if prefetching is retrofitted into the system. To even consider prefetching requires that there be a large surplus of bandwidth available. Having bandwidth that is merely adequate will not allow the prefetching to work well. In fact, it is more than likely that a processor will actually lose performance by prefetching – even when the standard prefetching metrics are auspicious – if there is not a large surplus of bandwidth.

On the other hand, if there is a large surplus of bandwidth, prefetching will work extraordinarily well. These two things (bandwidth and prefetching) work in strong synergy. Very large amounts of bandwidth enable prefetching to be sufficiently unconstrained so as to achieve very high coverage. Specifically, if accuracy is unimportant (which it is if the bandwidth is excessive), the coverage can be driven very high.

When it comes to the miss process, bandwidth and latency are not independent things. Since a deficiency in bandwidth manifests itself as latency, it is just as important to a system to provide high bandwidth as it is to keep the access times of the caches short.

11. Future Trends in Systems: Why Bandwidth Matters

What are the coming trends in computer systems, and will these trends stress the bandwidth even more? I touched on a few of those trends in Section 3. I will reiterate those, and talk about a few more trends. None of them is auspicious, and there is every reason to believe that bandwidth scaling is the next major systems roadblock after power management.

Figure 29 shows a general trend that has been progressing since the beginning of the electronic computing industry. The graph starts *circa* 1950, when

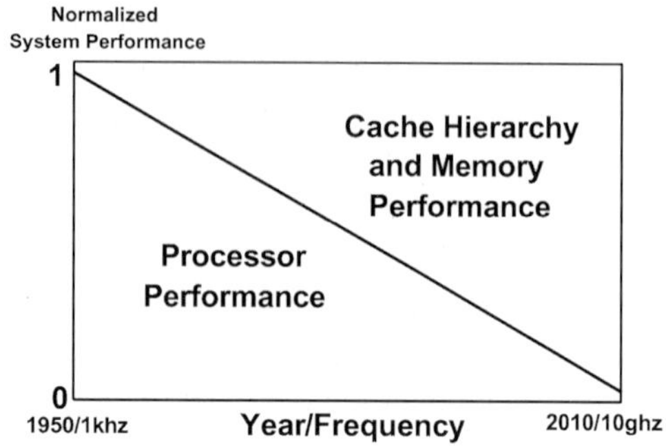

Figure 29. An abstract historical trend line showing that portion of system performance that is determined by the processor, and the (remaining) portion of system performance that is determined by the cache and memory hierarchy, starting from 1950, and projected to 2010.

the frequency was on the scale of 1 kilohertz [1], and it continues past today (into the next couple of technology generations) until 2010, where the frequency will not likely exceed 10 Gigahertz. This is a 7 order-of-magnitude increase over the 60-year span shown.

The graph shows that portion of the system performance that is determined by the processor core itself (the lower half plane), and that portion of the system performance that is determined by the cache and memory hierarchy (the upper half plane). The earliest computers did not have caches, or even hierarchies of memory. Programs were fairly self-contained, and were written to run within the single-level memory on the processor, which was very slow. Essentially all of the performance was dependent on the processor itself.

As processors got faster – mainly because of Moore's Law, but also through advances in microarchitecture – the processor portion of the performance shrunk. The memory portion of performance did not shrink at nearly the same rate, so when viewed proportionately, it grew *relative to the processor speed*, which was improving much faster. In attempting to scale the memory performance, caches were invented [18], and then they became hierarchical.

It is fundamental that the memory system could not scale at the same rate that Moore's Law made the processor scale. While Moore's Law made memory denser, as the processors became faster they needed more and more memory.

[1] In fact, there is no straightforward interpretation or notion of a "cycle time" in computers of this era. Pipelining had not yet been invented, and many operations were performed asynchronously. In addition, many operations were bit serial operations. ENIAC was working by 1946, and had an internal oscillator that produced pulses at 100 kilohertz that were phase shifted and used to produce other pulses which controlled the machine [19]. The processing yielded roughly 5000 arithmetic operations per second. Thus, the 1 kilohertz rate shown in Figure 29 is a very rough "equivalent" (order-of-magnitude) rate based on trying to translate the basic steps of performing an operation into time. The basic number used here comes from interpreting Figure 1.5-1 on page 37 in a book by Matick [20].

(There are various rules of thumb about the number of Megabytes per MIPS required to have a well-tuned system.)

Because each generation of systems required much more memory than the previous generation, although the memory technology became denser, the physical size of the memory system did not shrink nearly as fast as the processor did. This meant that the basic latencies up through the memory hierarchy and back expanded relative to the processor cycle time.

Further, since memory chips are sold by the billions, they are a commodity. The financial driver to any commodity item is cost, since margins are slim in any commodity market. The DRAM business is purely cost-driven; all of the emphasis is on density so as to make the margins viable. In DRAM, achieving the highest density precludes achieving the highest speed [21].

This is not true in logic technology, which is not (yet) a commodity technology, at least not in high-performance microprocessors. This is a second factor that causes the speed gap to widen. As we can see in the figure, these factors taken together cause the memory portion of the performance to become more and more significant. In large server systems today, more performance is lost to the cache and memory hierarchy than is lost to the processor; and this is getting worse. This problem is sometimes referred to as the "memory wall" [22].

The latencies in the cache and memory hierarchy cannot change very much. The only other lever to use to close this gap is bandwidth. Massive bandwidth can be used to do very aggressive prefetching (much more aggressive than the mechanisms generally discussed in the literature today); and remember that bandwidth and content are mutually fungible. If we can facilitate massive bandwidth, it can take the place of added capacity. This is the only real approach to smashing through the "memory wall."

In Section 3 I had pointed out that frequency scaling is slowing down considerably, at least in CMOS technology. While this offers some relief, there are three other trends in system design that (together) put even more pressure on the on-chip caches:

1. There is a concerted move to multiple cores per chip [2, 23]. This scales the on-chip MIPS as aggressively as frequency did, and it puts more pressure on the on-chip caches.
2. The cores are becoming multithreaded, and the degree of multithreading is growing [6]. This means that each core must have multiple working sets co-resident. This also puts more pressure on the on-chip caches.
3. With virtualization technology, the chips are being used to run as multiple independent systems simultaneously [7]. The on-chip caches must accommodate the storage associated with multiple virtual systems.

While all of these trends are advances, they all put more pressure on the on-chip caches, hence on the off-chip bandwidth. In addition to these trends, there are four other trends that merit comment.

First, off-chip drivers and receivers are much more complex, and are in a different field of design than CMOS logic. While the on-chip logic has been running at over 1 Gigahertz since the late 1990s, it is much harder to run off-chip signals at commensurate speeds. In the first place, at much above 1 Gigahertz, careful impedance control is needed in the packaging. The driver impedance must be matched to the wiring impedance, and the wiring must be terminated – which burns more power. Beyond a Gigahertz the signals may need special encoding, and may need special pre-distortion added to them. The circuits to do this take more area and burn even more power.

This is why as processor speeds have climbed into the 5 Gigahertz regime, the bus speeds have not kept pace. (And yes, it is known how to design off-chip drivers this fast, but doing it may be a poor choice because of area, power, and packaging costs.) When the processors run at these frequencies, the busses that service them run at a 2:1 or even a 3:1 reduction.

If logic speed continues to scale (say to 10 Gigahertz), the bus ratios will probably get worse. Recall that the trailing edge (hence the bus utilization) is directly proportional to this bus ratio. If processors get faster, the trailing edges will increase – probably almost in proportion.

Second, advances in microarchitecture have processor threads doing more speculation. This means executing past conditional branches (which may be guessed wrongly), doing more out-of-order processing, and running down multiple paths (knowing that not all of them are correct) [24, 25]. This increases the number of misses per useful instruction processed. Speculation fundamentally requires more bandwidth.

Third, as cache sizes increase, line sizes increase. The reason for this is that, even when the cache is made larger, hence slower, it is desirable to maintain the speed of the directory lookup. Since the directory has to have one entry per cache line, the only way to maintain the directory speed is to keep the number of entries (hence cache lines) constant. Thus, when the cache size is (say) doubled, the directory speed can be maintained only by doubling the line size.

Since the trailing edge is directly proportional to the line size, making the on-chip cache larger will drive the demand for more off-chip bandwidth. This is because, while doubling the cache will cut the miss rate (hence bandwidth) by roughly SQRT(2), doubling the line size will double the bandwidth needed for each miss. The net effect is to increase the bandwidth required by SQRT(2). Well if this is true, you might ask: "Then why are we making the caches larger?"

Very simply, the argument above assumes that nothing else on the chip is changing except for the cache size. This will not be the case. The reason that the caches are made larger is in response to the growing miss rates as other things on the chip continue to scale (speed of the processors, number of processors, degree of multithreading, and number of logical partitions). The caches are made larger so as to prevent the off-chip miss rate from growing too much. So the assertion that doubling the cache size will cut the miss rate is not necessarily true.

Were it true, the demand for more bandwidth would only grow by SQRT(2). The actual demand for bandwidth will grow by a lot more than this. One other method of increasing the line size without increasing the trailing edge (in direct proportion) is to sector the cache lines. While sectoring has not been used since the early 1960s [18], expect to see a resurgence in sectored caches in response to the bandwidth demand.

Fourth and finally, symmetric multiprocessor (SMP) sizes – meaning the number of processors in SMPs – are growing. Both the number of processors in a chip, and the number of chips, boards, etc., in the largest systems are increasing from generation to generation. This means that, when traffic leaves a processor chip, there is increasingly more other traffic in the system that it has to contend with. Thus the "value" of the off-chip bandwidth diminishes, even if its magnitude remains constant. It costs more in prioritization and queuing to move the bandwidth (leaving the chip) through the rest of the system.

All of these trends portend that bandwidth will be the next major bottleneck if scaling at any level (system, chip, processor, circuit, etc.) continues. In the final section I discuss the technology evolutions that are likely to occur in response to this demand.

12. Future Trends in Technology to Support Bandwidth Demands

Because bandwidth and content are mutually fungible, easing the growing demand for bandwidth requires technology innovation in two different areas. First, we need technology innovation that enables much denser storage for on-chip caches; and second, we need innovation in interconnection technology. I outline some likely directions in both areas below.

12.1. Dense On-chip Memory Technology

On-chip caches have always been made out of static random access memory (SRAM), and usually, the SRAM cell was a 6T (meaning 6 Transistor) cell [26, 27]. The 6T cell comprises two cross-coupled inverters, with a pass transistor on each pole to allow differential sensing and storing. As device sizes scale much below the 90 nanometer node, there appear to be stability issues with this type of cell because of the inherent variability in transistor strengths caused by localized dopant fluctuations in devices this small.

Thus, even if 6T SRAM were capable of handling the expanding working-sets of workloads (which is unlikely – as described in the previous section), as logic technology continues to scale, it appears that the density of 6T SRAM cannot continue to scale at the same rate as logic. Thus, there needs to be a denser technology solution to on-chip storage.

Today the densest semiconductor storage is dynamic random access memory (DRAM), which has a 1T1C (meaning 1 transistor, 1 capacitor) cell [28]. The problem with commodity DRAM is that it is too slow for cache applications. As previously explained, the reason that commodity DRAM is slow is that it is designed exclusively for density *because* it is a commodity. Embedded DRAM (EDRAM) is a relatively new technology that has been used in the ASICS domain (where density and low power was required), but it has not yet been deployed in the high-performance domain.

EDRAM is basically a different circuit-design point for DRAM that emphasizes performance as well as density [21]. It is less dense than DRAM, bus appears to be between $2\times$ and $4\times$ as dense as SRAM, depending on the speed required, which is in the range of $2\times - 4\times$ slower than SRAM. It represents a new {performance, density} point in on-chip storage. At the highest level of on-chip cache, the added latency is not nearly as significant a detractor as the added capacity is a benefit.

It is obvious that EDRAM will become ubiquitous in future high-performance processor chips. Beyond EDRAM there are two other dimensions in which to improve density further. The first is some new technology (perhaps magnetic) that is not yet known. The second is literally another dimension: three-dimensional (3D) circuit chips.

There has been lots of rudimentary work done in 3D processing technology. This work makes it fairly clear that 3D structures can be built. The most direct approach is to build two or more planes of 2D circuits, and laminate them together into a 3D stack that is interconnected with vias [29, 30]. The challenges that have not yet been ironed out are power distribution, cooling, design tools, and microarchitecture. These are things that will be solved, once they become important enough.

The real promise of these structures is that with dense vias, staggering amounts of bandwidth are possible between the planes. With cache levels stacked directly above each other, the number of connections can be extremely large (orders of magnitude more than today's bus widths), and the interconnection distance can be very short – a vertical via of a few hundred microns, with very little horizontal connection required. In addition to this making the transmissions wide and fast (thereby eliminating trailing edge), the short dataflow wires have the potential of being much lower power interconnections than what is achievable in horizontal on-chip busses today.

So it is clear that EDRAM will be the next step in on-chip storage. Several generations later, we expect to see 3D chips in which one or more planes will be EDRAM planes implemented as adjacent levels of cache. In addition to 3D enabling more on-chip capacity, it has the potential of offering staggering bandwidth between cache levels at very low power.

12.2. Interconnection Technology

Electronic signaling will not be sufficient to meet bandwidth requirements if systems continue to scale. To justify this statement I first use some approximate numbers, and then posit the parameters of a system in the near future to make an estimate of the required bandwidth. I discuss the implications of achieving this bandwidth electronically, and show that this particular case is actually understated, and that systems cannot evolve much further using only electrical interconnections.

With a 256 byte line size, a 1 Megabyte cache can have a miss rate in the range of 1 miss per 60 instructions in commercial business computing. If the largest on-chip cache is 16 Megabytes, this is four doublings in capacity (over 1 Megabyte), which drops the miss rate by a factor of 4 (assuming that the SQRT(2) rule holds). If the chip is being run as four virtual (disjoint) partitions, this quadruples the miss rate again, so the 16 Megabyte cache has a miss rate of 1 miss per 60 instructions when there are four virtual partitions.

If there are four processors on the chip that run these four partitions (that is, there are four virtual 4-way systems) against the 16 Megabyte cache, and the four processors effectively share half of the on-chip data, the effective miss rate doubles to 1 miss per 30 instructions. If each processor runs four threads, and the threads effectively share half of the on-chip data, the effective miss rate doubles again to 1 miss per 15 instructions.

If each thread runs at 4 CPI, the temporal miss rate is 1 miss per 60 cycles. If the processors run at 5 Gigahertz, this miss rate corresponds to 83.3 Megamisses per second. With a 256 byte line size, the total data moved is 21.3 Gigabytes per second. Since we showed that we need bandwidth of at least 3$\times$ the total data moved, the chip needs a minimum off-chip bandwidth of 64 Gigabytes per second. In bits (assuming 9 bits per byte with ECC) this is 576 Gigabits per second (Gbps).

Assuming that the off-chip bus is 64 bytes wide, this requires 576 signal pins, each running at 1 Gbps. Today, this is feasible. Note that, for good signal integrity, we require at least one power pin per signal pin. So this requires a packaging technology that supports well over 1200 pins per chip (there will be other signals besides this bus), and we need 576 on-chip receivers that run at 1 Gbps.

This calculation ignores the system topology, and assumes that there was one bus. Real SMP systems contain multiple processor chips. A pure hierarchical (tree) structure has unacceptable latency for maintaining coherency because many requests would require two round-trips up and back down the hierarchy. This is too many "hops," so many real systems have direct chip-to-chip busses (a fully connected topology) supplementing the hierarchical tree. The actual number of pins would be much larger than what is calculated above.

Continuing to ignore this major caveat, suppose we extrapolate this chip forward a few generations. Suppose we have 16 processors per chip running at 10 Gigahertz. We've quadrupled the number of processors, and doubled the frequency. The bandwidth has to increase by at least 8×; and remember that, to support these advances, the on-chip cache had to grow, which (as explained earlier) will have a deleterious effect on the bandwidth, perhaps another 2×. For the sake of argument and round numbers, let's assume that the bandwidth has to increase an order of magnitude, 10×.

Since bandwidth is equal to the number of channels times the data rate per channel, we can accomplish this by increasing the number of pins by 10× (to 12,000 pins). Or we can achieve this by running the signals 10× faster – at 10 Gbps. Or we can do something in between, say 6000 pins running at 5 Gbps. None of these solutions is likely in the electrical domain.

In the first place, escaping 6000 signals (that is, having wiring layers and vias in the package to transform the pin array escape pitch into the wiring pitch of the package) is very difficult, and requires many layers of metal. Further, because there are many layers of metal (and vias) in the signal paths, the paths have lots of discontinuities. This means that precise impedance control for this number of signals is practically impossible, and that lots of the transmitted power will be radiated into the package at high signaling rates. Having lots of signals will limit their speed.

On the other side of the equation, running well above 1–2 Gbps is difficult. As mentioned, in the escape path there are many discontinuities if the number of signals is large. Running much faster than this probably requires differential signaling (which doubles the number of pins), and some pre-distortion of the signals – which requires other circuity, hence power and area in the signal path. Above 3 Gbps the power increases quadratically as a function of the data rate.

Achieving 5 Gbps (as in this naive extrapolation) may require 100% of the chip power, and more than 100% of the chip area just to support the off-chip bandwidth. It is clear that, if systems continue to scale, we need a different interconnection technology that will scale with it. Further, we need transmission technology that runs with much less power.

Since we are discussing interconnection technology, we should first consider the system interconnection topology, and reconsider what is really required. Figure 30 shows how system topology has evolved. Multiprocessor systems of the 1970s (with the exception of IBM System 370) had the topology depicted on the left side of the figure. Multiple processors that required coherency were all connected to the same bus. This was simple because the number of channels was of order 1. Also, the coherency protocols were simple, since every entity in the system could see all of the traffic in the system.

These systems were modular, hence flexible, and moderately extendable. The processor frequencies were very slow by today's standards (below 1 Megahertz), and the shared bus structure was adequate at below 100 kilohertz. But the length of the bus, and all of the capacitive loading (due to the multiple drop points) prevented this kind of topology from scaling very far. You cannot put

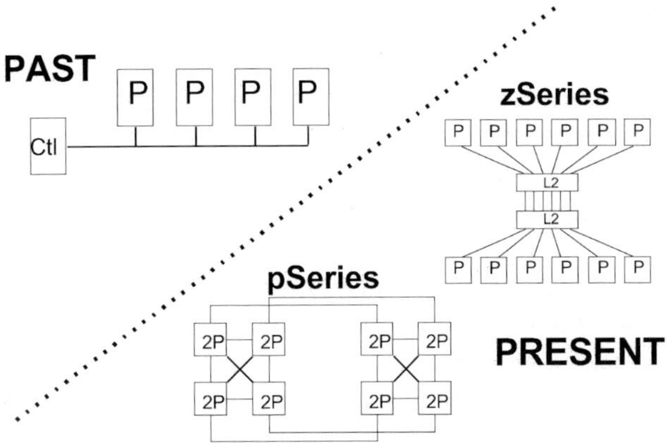

Figure 30. Physical system topology from the past, and physical system topology in the present - conceptually showing the IBM zSeries and IBM pSeries structures.

that many processors on a shared bus, nor can you run a multidrop structure up into the Gigahertz regime.

Instead, processor topologies like the ones on the right of Figure 30 evolved. The top figure is an IBM zSeries structure of the mid-1990s, which has continued to evolve [1]. It is a binodal structure, where each processor has its own bus to a shared L2 structure. The shared structure is the point of coherency. The bottom figure is an IBM pSeries structure of the same era. Physically, it looks like a hypercube, although logically, it is a "flat" SMP [2].

The main advantage of these structures is that the interconnections are all single-drop point-to-point wires. This allows the fastest possible electrical signaling to be used. However, the number of wiring channels required is massive (on the order of the square of the number of chips) and the packaging technology required to support this is niche, hence extremely expensive. Further, the coherency protocols have become quite complicated (because of the number of physical places that can have logically required information), which has a negative performance impact, and it makes design verification very complex.

Logically, these structures are all shared-bus, single-point-of-coherency systems. Physically, they have become widely distributed, "multiple-hop" systems with very expensive packaging and complicated coherency protocols. This evolution was entirely because of the requirement for point-to-point wiring, driven by the need to drive the busses fast. That is, this physically unnatural evolution was driven by the demand for bandwidth subject to the limitations of electrical interconnections.

It is clear that a good portion of the physical interconnection structure of large systems will become optical. However, for intra-system signaling the optics will *not* be a derivative of transport technology. Instead it will be a lower-speed, lower-power, *digital* technology that is silicon-CMOS compatible.

High-speed electrical signaling (below 3 Gbps) requires on the order of 5 milliwatts per Gbps. The digital optical technology to come will run at 5–10 milliwatts (frequency independent) up to 10 Gbps, i.e., it will run at an order of magnitude less power.

Transport optics evolved from a mindset of very high-speed bit-serial communications over large distances. The mindset was not a parallel bus (with parallel ECC) running a few centimeters, or at most a few feet. Thus, the transport mindset was one in which the communication channel (perhaps transcontinental, or transoceanic) is very expensive, so lots of complexity was put into the transceivers to make them extend to very high data rates (e.g. 80 Gbps).

A transport driver is an amplifier in which a laser – having a nonlinear transfer characteristic – is carefully biased, and then modulated over a short linear range in the small-signal domain. Because this must be very precise (and thermally controlled) to reach to very high speeds going long distances, the exact shape of the transfer characteristic of the laser is crucial, and becomes integral to the definition of laser "yield," hence cost. For this reason, linear arrays of lasers (as would be suited to parallel bus applications) are very expensive.

Digital optics will not modulate lasers this way. They will treat them as "on-off" (large signal) devices that run in the digital domain over short distances. The natural speed to use is the processor speed, or perhaps twice that (5–10 Gbps). It would be unnatural to run signals at 10× the processor speed, because the multiplexing would be complex, and would add latency to the path. Digital optics will not be bit serial. It will be parallel-bus optics (with ECC) just as intra-system electrical busses are today. This is natural. There is little notion of bit error rate (BER) in a parallel bus usage. Serial optics is also undesirable because the serialization process adds power and latency.

While (unless there is some new development in photonics) silicon cannot laze (because it has an indirect band-gap), and lasers will likely continue to be gallium arsenide, arrays of lasers can be grafted onto silicon, and connected to silicon CMOS drivers. These drivers will be simple on-off switches, perhaps with some passive bias [31]. Arrays of lasers will become cheaper both because they will become volumized, and because the definition of "yield" in this application is much more relaxed. Receivers will probably be metal-insulator-metal (MIM) structures in silicon CMOS.

In addition to easily providing a 5–10 Gbps data rate over relatively short distances with very low power (that is independent of the data rate), the principal advantage of optics is that optics can be multidropped without limiting its frequency. *This means that mutlidrop shared bus structures are enabled in the Gigahertz regime.*

Systems, or at least large subsystems, will revert to shared bus structures. This will obviate the massive number of wiring channels (order n-squared), and will greatly simplify the copper infrastructure. The copper infrastructure will still exist (for power distribution, service functions and diagnostics, and to support some of the control structure), but it will not require thousands of pins. The optical structure will provide the main bandwidth-intensive data flow within the system. The packaging will become cheaper (because of a drastic reduction in copper wiring channels), and the power associated with the bandwidth will be greatly reduced.

However, the packaging infrastructure will have to support multidrop optical channels. Since they will be multidrop, the optical "wiring" can be planar, i.e. it can be a single polymer layer within the copper infrastructure. Chips will have optical "pins" that will connect to the on-package optical channels with matched-index materials built as lens structures that will match the size of the solder balls on the chip [32].

Circuit boards will have optical connectors that will provide optical connections when the boards are plugged into their sockets. That is, board-to-board cabling (in addition to the copper backplane) is impractical, too complex, and error-prone. Backplanes will have a planar optical layer with "tap-point" gratings and couplings at the board-sites. Gratings can be designed to out-couple a small amount of the optical power in a channel, so that a backplane can support a large number of drops, say 16 or perhaps even 32.

Board-to-backplane connectors will likely have active components (optical to electrical to optical repeaters) so that precision mechanical alignment (which is very expensive) will not be required. Optical connections will be simple butt-couplings. With repeaters, the losses due to imprecise mechanical tolerances will be rendered irrelevant to the link budgets [33].

Finally, once a parallel-bus multidrop optical infrastructure is in place, small degrees of wavelength division multiplexing (WDM) will start to be incorporated. With a basic 850 nanometer laser, it is possible to shift the wavelength a few times by 50, or even 25 nanometers so as to reliably provide a few different colors (more than 2 but less than 10). This would enable a single channel (or chip connection) to carry multiple bits simultaneously; one bit per color.

Figure 31 shows the potential for the off-chip interconnection density of optics relative to copper. The large black circles depict 3-mil diameter solder balls on 6-mil centers, as is used in aggressive packaging infrastructures today. Interstitially between the solder balls, a number of 3×3 grids have been drawn. These grids show (to scale) the potential for placing groups of 10-micron diameter lasers. Each grid represents nine lasers. Assuming an optical connection that is the same size as a solder ball, each optical connector will support the nine lasers in its grid. Assuming that primitive WDM could extend to nine colors, this means that we can transmit a byte (with ECC) through each optical connector. The interstitial arrangement shows that associated with each copper solder ball, we can provide 3 bytes (27 lasers) of optical connections.

The digital-CMOS optical infrastructure described in this section is very different technology than transport optics. Attempts to scale transport optics into this domain will be unsatisfactory because the cost and the power will not

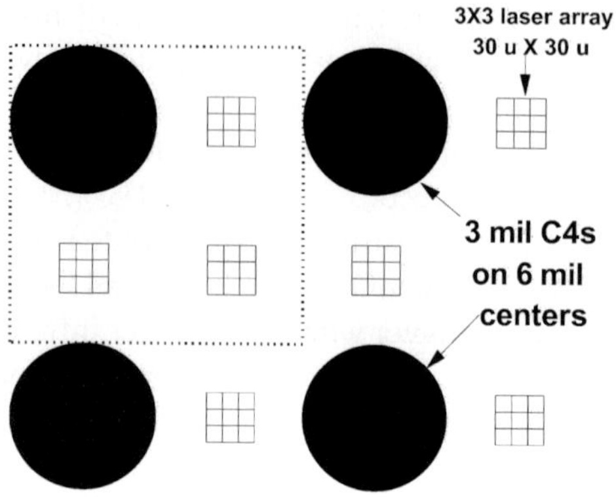

Figure 31. A scale drawing of an array of 3 mil solder balls placed on 6 mil centers, with 3 × 3 arrays of 10-micron diameter lasers placed interstitially between the solder balls.

scale sufficiently to be practical. But with a digital-CMOS optical infrastructure in place, system structures and costs will become simpler, will extend further, and will run with lower power.

13. Conclusions and Predictions

Bandwidth dominates all aspects of performance at the system level. This is apparent by looking at the unnatural physical structures that have evolved to support the bandwidth growth that accompanied that evolution. Bandwith is used primarily to service the cache misses in a system. An insufficiency of bandwidth manifests in the form of numerous trailing edge effects. There will be a resurgence in sectored caches so as to ameliorate some of these effects.

Bandwidth and content are mutually fungible entities. As systems continue to evolve, more pressure will be put on the on-chip caches, and on the off-chip bandwidth. I have explained why SRAM will not be able to provide the required on-chip storage capacity required, and why electrical signaling will not be able to provide the required off-chip bandwidth required as systems scale past the next few technology generations.

I have posited that on-chip cache hierarchies will eventually contain EDRAM, perhaps in the form of several planes in a 3D structure. In addition to providing more on-chip storage, 3D structures can be arranged so as to provide staggering amounts of bandwidth at very low power within the on-chip cache hierarchy.

I have also posited a new digital-CMOS optical technology that is *not* a derivative of optical transport technology. This technology will be all CMOS (except for the lasers), and will be designed to run at relatively (for optics) low data rates over short distances. Specifically, it will run in the 5–10 Gbps regime at distances up to a few feet. This will be much cheaper and run at much lower power than any transport derivative.

Further, the multidrop capability of optics will be used to facilitate a simple planar optical infrastructure that will replace the existing massive point-to-point infrastructure of copper today. This will simplify the copper infrastructure, and make packaging cheaper. It will further enable a resurgence to shared-bus system structures, which are simpler to operate and to verify. This will enable systems to evolve more easily and at lower power. While optics can be used to provide raw performance (and this is part of the puzzle) the real leverage of optics is that it can provide a real simplification to the packaging infrastructure and to the system topology. This is how the "optics card" will be played.

Bandwidth is the real key to system evolution. It is the reason that caches were invented and then made hierarchical. It is the reason that costly packaging evolved, and that system topologies (and their coherency protocols) have become complex. Continuing to evolve systems will require paradigm shifts in technology so as to enable much higher bandwidth at much lower power, and so as to provide much more on-chip storage.

Today, power is the main limitation to system evolution. In the past decade lots of work went into learning new techniques at all levels (device, circuit, microarchitecture, and system) to make power usage much more efficient. To the extent that this has been "solved," the next major hurdle will be bandwidth. This is the next "brick wall" that is standing in the road to systems evolution.

References

[1] Mak, P.; Strait, G.E.; Blake, M.A. *et al.* "Processor subsystem interconnect architecture for a large symmetric multiprocessor system", *IBM J. Res. Dev.*, **2004**, *48* (3/4).

[2] Tendler, J.M.; Dodson, J.S.; Fields, J.S.; Le, Jr. H.; Sinharoy, B. "POWER4 system microarchitecture", *IBM J. Res. Dev.*, **2002**, *46* (1).

[3] Winkel, T.M.; Becker, W.D.; Harrer, H. *et al.* "First and second-level packaging of the z990 processor cage", *IBM J. Res. Dev.*, **2004**, *48* (3/4).

[4] Srinivasan, V.; Brooks, D.; Gschwind, M.; Bose, P.; Zyuban, V.; Strenski, P.; Emma, P. "Optimizing pipelines for power and performance", *35th Annual IEEE/ACM Int. Symp. on Microarchitecture*, November **2002**, 333–334.

[5] Agarwal, V.; Hrishikesh, M.S.; Keckler, S.W.; Burger, D.; "Clock rate vs. IPC: the end of the road for conventional microarchitectures", *Proc. 27th Annu. Symp. Computer Architecture*", 10–14 June **2000**.

[6] Eggers, S.J.; Emer, J.S.; Levy, H.M.; Lo, J.L.; Stamm, R.L.; Tullsen, D.M. "Simultaneous multithreading: a platform for next-generation processors", *IEEE Micro*, **1997**, *17*(5), 12–19.

[7] Armstrong, W.J.; Arndt, R.L.; Boutcher, D.C. *et al.* "Advanced virtualization capabilities of POWER5 systems", *IBM J. Res. Dev.* **2005**, *49* (4/5).

[8] Emma, P.G. "Understanding some simple processor-performance limits", *IBM J. Res. Dev.*, **1997**, 215–232.

[9] Collier, W.W. *Reasoning About Parallel Architectures*, Prentice Hall, **1992**.

[10] Dubois, M.; Scheurich, C.; Briggs, F.A. "Synchronization, coherence, and event ordering in multiprocessors", *Computer*, **1988**, *21*(2), 9–21

[11] Chow, C.K. "On optimization of storage hierarchies", *IBM J. Res. Dev.* **1974**, *18*.

[12] Hill, M.D. Aspects of Cache Memory and Instruction Buffer Performance, PhD Thesis, University of California at Berkeley, **1987**.

[13] Emma, P.G. "Storage hierarchies", *Encyclopedia of Computer Science*, 3rd edn. Van Nostrand Reinhold, **1993**, p.1290.

[14] Jain, R. *The Art of Computer Systems Performance Analysis*, John Wiley & Sons, **1992**.

[15] Emma, P.G.; Hartstein, A.; Puzak, T.R.; Srinivasan, V. "Exploring the limits of prefetching", *IBM J. Res. Dev.* **2005**, *49*(1).

[16] Puzak, T.R.; Hartstein, A.; Emma, P.G.; Srinivasan, V. "When prefetching improves/degrades performance", *Proc. 2nd Conference on Computing Frontiers*, 4–6 May **2005**, 342–352.

[17] Dahlgren F.; Stenstrom, P. "Effectiveness of hardware-based stride and sequential prefetching in shared-memory multiprocessors", *1st IEEE Symposium on High-Performance Computer Architecture*, **1995**, 68.

[18] Liptay, J.S. "Structural aspects of the system/360 model 85, Part II: the cache", *IBM Sys. J.* **1968**, *7*(1).

[19] Lukoff, H. *From Dits to Bits*, Robotics Press, **1979**.

[20] Matick, R.E. *Computer Storage Systems and Technology*, John Wiley & Sons, **1977**.

[21] Matick R.E.; Schuster, S.E.; "Logic-based eDRAM: origins and rationale for use", *IBM J. Res. Dev.* **2005**, *49*(1).

[22] McKee, S.A. "Reflections on the memory wall", *Proc. First Conference on Computing Frontiers*, April **2004**, 162–167.

[23] Kalla, R.; Sinharoy, B.; Tendler. J.M. "IBM POWER5 chip: a dual-core multithreaded processor", *IEEE Micro*, **2004**, *4*(2).

[24] Chen, T.F. "Supporting highly speculative execution via adaptive branch trees", *Fourth International Symposium on High-Performance Computer Architecture*, 1–4 Feb. **1998**, 185–194.

[25] Uht, A.K.; Sindagi, V.; Hall, K. "Disjoint eager execution: an optimal form of speculative execution", *Proc. 28th Annual IEEE/ACM Int. Symp. on Microarchitecture*, 29 Nov.–1 Dec. **1995**, 313–325.

[26] Pleshko P.; Terman, L. "An investigation of the potential of MOS transistor memories", *Trans. Electronic Computers*, August **1966**.

[27] Schmidt, J. "MOS memory chip", *Solid State Design*, January **1965**.

[28] Terman, L. "MOSFET memory circuits", *IEEE Proc.*, July **1971**.

[29] Guarini K.W.; Wong, H.-S.P. "Wafer bonding for high-performance logic applications", *Wafer Bonding: Applications and Technology,* M. Alexe and U. Goesele (eds.), C.H.I.P.S. **2004**, 157–192.

[30] Guarini, K.W.; Topol, A.W.; Ieong M. *et al.* "Electrical integrity of state-of-the-art 0.13 μm SOI CMOS devices and circuits transferred for three-dimensional (3D) integrated circuit (IC) fabrication", *Tech. Dig. IEEE International Electron Devices Meeting*, San Francisco, CA, USA, December **2002**, 943–945.

[31] Bozso F.M.; Emma, P.G. "High speed data channel including a CMOS VCSEL driver and a high performance photodetector and CMOS photoreceiver", U.S. Patent # US20040101007A1, Assigned to IBM Corporation, Filed 27 Nov. 2002, Issued 27 May **2004**.

[32] Bozso, F.M.; Emma, P.G. "Optically connectable circuit board with optical components mounted thereon", U.S. Patent # US20040100781A1, Assigned to IBM Corporation, Filed 27 Nov. 2002, Issued 27 May **2004**.

[33] Bozso F.M.; Emma, P.G. "Backplane assembly with board-to-board optical interconnections and a method of continuity checking board connections", U.S. Patent # US20040100782A1, Assigned to IBM Corporation, Filed 27 Nov. 2002, Issued 27 May **2004**.

Chapter 11

HIGH-SPEED IO DESIGN

Warren R. Anderson
Intel Corporation

Abstract: This chapter explores common methods and circuit architectures used to transmit and receive data through off-chip links.

Key words: High-speed IO; off-chip links; serial data transfer; parallel data bus; FR-4; skin effect; dielectric loss; clock phase alignment; derived clocking; source synchronous; forwarded clocking; plesiochronous; mesochronous.

1. Introduction

In order for system performance to keep pace with the ever-increasing speed of the microprocessor, the bandwidth of the signaling into and out of the microprocessor must follow the trend in on-chip processing performance. However, the physical limitations imposed by the interconnect channel, which exhibits increasing amounts of signal loss and jitter amplification at higher frequencies, impedes the increase in off-chip signaling speed. Numerous strategies to cope with the interconnect properties have been developed, enabled by more sophisticated techniques to control and process the off-chip electrical signals.

This chapter discusses the most prevalent of these techniques, focusing on the chip-to-chip communication topologies common for microprocessors, namely access to memory, processor-to-processor communication for parallel computing, and processor-to-chipset communication. These links generally run over short distances of up to one or two meters and consist of buses of parallel data lanes carrying wide data words. Although serial communication shares many of the same properties and techniques as the parallel bus designs, serial

Vojin G. Oklobdzija and Ram K. Krishnamurthy (eds.),
High-Performance Energy-Efficient Microprocessor Design, 289–309.

communication, which generally takes place over much longer distances, will not be discussed explicitly here.

Our discussion begins with a comparison of the on-chip and off-chip data transmission environment, with an emphasis on the desired properties of the off-chip communication system. Several common signaling methods will be shown. We then explore the properties of the off-chip signaling medium, particularly those that dominate at high frequencies and therefore limit off-chip signaling speed. Techniques and example circuit topologies for adapting to these effects in both the time and voltage domains, as well as their limitations, are shown.

2. IO Signaling

The overall function of IO is to faithfully convey data from a transmitter chip to a receiver chip [5, 8]. Similar to on-chip communication, where data is passed from one section of the chip to another, off-chip communication must define the data representation for a "1" and a "0." It must also transmit the output data and capture the input data synchronously so that the input data stream can enter the synchronously clocked logic on the receiver side.

A typical configuration for microprocessor IO is shown in Figure 1. It depicts the processor in communication with a variety of external components, such as memory, a memory chipset, another processor, and a chipset communicating with external storage and networking devices. Off-chip communication takes place through a variety of parallel data buses. Each data bus consists of several parallel data lanes conveying information through an interconnection wire

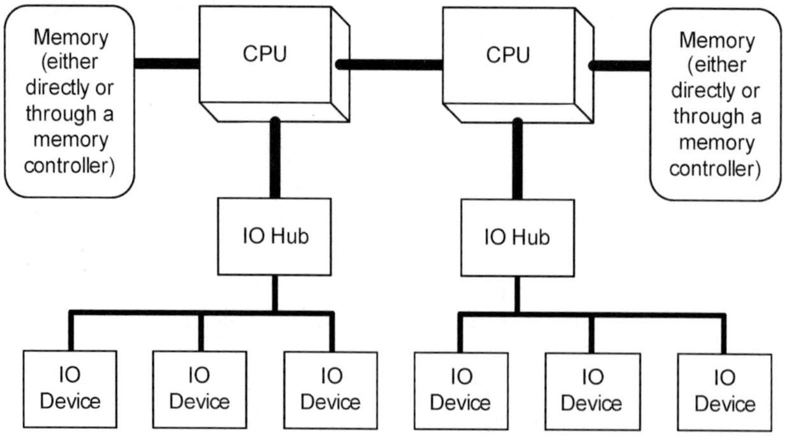

Figure 1. Typical components connected to a microprocessor.

or channel routing through the off-chip interconnection medium, which may consist of package wiring, board wiring, or cables.

Because the properties of the interconnect medium differ significantly from on-chip and off-chip connections, the optimal methods for communicating in these two environments also differ significantly. Off-chip communication must often take place between integrated circuits with different supply voltages, must compensate for losses in the interconnect medium, must traverse larger distances, and must take place in a noisy off-chip environment. Higher performance requires a robust representation in the voltage domain.

In the time domain, synchronization must be maintained across the parallel data bus. As shown in Figure 2, the larger off-chip communication distance creates skew across a parallel data bus, which, unless extremely well controlled, increases between transmitter and receiver. In addition, the receiver clock must fan out to all of the data lane receivers, which increases its jitter, and must be phase aligned to capture all of the data symbols in the valid data region. A data symbol is the representation for one bit of a "0" or a "1" on a data lane. Whereas on-chip clocks are designed with low skew between transmitting and receiving synchronization points, the IO transmitter and receiver reside on different pieces of silicon and often cannot be so well controlled. The IO architecture must compensate by aligning the receiver clock to sample the input data at the most optimum time.

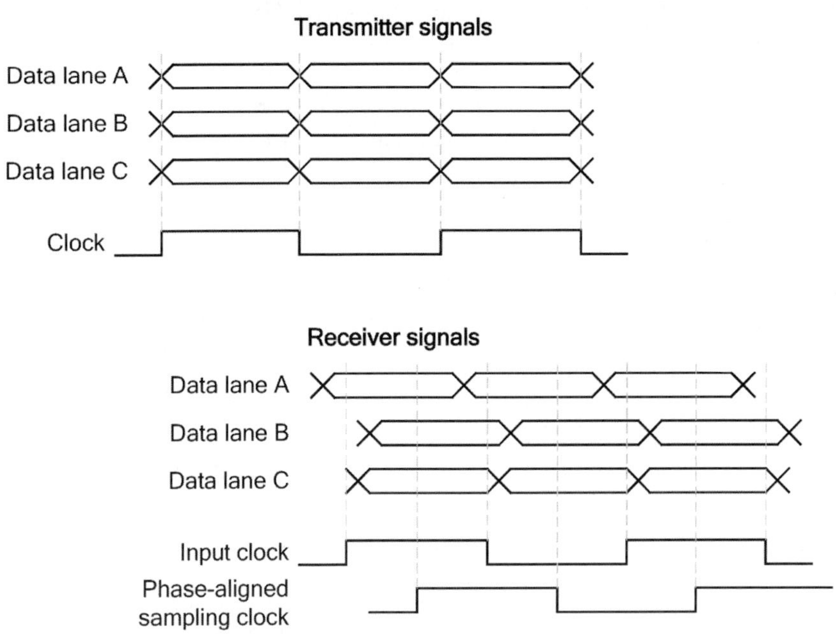

Figure 2. Typical timing of transmitter output and receiver input signals in a dual data-rate signaling IO system.

2.1. Single-ended Voltage-mode Signaling

Returning to the voltage domain, we first consider how the value of the data can be represented for off-chip signaling. As opposed to on-chip data propagation where a voltage representing a "1" must be near the positive supply rail Vdd and a voltage representing a "0" must be near the negative supply rail Vss, numerous schemes may be used for IO.

Some single-ended schemes work analogously to on-chip signaling, but must be standardized in order to enable interoperability among integrated circuits regardless of IC process technology. Off-chip signaling standards define explicit signaling voltage levels and tolerances independent from the on-chip supply, which can vary with process technology generation. At the input receiver, for example, a voltage below a maximum input voltage level V_{IL} represents a "0" and a voltage above a minimum input voltage level V_{IH} represents a "1." Minimum and maximum output levels are defined as V_{OL} and V_{OH}, usually with a slightly wider separation to allow for noise to degrade the signal between transmitter and receiver. Examples of voltage-mode signaling include the TTL and LVTTL standards. Such a scheme for a bi-directional interface is shown in Figure 3.

For high-speed communication, single-ended voltage-mode signaling exhibits numerous disadvantages. Since high-speed operation requires delivering as much energy as possible to the electrical pulse representing a symbol, any effect that removes energy from a symbol and combines it with another degrades operation speed. Transmission-line reflections represent one such effect.

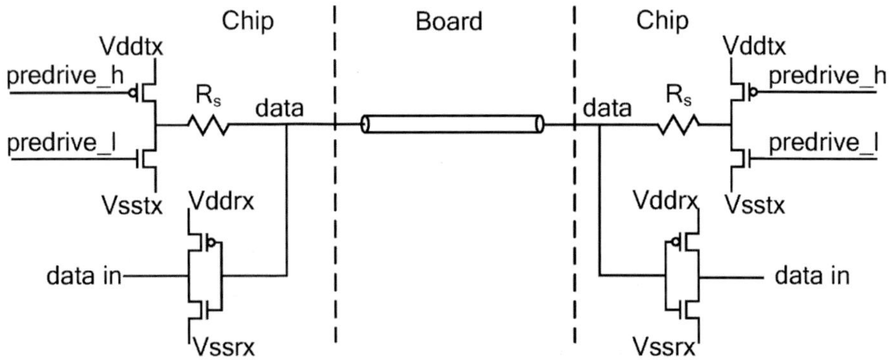

Figure 3. Bi-directional link using single-ended voltage mode signaling. The resistor in series with the output sets the V_{OH} and V_{OL} levels and also provides board impedance matching. The pre-driver outputs predrive_h and predrive-l are separated to tri-state the output driver when receiving input data. Separation of the transmitter and receiver supplies, as shown here, is often used to avoid simultaneous switching noise from the transmitter appearing on the supply of the more sensitive input receiver.

As will be discussed in Section 3.1.1, impedance discontinuities cause reflections that, if reflected back toward the receiver, create noise on un-related symbols. In order to absorb these reflections, high-speed transmitters must be impedance matched to the line. As shown in Figure 3, the driver itself becomes part of the termination network; therefore the driver must not only be controlled to launch the correct V_{OL} and V_{OH} levels, but it must also form the proper termination network, preferably at all voltage levels. This becomes extremely difficult in practice.

To overcome these difficulties and provide more noise immunity, the V_{OL}, V_{OH}, V_{IL}, and V_{IH} levels are often made wider than necessary, consuming more power. The wider swings are usually less well controlled, making higher-speed operation difficult.

2.2. Current Mode Signaling

Faster data rates can generally be obtained with current mode drivers. Signaling speeds can be further enhanced using current-mode signaling in pseudo-differential or differential mode, as will be described in Sections 2.3 and 2.4.

In current mode signaling, the output driver forces a current into the transmission line, using the natural impedance of the line to create a voltage. An example is shown in Figure 4, where the driver forces $I_{drive} = 20\,mA$ into the parallel combination of the $50\,\Omega$ interconnect impedance and the $R_{term} = 50\,\Omega$ near-end termination, generating a 500 mV signal. To avoid reflections, the line is terminated at the receiver end (far end). To absorb reflections from impedance discontinuities, the line is often terminated at the transmitter (near end) as well. The termination usually ties to ground to avoid depending on the value of the positive supply rail and to avoid coupling the signal to noise on the positive supply rail.

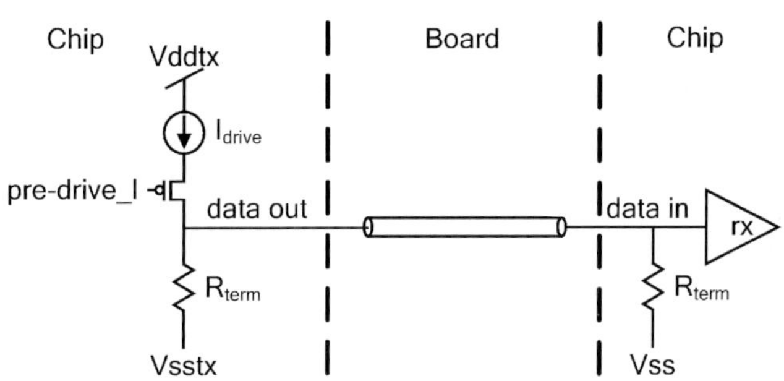

Figure 4. Current mode signaling.

Although a calibrated current-mode driver can launch the same signal on the line as a well-tuned voltage-mode driver, the current-mode driver decouples the driver function from the impedance matching section, here provided by the independent termination resistor. Since the current source portion of the current-mode driver has a high impedance, the actual impedance of the current source is not critical, as long as its impedance is much greater than that of the termination resistor. As a result, the driver is simpler to design and control than a voltage-mode driver, where the driver and impedance-matching functions are combined. Furthermore, current-mode signaling provides greater noise immunity since it decouples the driver from the positive supply rail.

Through the calibration of the current-mode output source with a known current or voltage, the output level can generally be driven within 10% or greater accuracy. Periodic calibration can compensate for voltage and temperature drifts on the transmitter die. The receiver, however, must still detect if the input voltage is above or below the input thresholds. As described in the next two sections, additional means are required to improve the accuracy of the input levels.

2.3. Pseudo-differential Signaling

The reduced swings usually found in current-mode signaling require a more sensitive means to discriminate between high and low input levels. One method for achieving this uses pseudo-differential signaling, which compares the signal at the receiver to a reference voltage. Since this reference voltage can be generated through matched resistor devices or other precision means, it will not depend strongly on process condition or temperature. Depending on the offset between the transmitter and receiver supplies and the receiver tolerance requirements, the reference voltage may be generated in the receiver, at the transmitter and routed to the receiver, or externally with precision, matched resistors. An example of a pseudo-differential bus is shown in Figure 5.

Pseudo-differential signaling has been used in both voltage mode and current mode standards, such as HSTL, GTL, and SSTL.

2.4. Differential Signaling

Although the explicit reference in pseudo-differential signaling provides for more robust signal detection, pseudo-differential signaling is still subject to several problems that must be eliminated for operation at higher speeds.

One such problem is simultaneous switching noise. Any signaling scheme using a single wire for each data lane creates simultaneous switching noise (SSO). SSO creates a transient decrease or increase in the transmitter supply

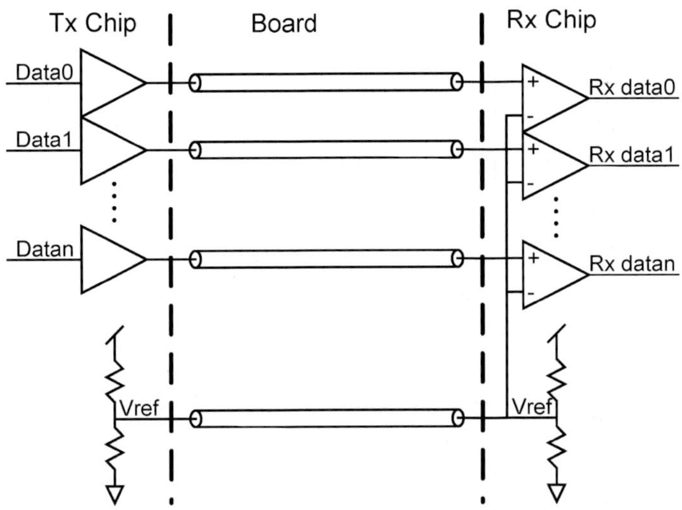

Figure 5. A parallel bus with pseudo-differential signaling.

voltage as it reacts to providing the dI/dt needed when switching the output [1, 9]. For example, consider the case of Figure 4 when the output drives low. In this condition no current flows in the transmission line. When the driver pulls the line high by sourcing a constant current from the positive supply rail into the output, this current creates a dI/dt in the circuit loop formed by the positive supply, the output, the interconnect signal's transmission line, its return path in the underlying ground plane, and the negative supply. Any inductance in this loop generates a transient voltage excursion when it experiences the dI/dt. This generally occurs where the driver's output pad and the supply rails interact with the package, particularly in bond wire designs. In our example, the inductance in the package between the positive supply on the board and on the die develops a voltage that temporarily decreases the on-die supply rail. Likewise, the inductance in the ground supply develops a voltage that temporarily raises the on-die ground voltage. Although the effect is also present on the signal itself, it is generally worse for the supplies since the ratio of signals to supply pairs is generally at least 2 to 1. Therefore, the effect is worst when all output drivers on a bus switch in the same direction simultaneously, creating the largest possible dI/dt. Although simultaneous switching can be tolerated through proper construction of the supply network [1, 9], it cannot be completely eliminated for single-ended signaling.

Differential signaling avoids this problem by transmitting the data and its complement on parallel interconnect wires to the receiver, where a positive or negative difference between the signal pair indicates a "1" or a "0," respectively. Provided the true and complement transmitters are balanced to draw

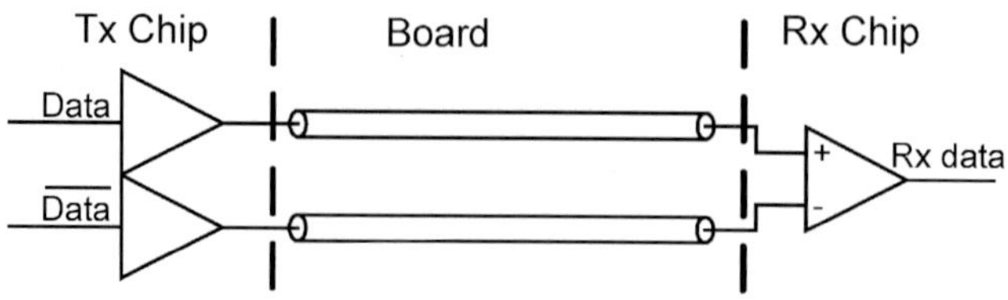

Figure 6. Differential signaling on one lane of an IO link.

equal currents even when switching, no simultaneous switching noise occurs. Furthermore, the voltage swing is effectively doubled. Rather than swinging $\pm \frac{1}{2}$ Vs around the pseudo-differential reference, for example, the signals can swing ±Vs between each other with the same single-ended output voltage. A final benefit is that noise that couples to both signals in the differential pair only alters the common mode and does not change the differential voltage. Therefore, differential signaling is more tolerant to certain types of noise than single-ended or pseudo-differential signaling.

The penalty for differential signaling is that each data lane now uses two interconnect wires, which reduces in half the number of lanes for a bus with a fixed number of interconnect wires. Furthermore, without optimization, differential signaling doubles the driver power. The factor of two decrease in bus width is justified if each lane in the link can run at double the single-ended data rate, which is often the case. Differential signaling is used in such standards as LVDS and PCI-Express. An example link topology is shown in Figure 6.

3. Coping with the Interconnect

The green epoxy glass resin printed wiring board material FR-4 is the workhorse of the electronics industry. A metal wiring trace over a power or ground plane with FR-4 dielectric in between forms a transmission line. This transmission line is far from ideal, however [2, 6, 7]. Several mechanisms create loss at high frequencies. Impedance discontinuities produce reflections. Adjacent signals and noisy supplies cause interference. Understanding these mechanisms through studying the properties of the interconnect is key to achieving higher speeds.

3.1. Interconnect Properties

An ideal transmission line consists of a distributed inductance per unit length l_0 and capacitance per unit length c_0, where both l_0 and c_0 depend on the geometry of the signal trace, the geometry of the dielectric surrounding the signal, and the dielectric constant. Signals injected into the line propagate with a velocity

$$v_0 = \frac{1}{\sqrt{l_0 c_0}} = \frac{c}{\sqrt{\varepsilon_r}}, \tag{1}$$

where c is the speed of light and ε_r is the relative permittivity of the dielectric. An ideal transmission line acts as a fixed impedance element, with impedance

$$Z_0 = \sqrt{\frac{l_0}{c_0}}. \tag{2}$$

3.1.1. Reflections and impedance discontinuities

Because the ideal line has no loss, an injected signal travels un-attenuated until it encounters a discontinuity, which may consist of a change in impedance or a load, such as the termination at the end of the line. When an incident wave of magnitude V_i propagating through a transmission line with impedance Z_0 encounters change to a new impedance Z_1, a reflection occurs. The ratio of the voltage of the reflected wave V_r to that of the incident wave V_i is given by

$$\frac{V_r}{V_i} = \frac{Z_1 - Z_0}{Z_1 + Z_0}. \tag{3}$$

Loads, stubs, and vias on the line create discontinuities as well. Capacitive loads or capacitive-like vias create a complex impedance. An incident wave into a capacitance initially sees a short, which decreases the impedance Z_1 to zero and causes a negative reflection by Eq. (3). The impedance rises to its steady-state value as the capacitor charges. Uniformly distributed loads add distributed capacitance per unit length and also alter the impedance through Eq. (2).

Since any reflection causes a loss of incident wave energy to the backwards-propagating pulse, only loss-less lines with uniform impedance allow all of the injected signal's energy to coherently propagate to the end of the line. Furthermore, termination that is not impedance-matched to the line causes reflections to occur which, if not completely absorbed, may combine with and distort other symbols. Therefore, operation at higher speeds requires minimizing all impedance discontinuities and terminating with a matched impedance.

3.1.2. Transmission line losses

Up to this point we have only considered loss-less lines. Unfortunately, transmission-line losses cause high-speed signals, even in a perfectly

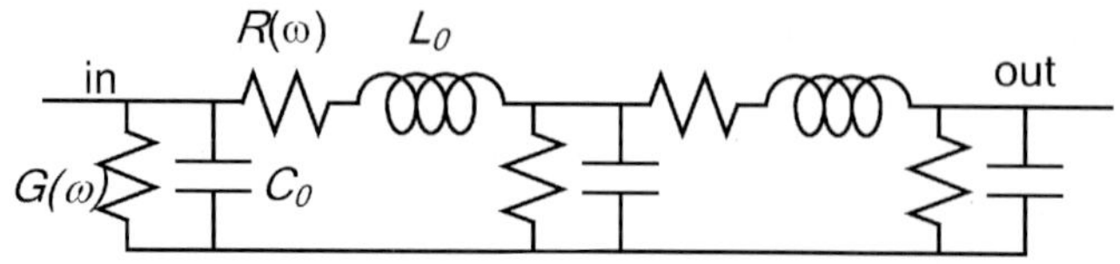

Figure 7. Transmission line with loss components.

impedance-matched and terminated line, to lose a portion of their energy through other means. A real transmission line exhibits a distributed resistance per unit length $R(\omega)$ in the signal trace and also contains, at high frequencies, a finite amount of conductance per unit length $G(\omega)$ through the dielectric between the signal and its return path, as shown in Figure 7.

For an injected signal $V_i(0, \omega)$ with angular frequency ω, the resulting signal $V(z, \omega)$ at any point z along the line is given by

$$V(z, \omega) = V_i(0, \omega)e^{-\gamma(\omega)z}, \tag{4}$$

where

$$\gamma(\omega) = \sqrt{(R(\omega) + j\omega L_0)(G(\omega) + j\omega C_0)}. \tag{5}$$

If both $R(\omega)$ and $G(\omega)$ are small, we can approximate $\gamma(\omega)$ by

$$\gamma(\omega) \cong \frac{R(\omega)}{2Z_0} + \frac{Z_0 G(\omega)}{2}. \tag{6}$$

As the notation indicates, both the series resistance $R(\omega)$ and dielectric loss $G(\omega)$ are frequency dependent. The frequency dependence of $R(\omega)$ arises from the skin effect, which confines the current ever closer to the surface of the conductor at higher frequencies. For a strip conductor of trace width w and resistivity ρ, the frequency dependence is given by [3]:

$$R(\omega) = \frac{1}{2w}\sqrt{\frac{\omega\mu\rho}{2}}. \tag{7}$$

The frequency dependence in the dielectric loss arises from the response of the medium to high-frequency electro-magnetic waves. From Figure 7, the admittance of the dielectric contains a real term $G(\omega)$ and an imaginary term $j\omega C$. Their ratio defines the loss tangent δ, given by

$$\tan\delta = \frac{G(\omega)}{\omega C}. \tag{8}$$

The loss tangent is nearly constant in frequency and is a fundamental property of the dielectric material. Rearranging Eq. (8) yields

$$G(\omega) = \omega C \tan\delta, \tag{9}$$

indicating that the conductance of the dielectric, and therefore the dielectric loss attenuation term from Eq. (6) is proportional to frequency. In typical FR-4 channels, dielectric loss dominates over skin-effect loss at frequencies greater than about 1GHz.

3.2. Inter-symbol Interference

Now consider a system sending symbols representing data from transmitter to receiver. For long lines at high data rates, the interconnect will carry many symbols in flight between the transmitter and the receiver. Because data carries information, an arbitrary pattern of "0" and "1" symbols will be present on the line, representing a variety of frequency components. Under these conditions, both reflections and frequency-dependent losses cause the output of the line at the receiver to depend strongly on the input data pattern.

Reflections not only take energy away from any given symbol, they also send energy from that symbol in the opposite direction. If the initial reflection reflects again, the reflected pulse joins other non-related symbols propagating in the direction of the receiver, interfering with and potentially corrupting the victim symbol. Furthermore, frequency-dependent loss causes data patterns with a high-frequency content to be attenuated while patterns with low frequency content are not. Both of these effects illustrate inter-symbol interference (ISI), where symbols can interfere with each other, resulting in strong data pattern-dependent characteristics for the signaling medium.

The frequency domain characteristics, represented by the ratio of the output signal to the input signal as shown in Figure 8, demonstrate where both of these effects occur. In the frequency domain, reflections cause dips and spikes at frequencies where the reflections result in destructive or constructive interference at the output. Loss is evident in the increasing attenuation of the output at high frequency. In the time domain, reflections diminish the amplitude of the initial step and also introduce delayed glitches at the output, potentially interfering with later symbols. Loss causes dispersion of the input step as well as a slowly-rising tail after the initial step.

These two effects, dispersion of the initial step and the slow, asymptotic approach to the steady-state condition, combine with the data pattern to create loss of margin on both the time and the voltage axes. Consider a lone symbol representing a "1" in field of "0" symbols, as shown by the pulse response characteristics of Figure 9. Prior to the "1," the line will have sat inactive for a period of time, allowing it to decay close to its steady-state condition. The interconnect losses disperse the rising edge of the "1" pulse and attenuate its peak. Likewise, losses do the same to the falling edge at the tail of the "1" pulse. As a result, the peak pulse amplitude, which must rise from the low steady-state level, is severely attenuated and, in some cases, may not cross the

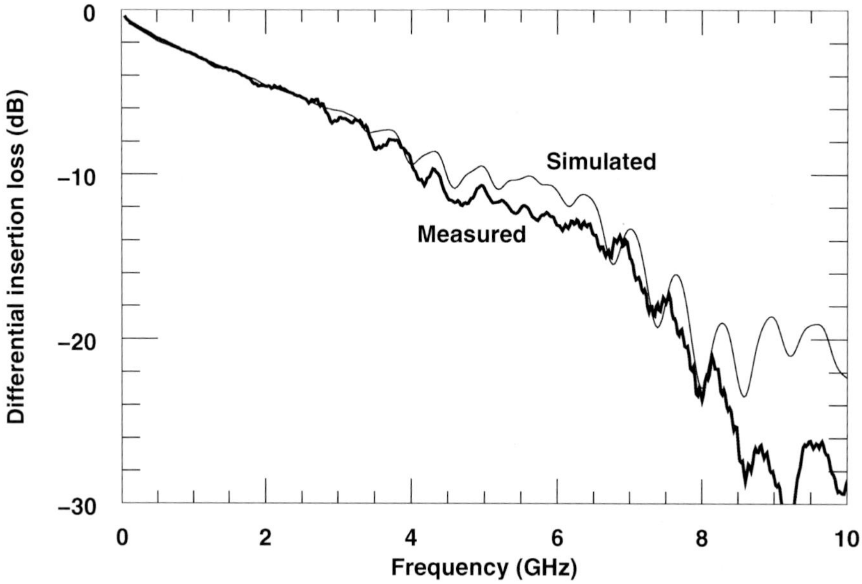

Figure 8. Frequency-domain characteristics of a differential transmission line consisting of two daughter cards and one baseboard, total length 15 inches (38 cm).

receiver threshold at all. Furthermore, the dispersion of the rising and falling edges compresses the symbol such that the pulse width is much narrower at the receiver. Without correction, noise in either voltage or time could corrupt the symbol.

Through superposition techniques, the pulse response can be extrapolated to provide the output characteristics for any input data pattern and to find the pattern yielding the worst-case minimum voltage and timing margin [10].

3.3. Equalization

Equalization techniques compensate for the frequency-dependent characteristics of the channel so that the combined frequency response of the system is nearly uniform over the frequencies of interest [4]. Imagine, for example, that the channel response is given by the transfer function $H(s)$. If we can process the input or output signal through another transfer function $G(s) = 1/H(s)$, the total transfer function of the system will be $H(s)G(s) = 1$.

In practice, it is difficult to cancel the channel response so accurately. However, even schemes that cancel the channel response at or near the highest operational frequency, where channel losses are greatest, can provide a significant benefit to IO performance.

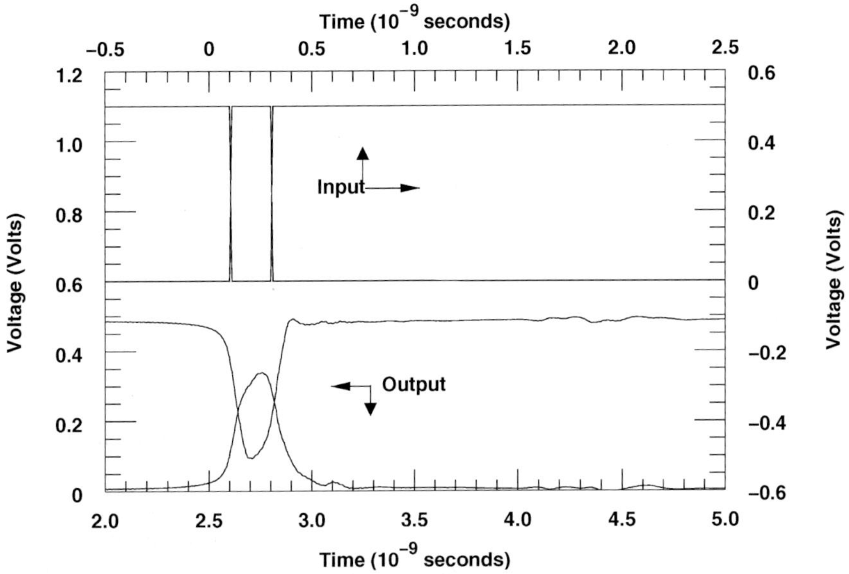

Figure 9. The pulse response of the differential transmission line of Figure 8. The input and output pulses are shown on different, translated time and voltage scales (left-bottom and top-right), that have been shifted for better comparison, with the arrows indicating the respective axes for each type.

Equalization can occur at the transmitter, the receiver, or both. At the transmitter, equalization is usually performed through pre-distortion of the input signal processed through on-chip logic [3, 4, 11]. Symbols with high-frequency components are injected into the line with higher amplitude while those of lower frequency are injected with lower amplitude. The example shown in Figure 10 uses a scheme where any symbol that is different from the previously transmitted symbol is sent with full amplitude. Symbols with the same value are attenuated. This is known as two-tap de-emphasis since only two bits of history are examined to decide the amplitude for the symbol entering the line, which is generally referred to as the cursor.

This scheme can be extended to any arbitrary number of symbols of history, either before or after the cursor, at the cost of power, die area, and potentially data latency. If we represent the data stream x_i with values of $+1$ and -1, equalization can be performed for the cursor symbol x_0 through a finite impulse response (FIR) filter as given by

$$y_0 = \alpha_{-m}x_{-m} + \cdots + \alpha_{-1}x_{-1} + \alpha_0 x_0 + \alpha_1 x_1 + \cdots + \alpha_m x_m, \tag{10}$$

where the α_i represent the coefficients of equalization required to cancel the channel response. This summation can be performed in digital logic for any

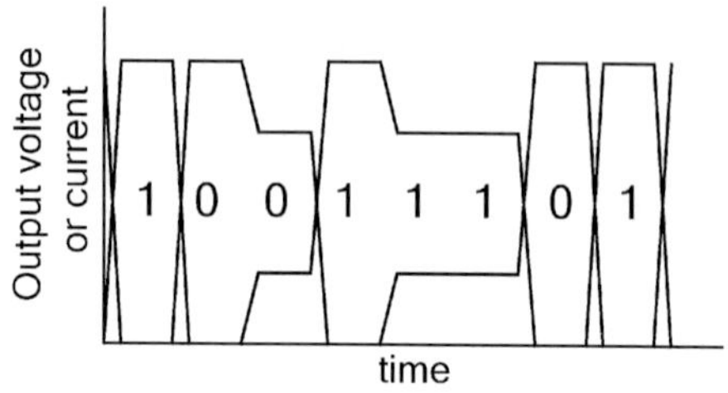

Figure 10. An example two-tap de-emphasis equalized waveform at the output of the transmitter.

arbitrary number of taps [11], but a typical 20-inch channel at 5 Gbits/s will require no or one tap prior to the cursor and one to four taps following the cursor.

Equalization can also be performed in the receiver through similar logic processing means. Receiver equalization requires an accurate capture of the initial portion of a data stream, usually through a training sequence, to provide knowledge of the history of the data stream. It also suffers from the amplification of the noise at the receiver along with the signal. However, the main advantage of receiver equalization is that it may apply a larger gain than the transmitter, which is limited in range between the maximum signal amplitude allowed out of the transmitter and the minimum signal needed to maintain an acceptable signal-to-noise ratio.

4. IO Clocking

Even if the link architecture can convey coherent symbols from transmitter to receiver through a high-loss channel, the symbols must be captured at the appropriate time and delivered synchronously to the receiver's processing unit. Figure 11(a) shows an ideal timing diagram for edge-triggered clocking of dual data-rate input data at the receiver's data sampling unit. Both rising and falling edges of the clock sample the input data. The clock is aligned to the center of the data symbol, which provides the greatest voltage in the sample and also the greatest timing margin. Any jitter of the data or the sampling clock with respect to one another results in timing margin loss, as shown in Figure 11(c). Jitter amounting to more than half of the symbol width causes the sampling clock to miss the data symbol entirely.

Two standard clocking topologies, derived clocking and source-synchronous or forwarded clocking, deliver the timing reference to the transmitter and receiver.

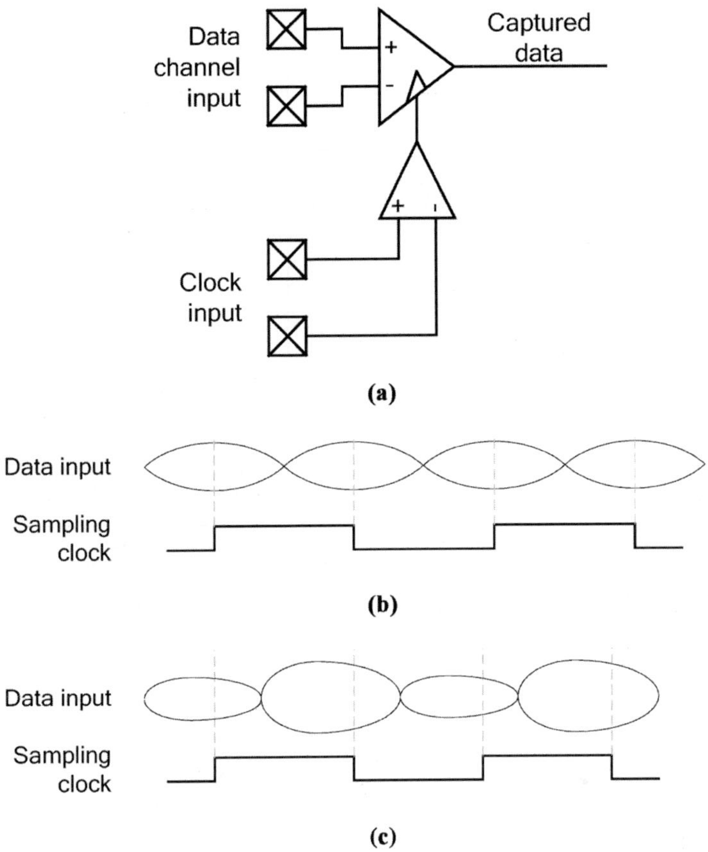

Figure 11. Synchronous capture of input data: **(a)** synchronous capture circuit, **(b)** ideal data and clock timing, **(c)** example timing with voltage noise and jitter.

4.1. Derived Clock Design

In derived clocking, a synchronization source is provided to both the transmitter and the receiver. This source may be common to both, as shown in Figure 12, or it may come from independent sources that are frequency matched to within a certain tolerance. Phase-locked loops (PLL) in both the transmitter and receiver multiply the input clock to the link frequency. In the transmitter, the link frequency clock is generally used without further phase adjustment to capture data from the internal processing unit and feed it onto the link. In the receiver, however, the PLL output clock must be phase aligned to coincide with the input data as shown in Figure 11(b).

The overall architecture to perform the phase alignment is shown in Figure 12. The receiver clock from the PLL enters two phase adjustment units, which deliver two clock phases to the input samplers. The first alignment unit

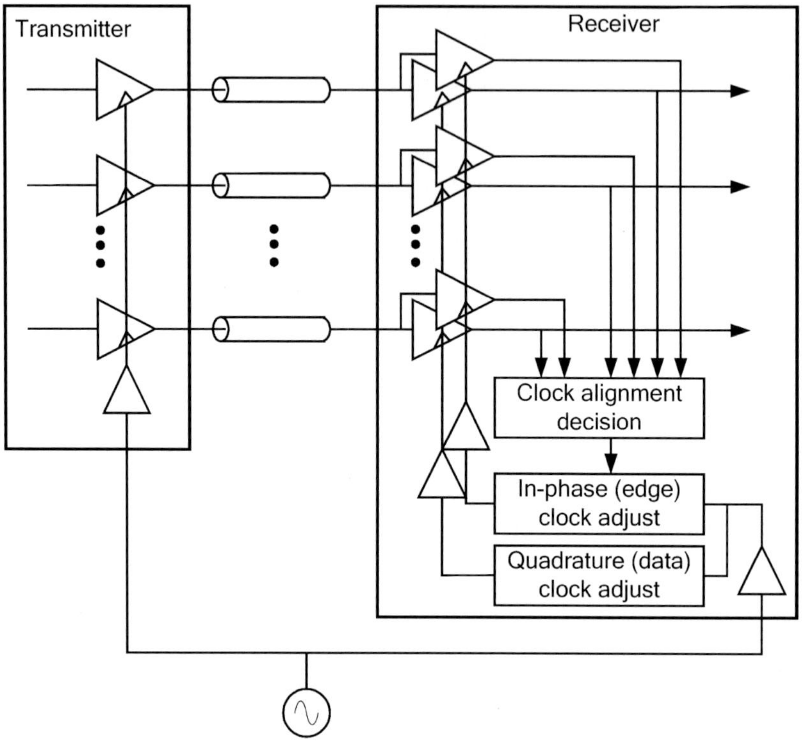

Figure 12. Clocking in a derived-clock architecture.

adjusts the phase of the sampling clock at the receiver to coincide with the data transitions at the symbol boundary. The second clock is shifted 90 degrees from the first clock and is used to sample the data at its midpoint. Following its capture, the data typically passes to an on-chip logic clock domain derived from the same input source but not adjusted in phase.

Figure 13 shows how over-sampling the input data in this manner provides phase alignment information through the clock alignment decision logic. If, as in the left two cases, the edge sampled data matches the earlier data sample, the clocks are early and should move later. In the right two cases, the edge samples match the later data sample, indicating that the clocks are late and should be moved earlier. By collecting this alignment information from the final data sampling point and feeding it back to the phase adjustment units, this architecture can also compensate for clock distribution delays in the system.

In fact, dynamic feedback also allows the alignment units to continuously track any drift between the clock and the data. Continuous alignment is often desirable for several reasons. In mesochronous systems, described in Section 4.3, the average frequency in the transmitter and in the receiver must be identical. However, voltage and temperature changes can cause a change

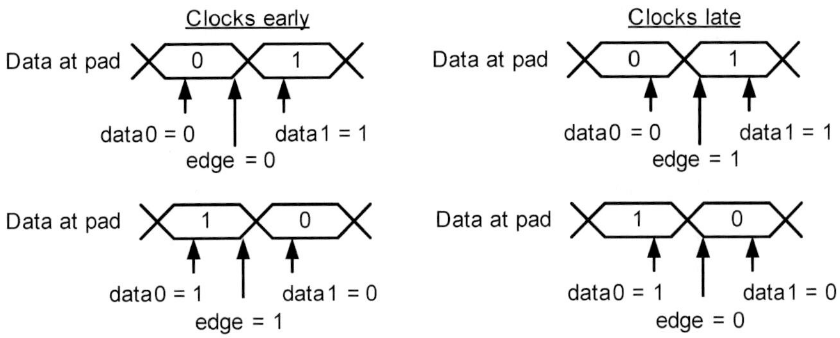

Figure 13. Over-sampling the input to lock the clock to the input data.

in circuit delay, shifting the data or clock in time. In plesiochronous systems, also described in Section 4.3, clocks may differ slightly in frequency, causing a continuous shift in phase between data and clock.

The scheme where the clock alignment is common across all data lanes, as illustrated in Figure 14, works only when the skew among data lanes is low enough to allow an adequate timing margin. When skew among data lanes becomes large, the clock alignment must be adjusted on a per-lane basis. This can be done by making the clock alignment decision and clock phase adjustment for every lane, at a cost of more circuits, area, and power to duplicate these circuit blocks in every lane.

4.1.1. Jitter in derived clock systems

Although dynamic tracking schemes follow slow drift, they are generally unable to correct for high-frequency jitter between clock and data. Such jitter causes misalignment of the sampling clock with respect to the data symbol and creates loss of timing margin which, when it exceeds half of the symbol width, causes data loss.

Timing misalignment occurs when differing jitter arises along the clock and data paths or when clock and data paths are of unmatched lengths so that they no longer share the common characteristics of the source clock. Specifically, the sources of jitter include

1. Jitter injected along one path that is not injected along the other, such as supply noise-induced delay variations in the transmitter or receiver.
2. Source clock jitter filtered by different PLL with differing characteristics.
3. Path length differences from the common point, which separates the original edge that creates the data from the original edge that creates the sampling clock.

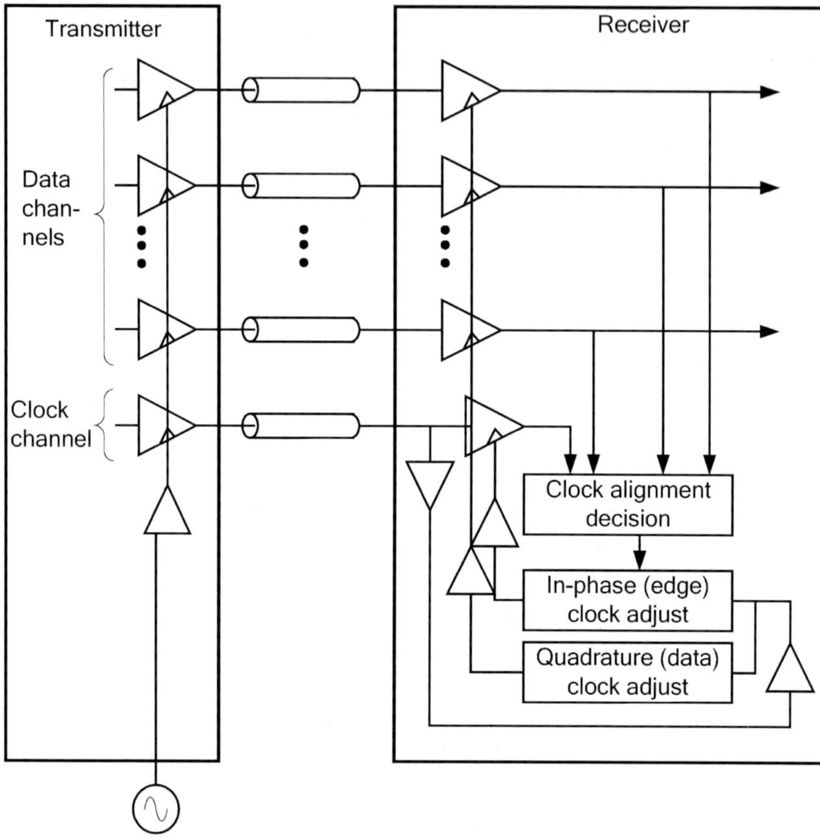

Figure 14. Clocking in a forwarded or source-synchronous clock architecture.

With enough knowledge of the system characteristics, the timing loss from these effects can be calculated [12] or measured [14].

4.2. Source Synchronous Design

An alternative clocking structure is source synchronous or forwarded clocking, shown in Figure 14. In this architecture, a clock lane is added in parallel with the data lanes and a clock is sent from transmitter to receiver. The forwarded clock at the receiver is amplified, phase aligned, and distributed to the input data samplers.

As long as the skew among data and clock lanes is within tolerable limits, clock alignment can be performed with the in-phase clock at the forwarded clock lane sampling input. Alignment is performed such that the in-phase clock is placed in the edge sampling position with respect to the forwarded clock input, as shown in Figure 13. The quadrature clock is shifted 90 degrees from this position so that it samples the data in the center of its valid region, providing the greatest margin for timing degradation through jitter. Alignment with the forwarded clock input eliminates the need for in-phase over-sampling at the data lane receivers.

Furthermore, because the forwarded clock shares the same source timing as the input data, both data and clock experience the same timing drift and jitter from the transmitter. This creates an inherent tracking between clock and data. In fact, dynamic tracking is often unnecessary in forwarded clock systems. Only periodic re-alignment is needed to compensate for temperature drift in the receiver clock path.

As with the derived clocking case, if skew among data lanes becomes too large, the clock alignment and phase adjustment can be pushed into every data lane to perform the clock alignment on a per-lane basis. The over-sampling receiver must be added back to the data lanes and the phase alignment overhead must be duplicated for every lane.

4.2.1. Jitter in source synchronous systems

Although in source synchronous systems the clock timing is common with the data timing at the point of transmission, clock and data paths are not identically matched. The clock traverses the amplifier, phase alignment, and distribution circuits in the receiver, each of which may add jitter to the clock that will not be seen in the data. Furthermore, these circuits plus any channel skew cause a delay mismatch between clock and data. Since the receiver clock path delay is fixed, jitter will accumulate up to the clock and data delay difference. Therefore, it is still critical to minimize jitter in the transmitter clock to achieve higher speeds.

4.3. Clock Drift Considerations

The construction of the source clocks into either a forwarded or a derived clock system also affects the gross synchronization between transmitter and receiver. Two situations are possible. In the first, the source clock to the transmitter and to the receiver may come from the same oscillator or from separate oscillators that are frequency matched such that the average frequency of the transmitter and the receiver clock is the same. This is known as a mesochronous system. In the second situation, known as a plesiochronous system, the average clock frequency of the source clock at the transmitter and at the receiver may differ by a small amount. The difference is usually constrained to parts per million. For both cases, the data path topology must comprehend the difference in data transfer rate arising from the clock system topology.

Although the rate-matching of a mesochronous system implies a straightforward one-for-one transfer along the data path, it is usually not so simple in actual systems. Although the average clock frequency must match between transmitter and receiver, short-term deviations may occur. These arise, for example, from differences in the response of the transmitter and the receiver PLL to phase noise from the reference oscillator or from voltage and temperature drift in the transmitter or receiver. These drifts cause the transmitter clock to run temporarily faster or slower than the receiver clock by a slight amount. To overcome the data-rate difference, we can buffer the data in a first-in first-out (FIFO) structure. This makes data available to the receiver in case its clock temporarily speeds up and also provides a buffer for received data in case the transmitter clock temporarily speeds up.

In a plesiochronous system, the clock frequency difference implies a continuous difference in the data rate between transmitter and receiver. Buffering the data does not provide a solution since any finite buffer will eventually run out. Possible solutions include either handshake mechanisms or skip characters. Hand-shake mechanisms work by transferring data only when it becomes available [15]. This can be done either per serial bit or by constructing the data into a parallel packet and transferring it when the packet is complete. Skip characters work by allowing the receiver to ignore or add in occasional null data sequences so that the same effective data transfer rate can be maintained.

5. Conclusions

Significant physical effects impede the operation of off-chip signaling at speeds higher than 1 to 2 Gbit/s. These include transmission-line effects such as dielectric loss and skin effect loss, inter-symbol interference, and clock- and data-skew loss. These effects can be overcome through the dedication of more on-chip computational resources to process the signal in a way that compensates for these effects. Such schemes include precision calibration of signaling levels, equalization, and clock phase adjustment. Further methods will be required to achieve speeds in excess of 5 Gbit/s.

Acknowledgements

The author is grateful to Xiaoxiong (Kevin) Gu and Mohiuddin Mazumder for providing the channel characteristics and simulation results. Thanks to Ken Drottar and Pascal Meier for valuable feedback.

References

[1] Bakoglu, H.B. *Circuits, Interconnections, and Packaging for VLSI.* Addison-Wesley, **1990**.

[2] Dabral, S.; Maloney, T.J. *Basic ESD and I/O Design,* John Wiley & Sons, **1998**.

[3] Dally, W.J.; Poulton, J.W. *Digital Systems Engineering,* Cambridge University Press, **1998**.

[4] Dally, W.J.; Poulton, J.W. "Transmitter equalization for 4-Gbps signaling", *IEEE Micro*, **1997**, 48–56.

[5] Horowitz, M.; Yang, C.-K.K.; Sidiropoulos, S. "High-speed electrical signaling: overview and limitations", *IEEE Micro*, **1998**, 12–24.

[6] Johnson, H.W.; Graham, M. *High-Speed Digital Design: A Handbook of Black Magic,* Prentice Hall, **1993**.

[7] Hall, S.H.; Hall, G.W.; McCall, J.A. *High-Speed Digital System Design: A Handbook of Interconnect Theory and Design Practices,* John Wiley & Sons, **2000**.

[8] Sidiropoulos, S.; Yang, C.-K.K.; Horowitz, M. "High-speed inter-chip signaling." In *Design of High-Performance Microprocessor Circuits,* Anantha Chandrakasan, William J. Bowhill, and Frank Fox, eds., IEEE Press, **2000**.

[9] Thierauf, S.C.; Anderson, W.R. "I/O and ESD circuit design." In *Design of High-Performance Microprocessor Circuits,* Anantha Chandrakasan, William J. Bowhill, and Frank Fox, eds., IEEE Press, **2000**.

[10] Casper, B.K.; Haycock, M.; Mooney, R. "An accurate and efficient analysis method for multi-Gb/s chip-to-chip signaling schemes", *Symposium on VLSI Circuits*, **2002**, 54–57.

[11] Erdogan, A.T.; Arslan, T.; Horrocks, D.H. "Low-power multiplication schemes for single multiplier CMOS based FIR digital filter implementations", *IEEE Int. Symp. Circuits Systems*, **1997**, 1940–1943.

[12] PCI Express Jitter Modeling (July 14, 2004), http://www.pcisig.com.

[13] Stojanovic, V.; Horowitz, M. "Modeling and analysis of high-speed links", *Custom Integrated Circuits Conference,* September **2003**.

[14] Kossel, M.A.; Schmatz, M.L. "Jitter measurements of high-speed serial links", *IEEE Design Test Comput.*, **2004**, 536–543.

[15] PCI Express Base Specification, Revision 1.1 (March 28, 2005), http://www.pcisig.com.

Chapter 12

PROCESSOR CORE AND LOW-POWER SOC DESIGN FOR EMBEDDED SYSTEMS

Naohiko Irie
Hitachi Ltd, Japan

Abstract: A processor core SH-X based on SuperH^{TM} architecture is described. It is implemented in a 130-nm CMOS process running at 400 MHz achieving 720 MIPS and 2.8 GFLOPS at a power of 250 mW under worst-case conditions. It has a dual-issue seven-stage pipeline architecture, but reaches 1.8 MIPS/MHz, which is equivalent to the previous five-stage processor. The processor meets the requirements of a wide range of applications, and is suitable for digital appliances aimed at the consumer market, such as cellular phones, digital still/video cameras, and car navigation systems. In this chapter a system-on-a-chip (SOC) implementation called SH-Mobile3 that uses SH-X core and low-power circuit technology's also described. The SOC applies power-switch circuit and low-leakage SRAM and achieves less than 100 μA in a stand-by mode.

Key words: processor core, embedded system, digital appliance, pipeline, floating-point unit (FPU), 3D graphics, system-on-a-chip (SOC), low power, power-switch, SRAM, leakage current

1. SH-X Processor Core for Wide Range Embedded Applications

1.1. Background and Issues for Embedded Systems

Processors for embedded systems are increasingly being used in digital appliances designed for the consumer market, such as cellular phones, digital still/video cameras, and car navigation systems. They must deliver high performance, while maintaining a reasonable size and low power consumption.

Vojin G. Oklobdzija and Ram K. Krishnamurthy (eds.),
High-Performance Energy-Efficient Microprocessor Design, 311–336.

Each digital consumer appliance has some unique features that have their own specific requirements for the processor core. Digital consumer appliances can be categorized into two types: one is a mobile type, such as cellular phones and digital still/video cameras, and the other is an equipped type, such as car navigation systems, DVD recorders, and game consoles.

Ideally, we should be able to provide an optimized chip for all appliances, to meet the wide range of requirements for various applications. For example, mobile-type appliances require lower power in average especially in stand-by mode. On the other hand, equipped-type appliances require performance headroom for functional extension under a certain peak power budget for thermal design. LSI implementation is required to satisfy the severe power and cost restrictions. As you can see, customization is an effective way to improve the cost-performance and power-performance ratio. Unfortunately, mask production costs increase for finer processes, and design and verification costs also increase with the integrated logic scale. Even though these initial costs increase, the mass-production costs decrease for the finer processes.

Figure 1 illustrates the cost structure model. The horizontal and vertical axes represent units and cost/sales, respectively, and the slope represents the unit cost of mass-production or sales price. The cost structure changes from cost #1 to cost #2 when using a finer process. Then, when costs and sales intersect, the break-even point increases. Therefore, the required market size for products using a customized chip increases when the finer processes are used. However, initial cost reduction changes the cost structure from #2 to #3, and decreases the break-even point.

Using common intellectual property (IP) and unifying design and verification of similar IPs are effective for reducing the initial costs. The processor core in particular requires enormous costs for design, verification, tool and software development, and using a common core for a lot of products contributes to the initial cost reduction. Therefore, it is important to make the core specifications flexible enough to expand to fit the required product range.

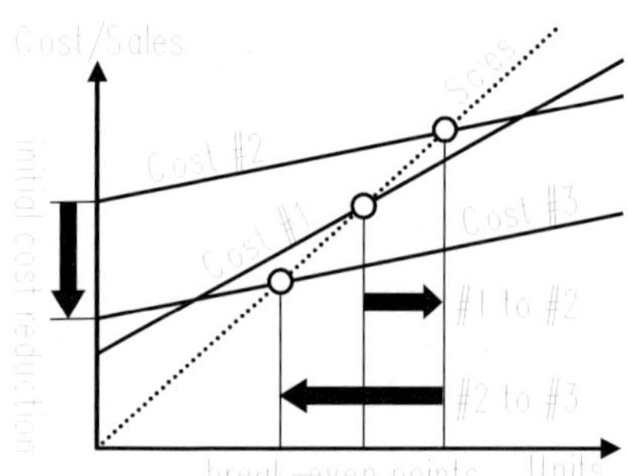

Figure 1. Initial costs of finer processes.

The flexibility can be enhanced by preparing optional modules, such as DSPs and floating-point units (FPUs), to augment the capabilities of a basic processing core. Optional DSPs can be added as execution units. These are much smaller than a full DSP core and can effectively accelerate standardized applications such as multimedia [1–3]. While PCs and game consoles may be able to justify the high cost of special graphics hardware that is much larger than a processor core [4, 5], other consumer appliances cannot. However, an optional FPU can handle a wide range of dynamic data, thus simplifying programming, especially for graphics acceleration. An optional FPU is therefore a good way of improving graphics performance with an embedded processor [6–8].

1.2. Specifications and Pipeline Structure

The flexible SuperH™ (SH) processor core SH-X was developed to meet the above requirements for the digital consumer appliances [9–11]. As mentioned above, we must reduce the initial costs including the design cost. Therefore, the SH-X is designed as the master design to implement various processor cores.

1.2.1. Processor core specifications

SH-4A and SH4AL-DSP are standard and low-power versions of the SH-X. The specifications of the cores are shown in Table 1.

The cores were fabricated using a 130-nm process. The SH-4A runs at 400 MHz with a supply voltage of 1.25 V in worst-case conditions, achieving

Table 1. SH-X processor core specifications

Core version	SH-4A	SH4AL-DSP
Process	130-nm CMOS	
Supply voltage	1.25 V	1.0 V
Optional modules	FPU	DSP
Frequency	400 MHz	200 MHz
Performance	720 MIPS[a]	360 MIPS[a]
	2.8 GFLOPS[b]	–
	36 M Polygons/s[c]	–
Power[a]	250 mW	80 mW
First-level caches	32 kB I- and D-caches	
First-level RAM	16 kB	
Second-level memory	256 kB cache or RAM	

[a] Measured using Dhrystone 2.1.
[b] Peak floating-point performance.
[c] Measured using a simple geometry benchmark program.

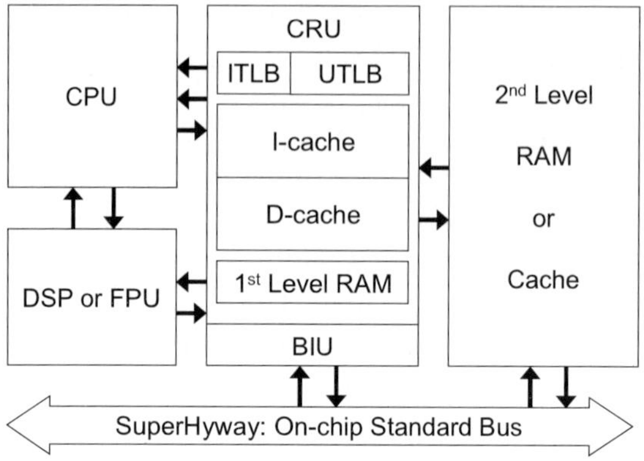

Figure 2. SH-X core block diagram.

720 MIPS and 250 mW (measured using Dhrystone 2.1), and 2.8 GFLOPS and 32 M polygons/s. It was first integrated into a product chip used for a car navigation system. The SH4AL-DSP runs at 200 MHz and 1.0 V, and achieves 360 MIPS and 80 mW resulting in 4500 MIPS/W. It was integrated into an application processor for cellular phones [12, 13]. The processor cores have 32-kB first-level instruction and data caches (I- and D-caches) and 16-kB RAM. They also have a 256-kB second-level cache or RAM, which is selectable. The size is flexible depending on requirements. The use of on-chip RAM ensures real-time response, which is a key feature of embedded processors.

Figure 2 illustrates the SH-X core block diagram. The SH-X core consists of a CPU, a cache RAM control unit (CRU), a bus interface unit (BIU), optional modules (DSP or FPU), and an optional second-level memory (RAM or cache). The SH-X core uses SuperHyway as the on-chip standard bus to interface with other on-chip intellectual properties (IPs).

1.2.2. Pipeline structure

The structure of the pipeline and its characteristics are illustrated in Figure 3. With dual-issue superscalar architecture, two instructions are issued to two of five execution pipelines, i.e. branch, CPU execution, load/store, DSP execution, and FPU execution. The FPU data transfer pipeline is categorized as being part of the load/store pipeline. A DSP instruction is a long instruction word (LIW) type and can specify both multiply and ALU operations, while the DSP execution pipeline can treat both of them simultaneously.

A seven-stage pipeline architecture is used to increase the clock rate. With this architecture the performance is typically about 20% lower than that of

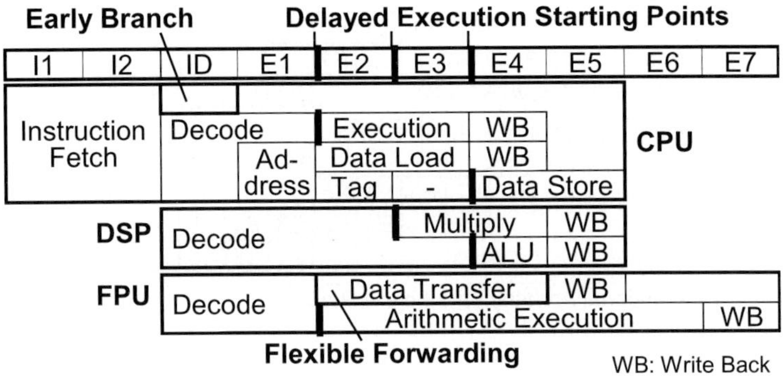

Figure 3. Pipeline structure.

a five-stage pipeline operating at the same frequency, mainly because of the long cycles needed to fetch instructions and to load and store data. Therefore, an early branch architecture, in which branch operations in the instruction queue are started out-of-order, is used. This reduces the longer branch penalty caused by longer instruction–fetch cycles. The delayed execution hides the load latency. However, it delays the timing for store data, and branch conditions. Store buffers are used to hide delays in store data, and branch prediction based on a branch history table hides delays due to branch conditions. Early register release and short latencies are useful for enhancing register availability. Flexible forwarding enables early release and simplifies programming.

Optimizing the pipeline by enhancing aspects of the microarchitecture enables the SH-X to compensate for the slower performance of the architecture. As a result it has achieved the 1.8 Dhrystone MIPS/MHz of the previous five-stage processor.

A conventional out-of-order architecture may achieve the same performance or better, but it requires a large amount of hardware. Although our architecture is less flexible than an out-of-order one, it is very efficient, achieving high performance with low-power and a small area.

1.2.3. Delayed executions

Figure 4 shows an example of the effects of delayed execution. Delayed execution accelerates multiple-cycle and dependent-instruction flows. The example shows a typical DSP instruction flow, i.e. a load, multiply, and store sequence. As shown in Figure 4, the load instruction calculates the load address at the E1 stage, and loads the data at the E2 and E3 stages. Then the multiply instruction multiplies the loaded data by other data in a register at the

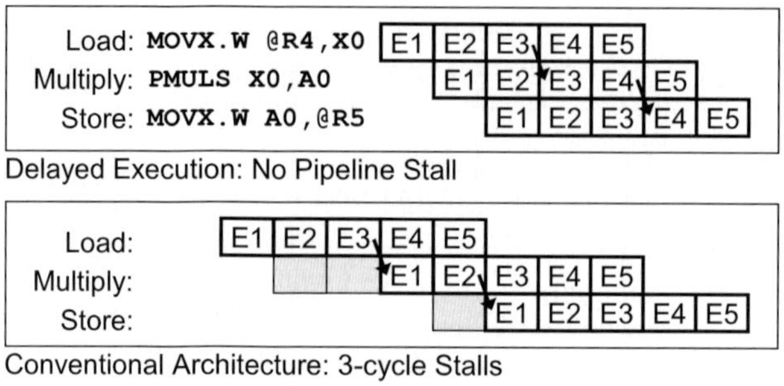

Figure 4. Example of effects of delayed execution.

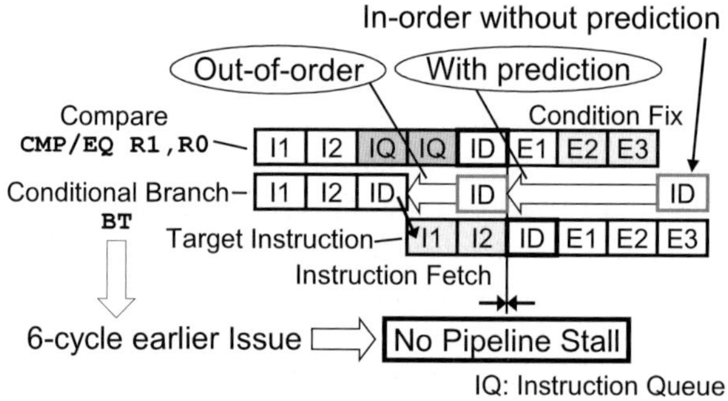

Figure 5. Example of effects of early branch architecture.

E3 and E4 stages. Finally, the store instruction calculates the store address at the E1 stage, and stores the multiplied data at the E4 and E5 stages. With delayed execution the sequence does not cause a pipeline stall. However, with conventional architecture, the same sequence would result in three-cycle stalls and a serious degradation in performance because conventional architecture would require all the source operands to be ready at the E1 stage.

1.2.4. Early branch architecture

Figure 5 shows an example of the effects of early branch architecture. The SH-X issues a branch out-of-order before a condition fix with a branch prediction. In this example the compare instruction stays in the instruction queue (IQ) for two cycles. During this time a conditional branch instruction initiates an instruction fetch out-of-order before a condition fix. The timing starts four cycles earlier with a prediction, and two cycles earlier when an instruction is issued out-of-order compared to the case when an instruction is issued in-order without a prediction. A conditional branch can then be issued six cycles earlier, and no pipeline stall occurs between the compare and target instructions.

The SH-X fetches four instructions per cycle. Even when the SH-X issues two instructions every cycle, the number of fetched instructions in the IQ increases until an instruction cache miss, a branch prediction miss, and so on. Therefore, instructions are likely stay in the IQ for several cycles, like the compare instruction in Figure 5, and out-of-order issuing of branches works effectively. The branch frequency of multimedia applications, which are the major applications of the SH-X, is relatively low as nested branch prediction works effectively, and we did not implement the nested prediction.

1.2.5. Flexible forwarding

Figure 6 shows an example of the effects of flexible forwarding. The FADD.S is a single precision add instruction with a latency of three, as shown in Table 2. The addition uses the pipeline stages from E2 to E4. In the case of early register release, the store instruction gets the add result FR1 forwarded at the E4 stage. The store instruction then stores the FR1 value immediately after it has been generated by the add instruction. In the case of late register allocation, the copy instruction gets the add result that the FR1 forwarded at the E1 stage. The copy destination register FR2 is then allocated late, and can be used for another purpose until it is allocated. Thus, flexible forwarding ensures that the timing of register release and allocation is flexible, easing programming constraints.

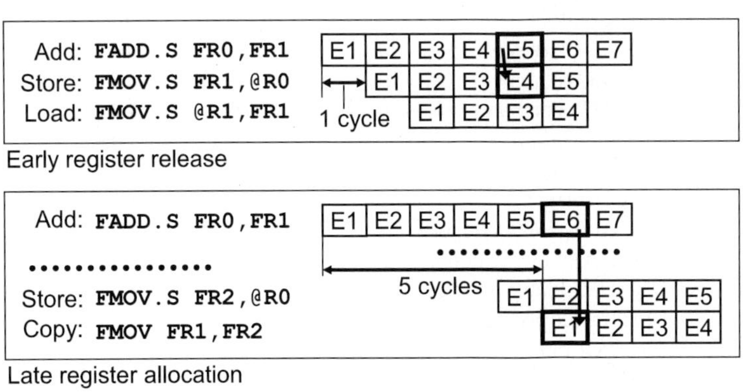

Figure 6. Example of effects of flexible forwarding.

1.3. Vector FPU Design

1.3.1. FPU arithmetic instructions

Since floating-point instructions tend to have long latencies, four special instructions – a four-element inner product (FIPR), transform four-element vector (FTRV), square-root reciprocal approximate (FSRRA), and sine and cosine approximate (FSCA) – were added to shorten the effective latencies. Table 2 shows the set of arithmetic instructions for the FPU. In the table the pitch represents resource-occupying cycles. The FDIV and FSQRT instructions occupy one dedicated resource, and the FSRRA and FSCA instructions occupy another. The values in parentheses indicate these dedicated resource-occupying cycles. An instruction using the same resources can be issued after the resource-occupying cycles. We call the cycles the pitch of the instruction.

Multiply–accumulate (MAC) is one of the most common operations in intensive computing applications. An FIPR instruction is implemented with an effective latency of one-quarter that of the equivalent sequence for one MUL and three MAC instructions. The use of four-way SIMD achieves the same throughput as an FIPR, but the latency is longer and the register file must be larger.

Table 2. FPU arithmetic instructions

Arithmetic instructions	Pitch/Latency	
	Single	Double
FADD (add)	1/3	1/5
FSUB (subtract)	1/3	1/5
FMUL (multiply)	1/5	3/7
FDIV (divide)	2(13)/17	2(28)/32
FSQRT (square root)	2(13)/17	2(28)/32
FCMP/EQ (compare)	1/1	1/1
FCMP/GT (compare)	1/1	1/1
FABS (absolute)	1/1	1/1
FNEG (negate)	1/1	1/1
FLOAT (integer to float)	1/3	1/5
FTRC (truncate to integer)	1/3	1/5
FCNVSD (single to double)	–	1/5
FCNVDS (double to single)	–	1/5
FMAC (multiply–accumulate)	1/5	–
FIPR (four-element inner product)	1/5	–
FTRV (transform four-element vector)	4/8	–
FSRRA (square-root reciprocal)	1(3)/5	–
FSCA (sine and cosine)	3(5)/7	–

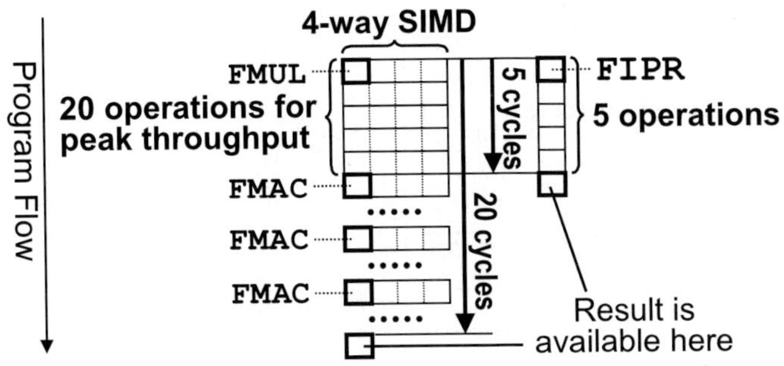

Figure 7. FIPR vs. four-way SIMD FMUL/FMAC.

Figure 7 illustrates the differences. In this example each box shows an operation issue slot. Since FMUL and FMAC have a five-cycle latency we must issue 20 independent operations for peak throughput in the case of four-way SIMD. The result is available 20 cycles after the FMUL issue. Therefore, FIPR requires one-quarter of the program's parallelism and latency. An FTRV instruction, which multiplies a four-by-four matrix and a four-element vector, is implemented using the FIPR hardware four times.

An FSRRA instruction is also used to accelerate vector normalization operations. It has an accuracy of 23 bits, which is similar to that of a series of square-root and divide instructions. Figure 8 compares the pitch and latency of the FSRRA and the equivalent sequence for an FSQRT and FDIV. There are two-cycle FSQRT and FDIV pitches because they require a one-cycle issue slot to initiate a special resource operation and a one-cycle post-process to normalize and round the result. The FSQRT and FDIV latencies take 17 cycles, and the result is available 34 cycles after the issue of the FSQRT. In contrast, the pitch and latency for the FSRRA are one and five; that is, only one-quarter and approximately one-fifth of the equivalent sequences, respectively. If we try to reduce these latencies we have to increase the special resources for FDIV/FSQRT. However, FSRRA is much faster using a similar amount of the resource.

An FSCA instruction is also defined to enhance graphics programmability. The FSCA generates the sine and cosine of the source operand with an error rate of less than 2^{-22}, which is sufficiently accurate for graphic applications.

1.3.2. Implementation

The FPU decodes FPU instructions, transfers FPU data as part of the load-store pipeline, and executes FPU arithmetic operations. This set of instructions consists of short- and long-latency instructions. The simple single-precision

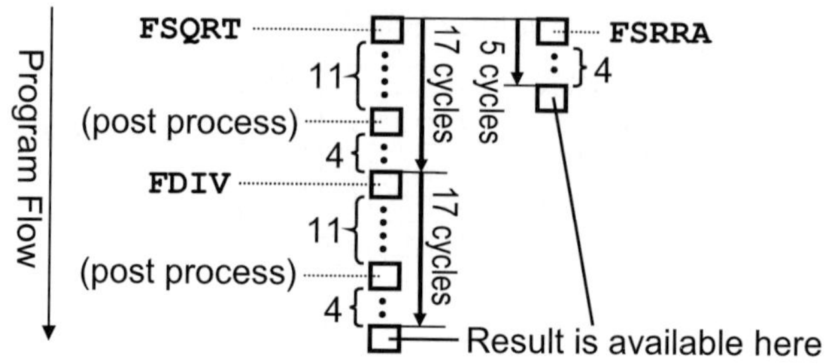

Figure 8. FSRRA vs. series of an FSQRT and an FDIV.

instructions, FADD, FSUB, FCMP, FLOAT, and FTRC, are categorized as short-latency instructions. They have three-cycle latency except FCMP, whose latency is one with the branch prediction. The FABS and FNEG instructions only treat the sign of the FPU data, and are executed by the load-store pipeline like transfer instructions.

Figure 9 illustrates the FPU arithmetic execution pipeline. With the delayed execution architecture, the register operand read and operand forwarding are done at the E1 stage, and the arithmetic operation starts at E2. The short arithmetic pipeline treats short-latency instructions. All the arithmetic pipelines share one register write port to reduce the number of ports. There are four forwarding source points to provide the specified latencies for any cycle distance of the define-and-use operations. The FDIV/FSQRT pipeline is occupied by 13/28 cycles to execute a single/double FDIV or FSQRT instruction, and these instructions cannot be issued frequently. The FSRRA/FSCA pipeline is three cycles long and is occupied by three times to execute an FSRRA or FSCA

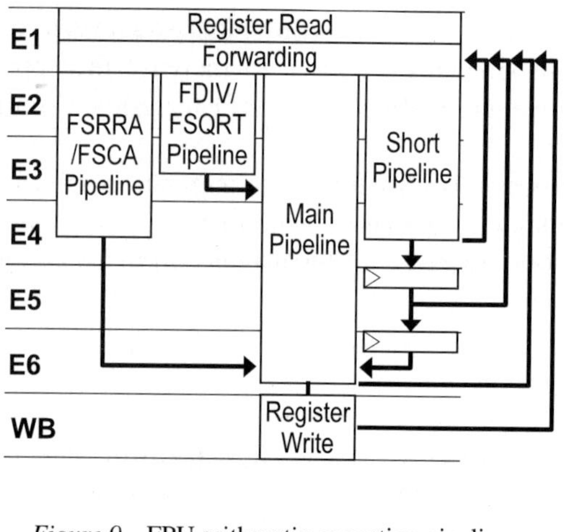

Figure 9. FPU arithmetic execution pipeline.

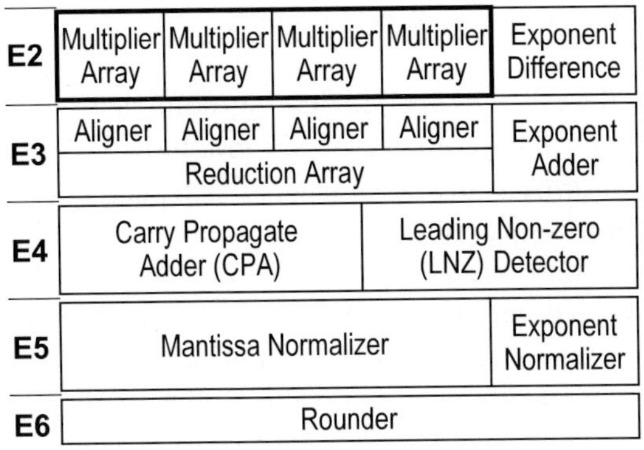

Figure 10. Structure of main FPU pipeline.

instruction. Therefore, the third E4 stage and E6 stage of the main pipeline are synchronized, and the FSRRA/FSCA pipeline output merges with the main pipeline at this point.

The FSRRA and FSCA are implemented by calculating the 3D polynomials of the properly divided periods. The width of the 3D term is eight bits, which adds a small area overhead, while enhancing accuracy and reducing latency.

Figure 10 illustrates the structure of the main FPU pipeline. There are four single-precision multiplier arrays at E2 to execute FIPR and FTRV and emulate double-precision multiplication. Their total area is less than that of a double-precision multiplier array. The calculation of exponent differences is also done at E2 for alignment operations by the four aligners at E3. The four aligners align eight terms, four sets of sum and carry pairs of four products generated by the four multiplier arrays, and reduce the eight terms to two using the reduction array at E3. The exponent value before normalization is also calculated by the exponent adder at E3. The carry-propagate adder (CPA) adds two terms from the reduction array, and the leading non-zero (LNZ) detector searches the LNZ position of the CPA result from the two CPA inputs precisely and with a speed comparable to that of the CPA at E4 [14]. Therefore, the result of the CPA can be normalized immediately after the CPA operation without the need to correct position errors, which is often necessary when using a conventional 1-bit error LNZ detector. Mantissa and exponent normalizers normalize the CPA and exponent-adder outputs at E5 controlled by the LNZ detector output. Finally, the rounder rounds the normalized results into the IEEE 754 format.

The extra hardware required for the special FPU instructions is about 30% of the original FPU hardware and the FPU area is about 10–20% of the SH-X core depending on the size of the first and second on-chip memories. Therefore, the extra hardware is about 3–6% of the SH-X core.

1.3.3. 3D graphics benchmark performance

We estimated the effect of special floating-point instructions using a simple benchmark based on the geometry of 3D graphics. Figure 11 illustrates the benchmark, which consists of coordinate and perspective transformations, and intensity calculations using parallel light. In general, a 3D object is divided into triangles or quadrangles. We used triangles and assumed a strip model with one vertex and one normal vector per polygon. Coordinate transformation can be used to make a polygon rotate, move in parallel, and increase or reduce in size by multiplying the transformation matrix by the vertex vector V of the polygon and getting a transformed vertex vector V″. Perspective transformation enables the x and y coordinates of the transformed vertex V″ to be projected onto a screen, and to get the coordinates S_x and S_y on the screen. Intensity calculation is used to calculate the surface intensity of the transformed polygon by transforming the polygon's normal vector N to get the cosine I of the transformed normal vector N′ and light vector L.

The formulas are shown in Figure 12. Since we need only V''_z of V″ to judge which object is closest to the viewing point, we do not need to calculate V''_x and V''_y. The coordinate and perspective transformation requires seven FMULs, 12 FMACs, and two FDIVs without the special floating-point instructions, and one FTRV, five FMULs, and two FSRRAs with them. Intensity calculation requires seven FMULs, twelve FMACs, one FSQRT, and one FDIV without the special floating-point instructions, and one FTRV, two FIPRs, one FSRRA, and one MUL with them.

Figure 13 shows the execution cycles for 3D geometry. After program optimization no register conflict occurs, and performance is restricted only by the arithmetic and FDIV/FSQRT special resource usage cycles. First, I will explain the coordinate and perspective transformation cycles. The gray areas of the graph represent the usage cycles. Without special instructions the FTRV, FIPR, FSRRA, and FDIV/FSQRT resources are occupied by the longest cycles, and these usage cycles determine the number of execution cycles, i.e. 26. Using the

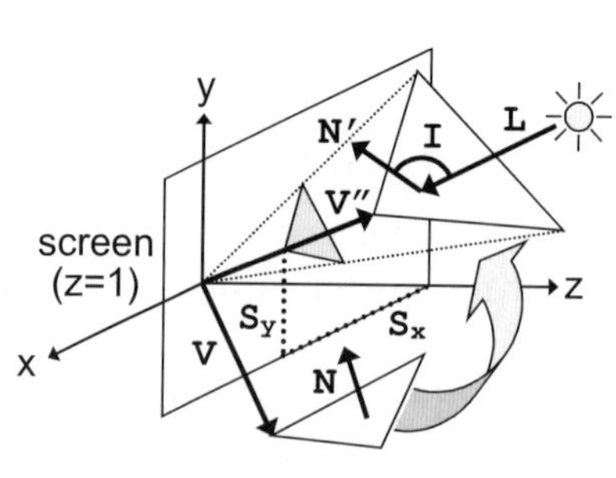

Figure 11. Simple 3D-geometry benchmark.

(1) Coordinate & Perspective Transformations

$$\begin{bmatrix} V_x' \\ V_y' \\ V_z' \\ V_w' \end{bmatrix} = \begin{bmatrix} T_{xx} & T_{xy} & T_{xz} & T_{xw} \\ T_{yx} & T_{yy} & T_{yz} & T_{yw} \\ T_{zx} & T_{zy} & T_{zz} & T_{zw} \\ T_{wx} & T_{wy} & T_{wz} & T_{ww} \end{bmatrix} \begin{bmatrix} V_x \\ V_y \\ V_z \\ 1 \end{bmatrix} \quad \begin{bmatrix} V_x'' \\ V_y'' \\ V_z'' \\ 1 \end{bmatrix} = \frac{1}{V_w'} \begin{bmatrix} V_x' \\ V_y' \\ V_z' \\ V_w' \end{bmatrix} \quad S_x = \frac{V_x''}{V_z''} \quad S_y = \frac{V_y''}{V_z''}$$

(2) Intensity Calculations (Parallel Light)

$$\begin{bmatrix} N_x' \\ N_y' \\ N_z' \\ N_w' \end{bmatrix} = \begin{bmatrix} T_{xx} & T_{xy} & T_{xz} & T_{xw} \\ T_{yx} & T_{yy} & T_{yz} & T_{yw} \\ T_{zx} & T_{zy} & T_{zz} & T_{zw} \\ T_{wx} & T_{wy} & T_{wz} & T_{ww} \end{bmatrix} \begin{bmatrix} N_x \\ N_y \\ N_z \\ 0 \end{bmatrix} \quad I = \frac{L_x N_x' + L_y N_y' + L_z N_z'}{\sqrt{N_x'^2 + N_y'^2 + N_z'^2}}$$

where:

$\begin{bmatrix} V_x \\ V_y \\ V_z \\ 1 \end{bmatrix}$: vertex vector $\begin{bmatrix} V_x' \\ V_y' \\ V_z' \\ V_w' \end{bmatrix}$: inter-mediate vector $\begin{bmatrix} V_x'' \\ V_y'' \\ V_z'' \\ 1 \end{bmatrix}$: trans-formed vertex vector $\begin{bmatrix} L_x \\ L_y \\ L_z \\ 0 \end{bmatrix}$: light vector

$\begin{bmatrix} N_x \\ N_y \\ N_z \\ 0 \end{bmatrix}$: normal vector $\begin{bmatrix} N_x' \\ N_y' \\ N_z' \\ N_w' \end{bmatrix}$: trans-formed normal vector $\begin{bmatrix} T_{xx} & T_{xy} & T_{xz} & T_{xw} \\ T_{yx} & T_{yy} & T_{yz} & T_{yw} \\ T_{zx} & T_{zy} & T_{zz} & T_{zw} \\ T_{wx} & T_{wy} & T_{wz} & T_{ww} \end{bmatrix}$: trans-form matrix

S_x, S_y: x,y coordinates on screen   I: surface intencity

Figure 12. Formulas for benchmark.

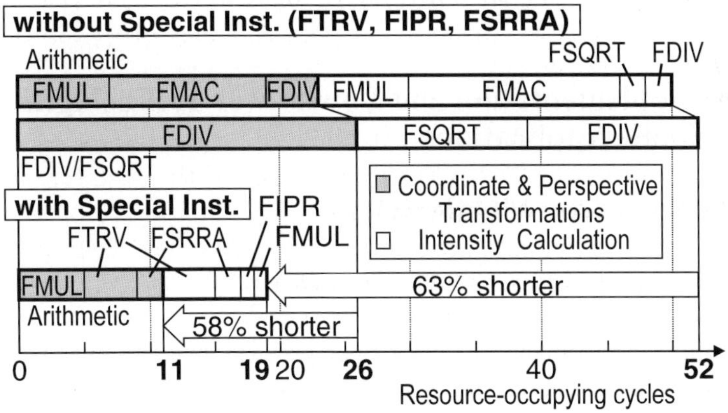

Figure 13. 3D Geometry execution cycles.

special instructions enables some of these instructions to be replaced. In this case the arithmetic resource usage cycles determine the number of execution cycles, i.e. 11, which are 58% shorter than when special instructions are not used. Similarly, when intensity is also calculated, the execution cycles are 19 and 52 with and without special instructions, respectively, and 63% shorter using special instructions compared to not using them.

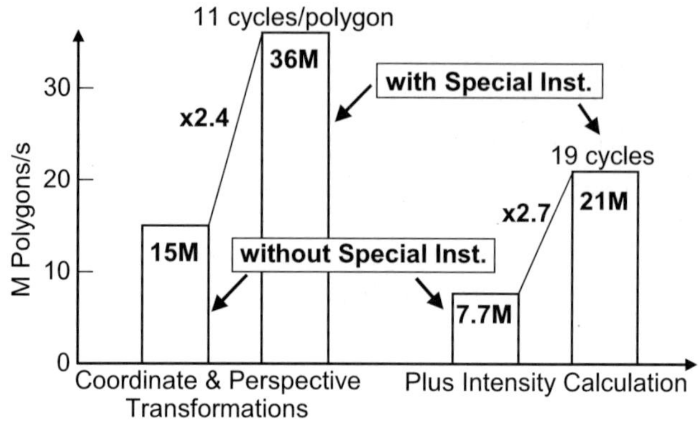

Figure 14. 3D geometry benchmark performance at 400 MHz.

Figure 14 shows the 3D-geometry benchmark performance at 400 MHz, according to the cycles shown in Table 3. Without special instructions the coordinate and perspective transformation performance is 15 M polygons/s. With special instructions, the performance is accelerated 2.4 times, increasing to 36 M polygons/s. Similarly, with intensity calculation, but without any special instructions 7.7 M polygons/s is achieved. Using special instructions the performance is accelerated 2.7 times, increasing to 21 M polygons/s.

2. SOC Implementation and Low-Power Technologies for Mobile Applications

2.1. Outline of the Mobile SOC

2.1.1. Background and issues for mobile SOCs

To get high performance using a finer process, leakage current control to meet power budget is necessary especially for embedded systems. In this Section 1 describe a method of reducing leakage power in an SH-Mobile3, application processor in 3G cellular phones as an example of SH-based mobile SOCs [12]. 3G phones are used not only for voice communication, email, and web browsing, but also for more advanced functions such as videophone and 3D games. The phones have an application processor embedded in them in addition to a base-band processor to achieve multimedia performance on demand without compromising standby or talk-time capacity [15].

The challenge for chip designers is to maintain a long enough battery life to support these applications. A common solution is to provide several low-power standby modes in the microprocessors. An important aspect of these standby modes is not only the power consumption in each mode, but also the transition time from the standby to the active mode. A long transition time may cause a significant speed overhead and this prevents using the standby mode, resulting in high leakage-power consumption. Minimizing the leakage current for various phone operating scenarios is therefore important.

This section describes the techniques used to reduce the leakage power used in the SH-Mobile3 processor. A notable feature is the implementation of on-chip power switches, which enables two new hierarchical standby modes: resume standby (R-standby) and ultra standby (U-standby) modes.

2.1.2. Chip overview

The SH-Mobile3 is a system-on-a-chip (SOC) device that is implemented on a low-power SOC design platform. This platform enables advanced circuit techniques, including on-chip power switches, to be used. These include thick-tox on-chip power switches (PSWs) for plugging leakage currents, μI/O for supporting multiple power domains with a wide range of conversion functions, and a low leakage data-retention RAM. (Details of the low leakage data-retention RAM are described in Section 2.)

Figure 15 shows a chip micrograph of the SH-Mobile3. A 130-nm five-layer-Cu dual-Vth dual-tox CMOS technology is used. The supply voltage for the core is 1.2 V with 1.8/3.3 V for the I/O. The operating frequency is 216 MHz under the worst PVT conditions. The chip size is $7.7 \times 7.6\,mm^2$. The SH-Mobile3 integrates one SH-X core with DSP, 32-kB four-way set-associative

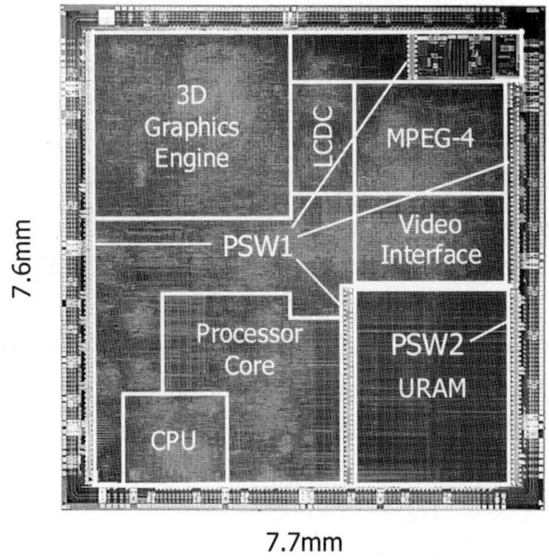

Figure 15. Chip micrograph of SH-Mobile3.

instruction and data caches, a four-entry instruction TLB, a 64-entry unified TLB, a 16-kB local RAM (XYRAM), a 256-kB user RAM (URAM), several media-processing IPs, such as MPEG-4 and 3D graphics accelerators, and other peripheral modules. The on-chip power switches are called PSW1 and PSW2.

2.1.3. Basic concept for lowering power

A key hint to achieving low power in cellular phones is that applications that run on phones are more limited than on PCs. Integrating a dedicated computation engine and providing sufficient performance at a minimum operating frequency is an effective way of improving overall power efficiency. Accordingly, the SH-Mobile3 includes advanced CMOS technology and integration of high-performance-per-clock dedicated multiple computation engines such as a 1.8-MIPS/MHz embedded processor core described the previous section.

Lowering the operating frequency also enables the threshold voltage to be raised. The operating frequency of the SH-Mobile3 is thus successively reduced to 200 MHz, and leakage power consumption is limited to about 1% of the total power consumption. However, this leakage current is not low enough for a cellular phone in standby mode. It should be noted that the leakage budget for the application processor is about 10–100 μA.

A back-biasing technique [16] is an effective way of reducing the leakage. However, this is unsuitable for high-Vth thin-tox circuits because it cannot plug the gate-tunneling leakage current and the gate-induced drain leakage (GIDL) current is not negligible [17, 18]. Back biasing is also less effective in advanced process technology. Power gating, i.e. using off-chip regulators to cut off the power supply to the chip externally, is another solution. The SH-Mobile1 uses this method [1], but it requires multiple power supply channels. It is also difficult to shorten the transition time from standby mode due to the large C or L components on the power line between the chip and off-chip regulators. The SH-Mobile3 therefore uses on-chip power switches.

Using power cut scheme with on-chip power switches, we need to consider which area can be turned on in stand-by mode. The straightforward way of cutting power is to turn on only the I/O module and wake-up control part. This mode is implemented as ultra-standby (U-standby) mode [1] targeting under 10 μA leakage current. But it takes a long recovery time from U-standby to normal mode and it causes a limited opportunity to enter U-standby mode. To solve this problem we also implemented resume-standby (R-standby) mode. In this mode data for resumption is kept in a back-up area which is turned on in R-standby mode. To realize this mode, optimized on-chip power switches and low-leakage SRAM for back-up data are required.

2.2. Circuit Technology for Mobile SOCs

2.2.1. On-chip power switches

Figure 16 shows the basic configuration of the two power domains and on-chip power switches. The on-chip power switches, PSW1 and PSW2, provide local ground level (vssm1 and vssm2). PSWC is an on-chip power switch controller. The interface circuitry between the two power domains is based on a μI/O, which is described below. The critical factors in implementing the on-chip power switches are as follows:

1. The configuration of the switches (polarity of MOS, threshold voltage, gate-oxide thickness).
2. The size of the switches.
3. Preventions of invalid signal transmission while the power supply on one side is cut off.
4. Prevention of rush current when power switches are turned on.

2.2.1.1. Configuration of on-chip power switch. A high-Vth thick-tox NMOS transistor is used for the power switch for three reasons. First, an NMOS transistor has a g_m more than two times larger than that of a PMOS. (The on-resistance of the power switches is expressed by $1/g_m$.) Low on-resistance is essential to minimize area and speed overheads.

Secondly, there are two options for the gate-oxide thickness of the power switches: a thin-tox or a thick-tox MOS can be used in the I/O circuitry. The crucial factor is the gate-tunneling current. The gate-tunneling current in the power switches is not negligible because it flows even when the switches are turned off and it is therefore essential that they are large enough to minimize speed overheads. A thin-tox MOS has too much gate-tunneling current for the power switches to clear the leakage budget so a thick-tox MOS is the only solution possible.

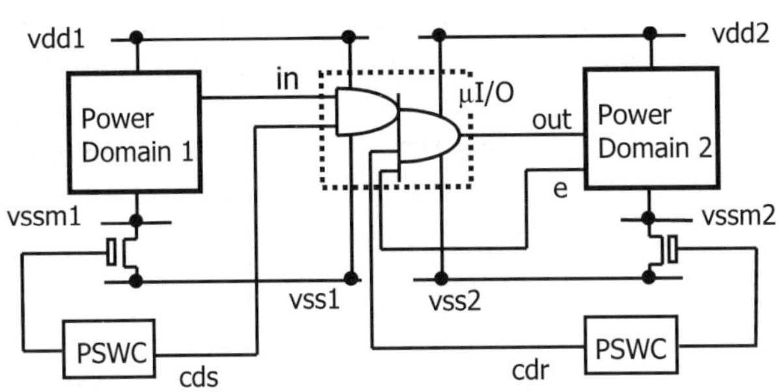

Figure 16. Basic configuration of two power domains and on-chip power switches.

Lastly, a low-Vth NMOS transistor provides low on-resistance, but it needs negative voltage to turn it off completely, thus requiring an additional power supply or on-chip voltage generator. The on-resistance of a high-Vth thin-tox NMOS may be insufficient for the power switches. However, a high-Vth thick-tox NMOS, which applies an I/O supply voltage (3.3 V) to the gates, provides sufficient on-resistance.

2.2.1.2. Size of on-chip power switch. The size of the on-chip power switch should be determined so as to ensure that the speed overhead is negligible when the power switch is implemented. Large on-resistance of the power switch (R_{sw}) causes a voltage drop across the power switch, resulting in speed degradation. In the SH-Mobile3, the size of the power switches is designed so that the R_{sw} is less than 0.1% of the equivalent resistance (R_{eq}) of a circuit to which power is supplied via a power switch. The R_{eq} is defined as:

$$R_{eq} = V/I_{max},$$

where V is the supply voltage applied to the circuit, and I_{max} is the maximum current in the time resolution of the t_c of the decoupling and/or parasitic capacitors on the terminal between the power switch and the circuit. For example, when $R_{sw} = 2k\Omega \cdot \mu m$, $I_{max} = 1\text{A}$, $V = 1.2\,\text{V}$, and $R_{eq} = 1.2\,\Omega$, the R_{sw} should be less than $12\,\text{m}\Omega$. In total, the power switches should be more than 166-mm wide.

2.2.1.3. Prevention of invalid signal transmission. It is essential to prevent invalid signal transmission between the power-on and power-off domains because this causes a significant increase in power due to short-circuit currents. The μI/O in Figure 16 prevents this by using a four-input AND function [19]. The μI/O may require a signal level-shift function when the sender's supply voltage is different from the receiver's one. Figure 17 shows an μI/O with a signal level-shift function. LC is the level-shifter circuitry, where a dual-rail input signal (n1 and /n1) with a signal swing of vdd1 is converted to a single-rail output signal (n2) with a signal swing of vdd2.

The four-input AND function in the μI/O supports both internal power shutdown by the on-chip power switches and external shutdown by off-chip

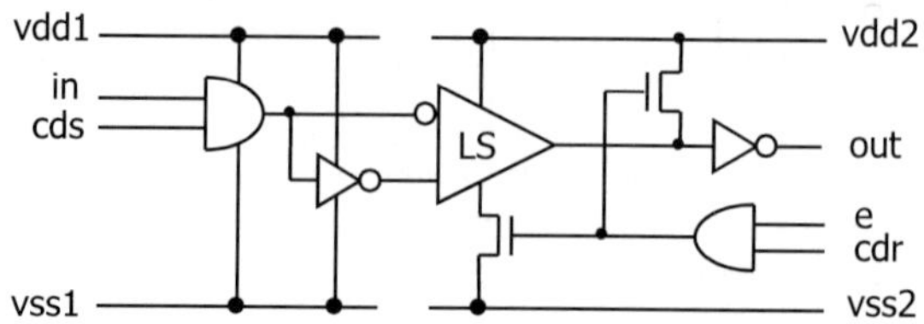

Figure 17. μI/O with level shifting function.

regulators. Three input signals (cds, cdr, and e) are used as control signals. The cds and cdr signals are automatically controlled by a power switch controller (PSWC1 or 2); the cds and cdr are driven to low when the sender and receiver domains, respectively, are turned off internally by the power switches. All the designers have to do is to drive "e" to low when the sender domain is turned off externally.

2.2.1.4. Preventing rush current. Turning on the power switches may cause a large rush current on power lines because large parasitic capacitances (tens of n-F) are required to charge the initial voltages [15]. This may result in a large drop in the supply voltage. When power lines are shared with other domains, this drop in supply may be critical for other domains. Supply variation due to a supply drop can lead to timing violations. The power-switch controller implemented in the SH-Mobile3 therefore uses a slew-rate control scheme to prevent rush currents.

Figure 18(a) shows a block diagram of a power-switch controller (PSWC). The gate of the power switch (g) is driven at a low slew rate using two drivers (C1 and C2). Figure 18(b) shows simulated waveforms. In the first stage (at t1),

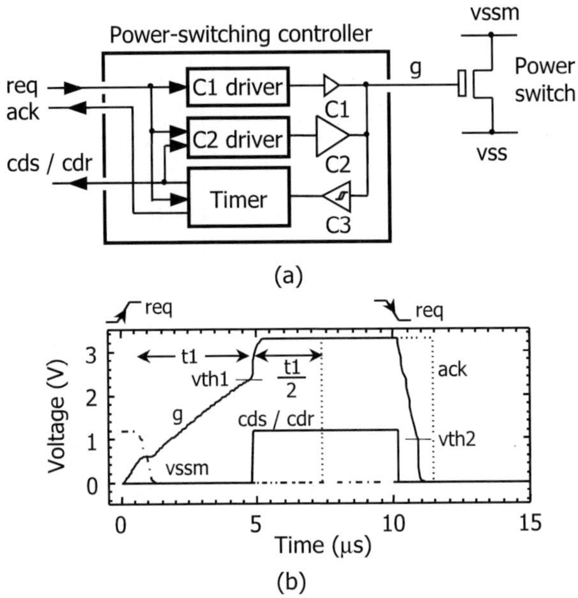

Figure 18. **(a)** Block diagram of the power-switch controller (PSWC) and **(b)** simulated waveforms.

g is driven by a small driver (C1). After C3 detects that g is above the threshold voltage (Vth1) at t1a, a large driver (C2) drives g again to keep it at low impedance. A simulated rush current showed that this system reduced the rush current at a rate of over 20 dB.

The slew-rate control scheme not only prevents rush current but also provides a req/ack handshake interface so that the domain can assess the condition of the power-switch state. A timer in the power-switch controller counts from t1 to t1a (see Figure 18(b)), and drives ack to high at t1b, satisfying $t1b = (t1a - t1)/2$. Thus, when the ack signal is high, the power switch is guaranteed to be completely on. This decreases the overhead for the transition time from the standby to the active mode.

2.2.2. *Implementation of power switches*

Figure 19 shows a block diagram of the SH-Mobile3. The chip is divided into four power domains. The power supply for power domains 1 and 2 can be cut off independently by the on-chip power switches (PSW1 and PSW2). Signals across the power-domain boundary are routed via the μI/O. The power switches and μI/O are controlled by a power-switch controller (PSC) located in power domain 3. The total gate width of PSW1 and PSW2 are 794 and 221 mm, respectively.

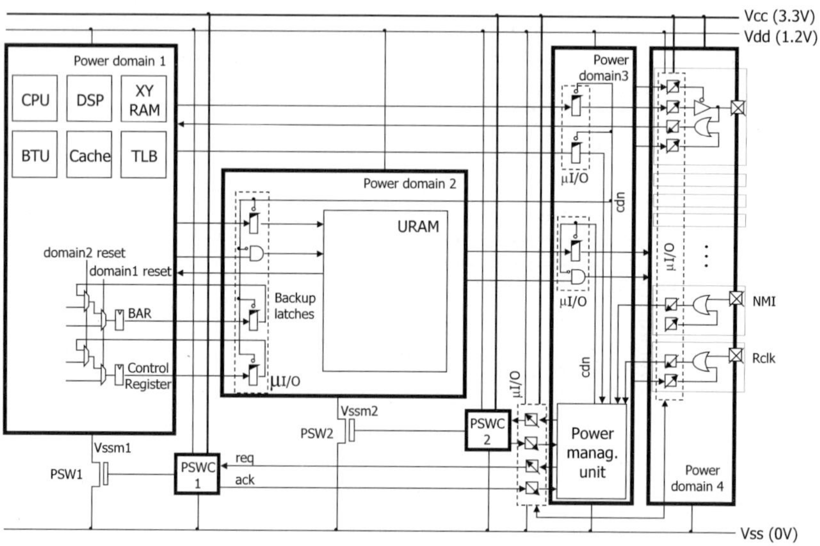

Figure 19. Block diagram of the SH-Mobile3.

2.2.3. Leakage reduction in SRAM module

In R-standby mode, a large capacity SRAM, URAM, in power domain 2 has to retain its data (see Figure 19). For data retention the power of the SRAM module cannot be cut off and the leakage current of the SRAM module has a large impact on the whole leakage power. To reduce the leakage current in R-standby mode, the SRAM module supports a standby mode, in which the SRAM module retains its data with lower leakage power. Besides, increasing leakage current in the active state will be a problem for saving power consumption. One of the solutions for this problem is reducing leakage current during low-speed operation. A low-leakage active mode is implemented in the SRAM module. This mode is not supported in the SH-Mobile3 processor, but this mode will be important when the active leakage current becomes a critical issue.

We first consider a SRAM that fits these operating modes of the processor adaptively. We developed a on-chip SRAM [13] with three operating modes: a high-speed mode, low-leakage active mode, and standby mode (R-standby in the SH-Mobile3). Note that this SRAM features leakage-current reduction for word-drivers even in high-speed active mode. To the best of our knowledge this is the first trial in which a leakage-reduction circuit technique has been implemented in an actual production chip to reduce leakage current in a high-speed active mode.

2.2.3.1. Source-line self-bias technique. The SRAM has to retain its data during the standby state. Therefore, the leakage current of the SRAM memory cell has to be reduced with data retention. In this SRAM, the Vssm, the voltage of the source power line of the memory cells (Figure 20(a)), is increased to reduce the leakage current. As the Vssm voltage increases, the leakage current is further reduced. However, the memory cells then become

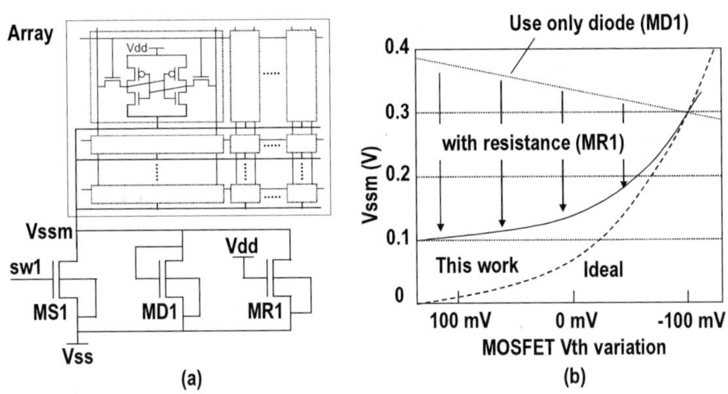

Figure 20. Self-biased source line voltage control [13].

less stable. To satisfy the requirements of both low leakage and high stability, the lowest Vssm that satisfies the leakage target must be generated. Therefore, we developed a new Vssm controller, the PLVC1, which consists of three nMOSFETs, MS1, MD1, and MR1. The MS1 works as a power switch between the Vssm and Vss, MD1 as a diode, and MR1 as a resistor. The MR1 has a long gate length and is normally on.

When a manufactured MOSFET has a high Vth, memory-cell leakage becomes smaller and the current through the MR1 is greater than the memory-cell leakage, so the Vssm voltage becomes lower. However, when the Vth is low, the Vssm voltage is high, but the MD1 restricts the rise in the Vssm voltage. This keeps the voltage low enough to retain stored data. Figure 20(b) shows the Vth of a manufactured MOSFET vs. the Vssm voltage. The horizontal axis shows the Vth difference in the Vth between the designed value and the value of the manufactured MOSFET, which varies according to process variations. When the Vth of the manufactured transistor is low and the leakage current is high, the Vssm voltage has to be high to significantly reduce the leakage current. The broken line indicates the ideal values that just satisfy the leakage target and ensure the highest memory-cell stability. If the voltage controller is composed of only a power switch and a diode, the Vssm voltage is indicated by the "use only diode" line in the graph. This line satisfies the leakage target, but with low operation stability.

Using three types of MOSFETs, MS1, MD1, and MR1, for the voltage controller, ensures that the Vssm voltage is closer to the ideal value. These values satisfy the leakage target and maintain better stability.

2.2.3.2. Leakage-current reduction of array-associated circuits. Figure 21 shows the leakage-current reduction scheme for an array of associated circuits. The memory array is divided into four parts according to their structure, and the leakage current of each part is reduced by controlling each voltage line.

One part is word drivers, and the leakage current is reduced by a cut in its Vdd line (Vddw) by a pMOSFET power switch controlled by SW2. Since the Vddw is only connected to word drivers, and its parasitic load is small, the Vddw can be charged to Vdd level even if the Vddw is activated after the bank selection. The Vddw can therefore be controlled every cycle, and be in an active state when the bank is accessed and be inactive when the bank is not accessed.

One part is a memory cell array, and the leakage current of the cell array can be reduced by the source-level control (see Section 2.1), and its state is controlled by the SW1 signal. One part is a peripheral circuit in the bank, for example the controller and sense/write a circuit, and the leakage current is reduced by cutting its Vss line (Vssa) by a nMOSFET power switch controlled by SW3. The Vssm and Vssa have a large parasitic load and it takes a long time

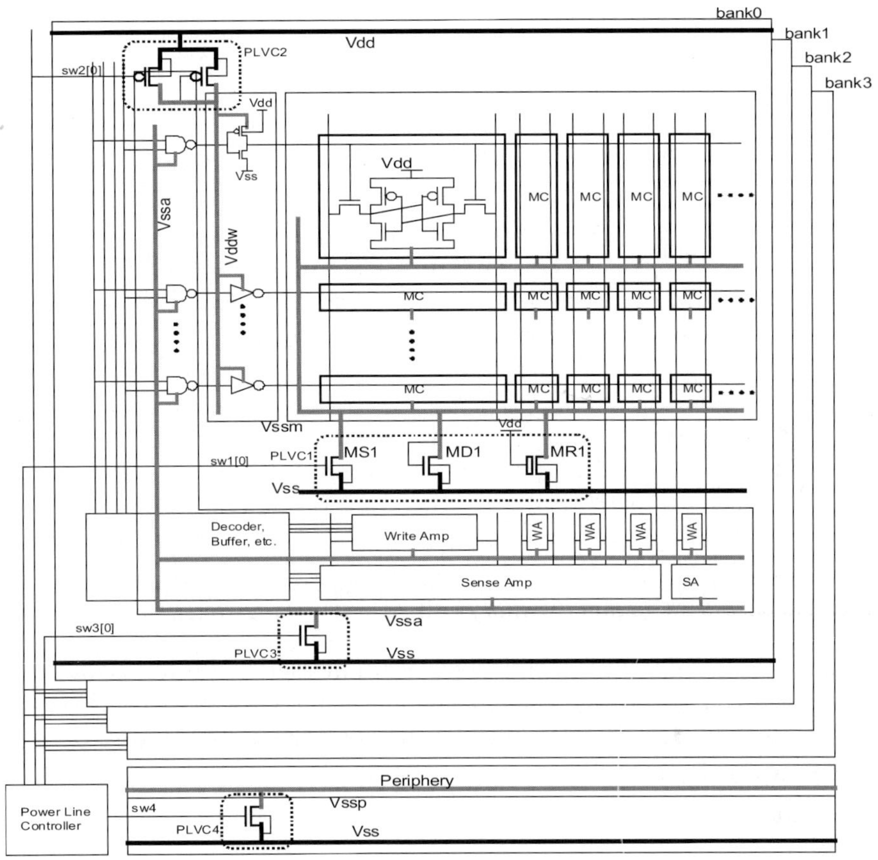

Figure 21. Low leakage SRAM module structure [13].

to be discharged to the Vss level, so the Vssm and Vssa cannot be activated in time in high-speed mode. Therefore Vssm and Vssa are always Vss level in the high-speed mode, and Vssm and Vssa of the non-accessed bank can be controlled to a low-leakage state only in low-speed low-leakage mode. The remaining part is the bank-control circuits and its leakage current is controlled by an SW4 signal. In active mode (high speed and low speed), the bank-control circuit has to operate; therefore SW4 is only deactivated in standby mode. Table 3 shows the state of the switches.

2.2.3.3. Leakage-current reduction of 1-Mbit SRAM module. Figure **??** shows the measurement results for a worst-leakage sample of the 1-Mbit SRAM module, which is the equivalent circuit of the SH-Mobile3 shown in Figure 15. The temperature was 45°C. The figure shows the leakage current in each operating mode. The values of conventional circuits were estimated

Table 3. Switch control

	SW1	SW2	SW3	SW4
Active mode	On	On/Off	On	On
Low-leakage active	On/Off	On/Off	On/Off	On
Standby	Off	Off	Off	Off

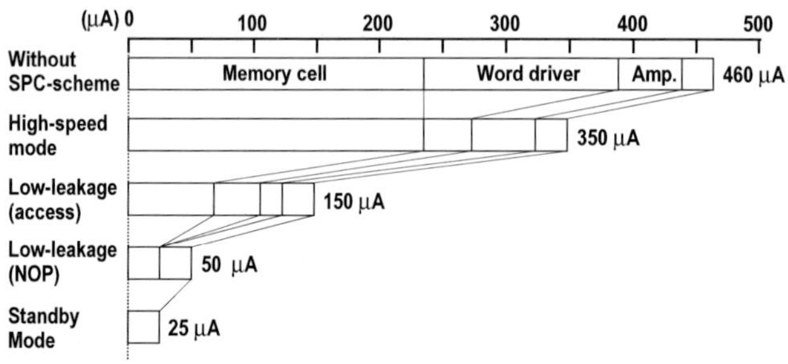

Figure 22. Leakage currents in 1-Mb SRAM module [13].

using the results of leakage simulations and of measuring this prototype chip. The technologies described in this chapter reduced active leakage by 25% even in high-speed mode (150 MHz), by 90% in low-leakage mode (10 MHz), and by 95% in standby mode.

2.3. Measurement Result of Stand-by Modes

The use of on-chip power-switch and low-leakage SRAM technology enables two new standby modes in the SH-Mobile3: resume standby (R-standby) and ultra standby (U-standby) modes. Figure 23 shows the standby power consumption, or leakage current, measured at room temperature at a power supply voltage of 1.2 V. In standby mode, without the power being cut off, the leakage current is 2.2 mA.

In the U-standby mode, both PSW1 and PSW2 are turned off (Figure 19), producing ultra-low leakage of only 11 μA. In this case, however, the transition to the active mode takes longer because most of the information on the chip is lost and the system requires a boot sequence. This standby mode is used when a flip-type cellular phone is closed.

In the R-standby mode, only PSW1 is turned off. This cuts off the power supply to the CPU core, MPEG and 3DG hardware IPs, and peripherals. The power supply to the low-leakage SRAM and back-up registers is kept on. As a result, recovery from the R-standby mode is much more rapid than from the U-standby mode. The leakage current in this mode is 86 μA.

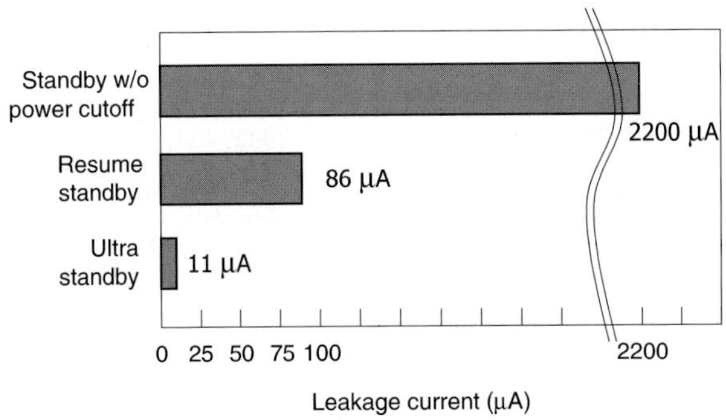

Figure 23. Leakage current consumption in each standby mode.

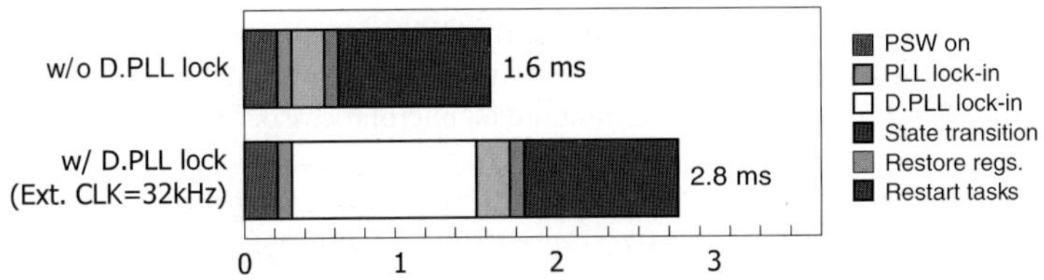

Figure 24. Transition time from the resume standby mode.

To ensure a quick transition from the R-standby mode with a minimum hardware cost, data backup using a backup latch (see Figure 19) is used. If all the information in the flip-flops is cleared by turning off the power switches, a longer recovery time is required. To avoid this, backup latches are implemented in the same power domain as the memory, and the power supply for this domain is kept on in the resume-standby mode. Key information for achieving quick recovery is stored in the backup latches before the transition to the resume standby mode, and this is restored to the original flip-flops during the recovery operation. The control-register contents that are needed immediately after wake-up, such as clock and interrupt settings, are saved to the backup latches. The boot address register (BAR) that holds the restart address is also backed up in the backup latches.

The transition time from the resume standby mode is plotted in Figure 24. Transition is triggered by an interrupt signal. When an interruption occurs, hardware recovery operations such as power-switch control and a PLL lock are activated. Software recovery operations then start from the BAR address and the OS recovery routine is executed. This bar graph shows a breakdown of the recovery time. A digital PLL generates 27 MHz from a 32-kHz clock. This needs a long lock time, so if a 27-MHz clock is available for an external clock input, the recovery time is only 1.6 ms. Otherwise, the recovery time is 2.8 ms.

References

[1] Yamada, T. *et al.* "A 133MHz 170mW 10μA standby application processor for 3G cellular phones", *ISSCC Dig. Tech. Papers*, **2002**, 474, 370–371.

[2] Yamada, T. *et al.* "A low-power embedded RISC microprocessor with an integrated DSP for mobile applications", *IEICE Trans.*, *E85-C*(2), 253–262.

[3] Tsunoda, T. *et al.* "Application processor for 3G cellular phones", *COOL Chips* V *Proc.*, April **2002**, I, 102–111.

[4] Kutaragi, K. *et al.* "A microprocessor with a 128b CPU, 10 floating-point MACs, 4 floating-point dividers, and an MPEG2 decoder", *ISSCC Dig. Tech. Papers*, February **1999**, 256–257.

[5] Rogenmoser, R. *et al.* "A dual-issue floating-point coprocessor with SIMD architecture and fast 3D functions", *ISSCC Dig. Tech. Papers*, February **2002**, 414–415.

[6] Arakawa, F. *et al.* "SH4 RISC multimedia microprocessor" *HOT Chips IX Symp. Rec.*, August **1997**, 165–176.

[7] Nishii, O. *et al.* "A 200MHz 1.2W 1.4GFLOPS microprocessor with graphic operation unit", *ISSCC Dig. Tech. Papers*, February **1998**, *447*, 288–289.

[8] Arakawa, F. *et al.* "SH4 RISC multimedia microprocessor" *IEEE Micro*, **1998**, *18*(2), 26–34.

[9] Yoshioka, S.; Hattori, T. "SH-X 4500MIPS/W 2 2-way superscalar CPU core and its SoC products", *Microprocessor Forum 2003 Conf. Program*, Session 4: Low-Power Processors, San Jose, USA, October **2003**.

[10] Arakawa, F. *et al.* "An embedded processor core for consumer appliances with 2.8GFLOPS and 36M polygons/s FPU", *ISSCC Dig. Tech. Papers*, February **2004**, *531*, 334–335.

[11] Arakawa, F. *et al.* "An embedded processor core for consumer appliances with 2.8 GFLOPS and 36M polygons/s FPU", *IEICE Trans. Fund.*, **2004**, *E87*A(12), 3068–3074.

[12] Kamei, T. *et al.* "A resume-standby application processor for 3G cellular phones", *ISSCC Dig. Tech. Papers*, February **2004**, *531*, 336–337.

[13] Yamaoka, M. *et al.* "A 300MHz 25μA/Mb leakage on-chip SRAM module featuring process-variation immunity and low-leakage-active mode for mobile-phone application processor" *ISSCC Dig. Tech. Papers*, February **2004**, *542*, 494–495.

[14] Arakawa, F. *et al.* "An exact leading non-zero detector for a floating-point unit", *IEICE Trans. Electron.*, **2005**, *E88*C(4), 570–571.

[15] Royannez, P. *et al.* "9nm low leakage SoC design techniques for wireless applications", *ISSCC Dig. Tech. Papers*, February **2005**, 138–139.

[16] Kuroda, T. *et al.* "A 0.9V 150MHz 10mW 4 mm2 2-D discrete cosine transform core processor with variable-threshold scheme", *ISSCC Dig. Tech. Papers*, February **1996**, 166–167.

[17] Soden, J.M. *et al.* "Identifying defects in deep-submicron CMOS ICs", *IEEE Spectrum*, September **1996**, 66–71.

[18] Chan, T.Y. *et al.* "The impact of gate-induced drain current on MOSFET scaling", *IEDM*, December **1987**, 718–721.

[19] Kanno, Y.; Mizuno, H.; Oodaira, N.; Yasu, Y.; Yanagisawa, K. "μI/O architecture for 0.13-μm wide-voltage-range system-on-a-package (SoP) designs", *Symp. on* VLSI Circuits, *Digest of Technical Papers*, **2002**, 168.

INDEX